Short-Term Bioassays in the Analysis of Complex Environmental Mixtures IV

ENVIRONMENTAL SCIENCE RESEARCH

Recent Volumes in this Series

Volume 24 – COMPARATIVE CHEMICAL MUTAGENESIS
Edited by Frederick J. de Serres and Michael D. Shelby

Volume 25 – GENOTOXIC EFFECTS OF AIRBORNE AGENTS
Edited by Raymond R. Tice,
Daniel L. Costa, and Karen Schaich

Volume 26 – HUMAN AND ENVIRONMENTAL RISKS OF CHLORINATED DIOXINS AND RELATED COMPOUNDS
Edited by Richard E. Tucker, Alvin L. Young, and
Allan P. Gray

Volume 27 – SHORT-TERM BIOASSAYS IN THE ANALYSIS OF COMPLEX ENVIRONMENTAL MIXTURES III
Edited by Michael D. Waters, Shahbeg S. Sandhu,
Joellen Lewtas, Larry Claxton, Neil Chernoff,
and Stephen Nesnow

Volume 28 – UTILIZATION OF MAMMALIAN SPECIFIC LOCUS STUDIES IN HAZARD EVALUATION AND ESTIMATION OF GENETIC RISK
Edited by Frederick J. de Serres and William Sheridan

Volume 29 – APPLICATION OF BIOLOGICAL MARKERS TO CARCINOGEN TESTING
Edited by Harry A. Milman and Stewart Sell

Volume 30 – INDIVIDUAL SUSCEPTIBILITY TO GENOTOXIC AGENTS IN THE HUMAN POPULATION
Edited by Frederick J. de Serres and Ronald W. Pero

Volume 31 – MUTATION, CANCER, AND MALFORMATION
Edited by Ernest H. Y. Chu and Walderico M. Generoso

Volume 32 – SHORT-TERM BIOASSAYS IN THE ANALYSIS OF COMPLEX ENVIRONMENTAL MIXTURES IV
Edited by Michael D. Waters, Shahbeg S. Sandhu, Joellen Lewtas,
Larry Claxton, Gary Strauss, and Stephen Nesnow

Short-Term Bioassays in the Analysis of Complex Environmental Mixtures IV

Edited by

MICHAEL D. WATERS
SHAHBEG S. SANDHU
JOELLEN LEWTAS
LARRY CLAXTON
GARY STRAUSS
and
STEPHEN NESNOW

U.S. Environmental Protection Agency
Research Triangle Park, North Carolina

PLENUM PRESS • NEW YORK AND LONDON

Library of Congress Cataloging in Publication Data

Symposium on Short-Term Genetic Bioassays in the Evaluation of Complex Environmental Mixtures (1984: Chapel Hill, N.C.)
Short-term bioassays in the analysis of complex environmental mixtures IV.

(Environmental science research; v. 32)
Bibliography: p.
Includes index.
1. Genetic toxicology—Technique—Congresses. 2. Toxicity testing—Congresses. 3. Pollutants—Toxicology—Congresses. 4. Biological assay—Congresses. I. Waters, Michael D. II. Title. III. Series.
RA1224.3.S96 1984 615.9′07 85-9505
ISBN 0-306-42015-5

The research described in this volume has been reviewed by the Health Effects Research Laboratory, U.S. Environmental Protection Agency, and approved for publication. Approval does not signify that the contents necessarily reflect the views and policies of the U.S. Environmental Protection Agency, nor does mention of trade names or commercial products constitute endorsement or recommendation for use.

Proceedings of a Symposium on Short-Term Genetic Bioassays in the Evaluation of Complex Environmental Mixtures, held March 27–29, 1984, in Chapel Hill, North Carolina

First Plenum printing 1985

Plenum Press is
A Division of Plenum Publishing Corporation
233 Spring Street, New York, N.Y. 10013

Printed in the United States of America

FOREWORD

With this proceedings of the fourth symposium on complex mixtures, we continue to revise and extend our knowledge of genetic methods for the evaluation of chemical mixtures in the environment. The early chapters of this volume are devoted to new bioassay techniques that are directly applicable to the monitoring of environments contaminated with genotoxic chemicals. Microbiological methods have been further refined to meet the special needs of atmospheric monitoring so that very small samples may now be efficiently tested. New *in situ* methods utilizing green plants actually avoid many of the usual difficulties of sample collection and preparation and offer special advantages in monitoring wastewater, sludges, and hazardous wastes. Insects also are being employed very effectively in the evaluation of gaseous air pollutants in controlled laboratory investigations.

Increased emphasis has been placed on a comprehensive assessment of the potential of complex mixtures to cause various kinds of genetic damage. New assays for chromosome structural and numerical aberrations in mammalian cells *in vitro* have been developed and are being applied in laboratory studies. Efforts to link tests for gene mutation and cell transformation *in vitro* with assays for tumorigenesis *in vivo* are contributing to the validation of the short-term testing approach. Studies comparing *in vitro* and *in vivo* data on a coal conversion by-product, on polycyclic aromatic hydrocarbons, and on mineral fibers are reported in separate papers.

Later chapters are devoted to investigations on the fractionation and biological evaluation of specific chemical components within complex mixtures. Toward this end, it has been possible to generate complex atmospheres in which individual components are varied in concentration so that the biological consequences may be observed using microbial assays. Similar techniques have been used to demonstrate the formation of potent mutagens, the nitropyrenes, in environmental mixtures as well as in foods. The compound 1-nitropyrene was detected in cooked meat, but was not found to be carcinogenic in mice. On the other hand, the combustion emission constituent 1,6-dinitropyrene was clearly demonstrated to be a carcinogen in mice.

The effect of atmospheric transformation of automotive exhaust organics and of woodstove emissions has received much needed attention. Chamber studies showed that atmospheric processes can both increase and decrease the mutagenicity of emitted organic compounds. Atmospheric gases such as ozone and nitrogen oxides and physical factors such as light intensity and humidity dramatically modify the chemical transformation processes as monitored by analytical techniques and microbial assays.

Fractions of ambient air particulate extracts have been extensively studied in a major statewide sampling and analysis program. The results will provide reference points for future sampling and mutagenicity studies in other parts of the country. In another study involving five cities, inter-urban variations in the mutagenic activity of ambient aerosols were investigated in reference to fuel use patterns and seasonal variables. Mutagenic activities reflected differences in the chemical composition of the aerosols from the five cities studied; these differences, in turn, could be related to fuel use patterns. Similarly, the wastewaters and sludges of a major municipal treatment plant were impacted by industrial effluents and, again, microbial systems demonstrated their utility in municipal waste management.

The assessment of human health risk from complex mixtures of environmental chemicals remains a difficult problem. Definite progress has been made in the development and implementation of short-term tests that are directly applicable to the monitoring of individuals exposed to complex mixtures, especially in the workplace. The mutagenicity of urine has been studied as a non-specific indicator of exposure to genotoxic agents. Cytogenetic analyses of peripheral blood lymphocytes appear to provide a more direct biological measure of induced genetic damage in somatic cells. Studies are reported on pharmaceutical operations with cytostatic drugs, on formaldehyde in mobile homes, and on ethylene oxide in sterilization facilities. The latter study on ethylene oxide is the largest to date and clearly demonstrates that exposure to the compound is associated with an increase in sister chromatid exchanges in peripheral lymphocytes. Future investigations may be expected to monitor other types of genetic damage in lymphocytes, including specific gene mutation. The course of development of a clonal assay for gene mutation in human lymphocytes is described in detail. There has as yet been only limited use of such direct human monitoring approaches in mutagenesis or carcinogenesis risk assessment. Short-term genetic bioassay data, in the context of risk assessment, has been used principally to estimate comparative carcinogenic potency, as described in the last paper in the volume. This approach will be applied in the future to a wide variety of test data on complex mixtures.

Additional research must be performed to further characterize the nature and extent of genetic hazards posed by human exposure to complex chemical mixtures. Future volumes may be expected to provide a more exhaustive discussion of methods to integrate chemical analytical data with short-term test data and data obtained directly from human monitoring. In this way, methods for hazard identification may be extended in their utility, and risk assessment techniques may be refined by direct measurements of genetic damage in humans.

Michael D. Waters

CONTENTS

---

*Invited paper.
†Based on poster presentation.

KEYNOTE ADDRESS

KEYNOTE ADDRESS

Scott Baker

Office of Research and Development, U.S. Environmental Protection Agency, Washington, D.C. 20460

Our view of the environment has evolved from simplistic impressions of global systems to a realization of the intricate complexities that pervade them. We began a decade ago with a single chemical-single medium mentality and conducted excellent research in this mode. Until the mid-1970's, we thought that the environment could be protected if we controlled a few air pollutants and a few water pollutants. Until recently, we thought the effects of a pollutant were confined to a fairly small area surrounding its source and to a single medium--air, land, or water. Until recently, we believed that the principal dangers posed by pollutants were their acute effects on humans and wildlife. The growing recognition of the hazards presented by toxic substances has made us realize that, in fact, we must cope with thousands of pollutants. We constantly strive for better ways to determine the health effects of these pollutants, to monitor their presence in the environment, and to control them. Now we know that air pollutants may be transported hundreds of miles through the atmosphere and end up in the water. Yet we do not have the scientific understanding or information that is needed to cope with remote impacts. We now realize that chronic health problems may be more important than acute effects, and we have learned that the adverse effects of some pollutants may not be manifested for decades.

In ten years, we have undertaken an accelerated mission of exploration and discovery in environmental science that has led us to stand now in awe of our own realization that the answers we thought we had were just way points. This symposium is one of many new signals that we are now striking new commitments to deal with issues in a set of real-world circumstances--where multipollutants interact in multimedia and exert collective health and environmental effects.

We must be concerned not only about combinations of chemicals identified in effluents and emissions, but also about the dynamics of their transformation and transport. Chemicals released into the air, water, and soil will generally react with other chemicals in those media. A complex series of reactions may continue along extended physical transport pathways before the materials find semipermanent storage sites in one or more environmental compartments. The pathways for some pollutant chemicals are relatively simple and reasonably well understood. For example, carbon monoxide reacts with oxygen to form carbon dioxide, or it is metabolized in the biosphere. Sulfur dioxide, another combustion product, has a much more complex series of atmospheric oxidative interactions that ultimately result in its conversion to sulfuric acid and sulfate salts. In contrast to carbon monoxide, this complex activity produces greater impacts on downwind populations as atmospheric conversion proceeds.

On the other hand, chemical interactions in water and soil are usually site specific, and there are few situations where sufficient information exists for the development of good case histories. Cases in which the environment is contaminated by a poorly characterized mixture of chemicals (for instance, from industrial effluents or chemical landfills) raise special problems of adequate sampling and mixture characterization. These problems complicate our ability to characterize exposure accurately.

An important consideration in exposure is that mixtures of chemicals may be separated as they are transported through the various media of the environment. For example, volatile components of mixtures are differentially transported across water-air or soil-air interfaces. Some parts of the natural environment act as giant chromatographs, and the principles of chromatography can be applied to estimate the magnitudes of exposures.

In some cases, the environmental behavior of chemicals may be significantly influenced by the presence of other chemicals in complex mixtures. The presence of one chemical may strongly influence the biological uptake and retention of others. Few such cases have been documented, but they are potentially significant in assessing human exposure to complex mixtures.

Concomitant exposure to multiple organic and inorganic chemicals, particulate matter, and radiochemicals may cause a variety of acute and chronic effects in humans, and appropriate tests must be developed to search for these effects. In the future, *in vivo* tests will become increasingly difficult as we try to interpret synergistic, antagonistic, or otherwise combined effects of exposure to multiple chemicals. This will make our reliance on simpler, more controllable *in vitro* systems more critical. In the recent past, a series of mutagenicity screens was conducted at EPA to compare the potency of diesel exhaust fractions with other mixtures. That was the beginning of serious efforts in EPA to untangle difficult issues concerning complex mixtures. The experience provided some surprising conclusions that could not have been disclosed by single-substance experimentation.

The challenges for the future lie in areas that may not be readily apparent. In the past, our predictive capability in determining carcinogenic potential from mutagenic screens has met with moderate success.

Because of the variability present in experiments with complex mixtures, we can anticipate a diminishing return in predictive capability of mutagenic screens. We therefore have to find new ways of flagging genetic effects. As a windfall from the burgeoning technology of genetic engineering, new strains of organisms with different specific indicators of genetic alteration will have to be developed for comparative testing, particularly since in the future we will be dealing with mixtures of increasingly more complex and troublesome character. Mutagenicity testing will have to be more closely integrated with other existing and emerging indicators of toxicity, such as molecular markers and other biomarkers, immunologic complexes, and biochemical kinetic indices. We have indications today of the importance that these concepts will have in the future. Some of EPA's current regulations are based on *biological* end points rather than *chemical* end points. The water spider is proving to be a sensitive indicator of environmental toxicity--important for implications of drinking water contaminants and human health. One can readily see the utility of these advances in studies with complex mixtures.

Risk assessment has become an integral part of the regulatory decision-making process in the government. As we continually refine our risk assessment capabilities to address such emotionally charged issues as linearity, threshold, weight of evidence, upper-bound estimation, and the value of human life, greater demands will be placed on the power of short-term tests, such as mutagenicity, to be predictive of health insult. Complex mixtures will more likely give rise to higher false positive and false negative rates at a time when techniques in risk assessment will be attempting to reduce them. The challenge here for researchers in mutagenesis will be to design integrated systems that permit us to recognize false results.

Perhaps our greatest challenge will be the continued enhancement of public understanding. The public thinks we know what *all* of the bad pollutants are, *exactly* what adverse health or environmental effects they cause, how to measure them *precisely*, and how to control them *absolutely*. In the future, we can look forward to providing more complex explanations for the effects of more complex mixtures. Yet we must continue to strive for a better informed public. We, as scientists, have an obligation to contribute to this goal.

Our understanding of the nature of environmental problems requires a new, innovative research agenda that incorporates the interrelationships of toxic substances. We must continue to look beyond genetics to other end points and then determine, with these end points, how genetics can best fortify our understanding. We must find ways to parlay effective research programs on complex mixtures into better estimates of human risk, better policy decisions, and, ultimately, better protection of the environment and human health.

SESSION 1

NEW DEVELOPMENTS AND TECHNIQUES

USE OF THE MICROSCREEN ASSAY FOR AIRBORNE PARTICULATE ORGANIC MATTER

Toby G. Rossman, LeRoy W. Meyer, James P. Butler, and Joan M. Daisey

Institute of Environmental Medicine, New York University Medical Center, New York, New York 10016

INTRODUCTION

The value of short-term *in vitro* assays to screen for potential human mutagens and carcinogens is well established. The most extensively validated short-term assay is the Ames test (Ames et al., 1975). This assay is usually included in the "battery" of tests used both in industry and in government screening programs. Other assays are usually included in the battery in order to increase the probability of detecting all genotoxic agents.

The Ames test has evolved from an assay of single chemicals into an assay of such complex mixtures as human fluids, metabolites of carcinogens, and environmental samples such as airborne extractable organic material, diesel and automobile exhaust, and cigarette smoke condensate (Talcott and Wei, 1977; Claxton, 1983; Kier et al., 1974). This assay requires only a fraction of the material necessary for mouse skin carcinogenicity studies.

A major role for *in vitro* assays is to aid in the fractionation of complex environmental mixtures. However, a problem in assessing the biologically significant chemical classes from environmental samples has been the lack of a screening method suitable for testing small samples. A U.S. Department of Energy sponsored workshop concerning organic air pollutant research priorities concluded "Bacterial mutagen assays need to be developed or improved to afford a higher probability of fast and reliable identification of most biologically active agents" (USDOE, 1982).

Although five different tester strains are recommended for general screening purposes in the Ames test (de Serres and Shelby, 1979), many workers studying environmental samples use only one or two strains, due to lack of material. Claxton (1983) points out that many workers studying automotive emissions work exclusively with TA98 and TA100 largely because of limitations in the amount of sample, and states that "investigators have questioned whether the *Salmonella* bioassay correlates well enough with other bioassys to use as a routine screen. This knowledge would be useful in the development of new combustion and control techniques."

In the Ames test, reversion of the tester strain requires specific changes in base sequence (Ames et al., 1975). Additional tester strains have been developed to detect agents (e.g., reactive oxygen species) that cause damage at A·T sites and are not detectable in the "standard" strains (Levin et al., 1982). While this approach yields much useful information as to the mutagenic mechanism of a particular agent, the inclusion of many tester strains in screening assays is a disadvantage in cases where only small amounts of the test agent are available. In addition, agents that cause deletions would not be detectable in any reversion assay. The correspondence between carcinogenicity and mutagenicity in the Ames test depends very much upon the group of compounds chosen for a validation study, and can vary from 50% to about 90% (Purchase, 1982). Certain classes of compounds (e.g., N-nitrosamines) are not well identified, and other classes (e.g., bifunctional alkylating agents) are not identified in the "standard" Ames strains but require other strains not usually included in the assay (e.g., uvr$^+$ strains). These and other drawbacks of a screening system based solely on reversions are discussed at greater length elsewhere (Skopek et al., 1978; Mohn, 1981).

What appears to be needed at this point is an assay of broad specificity that is rapid, inexpensive, and very sensitive. In screening for agents of unknown genotoxic effect, an assay with an end point capable of detecting almost any kind of DNA damage would seem desirable.

The Microscreen system is a multiple end point system that utilizes *E. coli* $WP2_s(\lambda)$, a strain developed in this laboratory, carrying the *uvrA* mutation (which renders the bacteria unable to perform excision repair of most lesions), the *trpE* mutation (which enables Trp$^+$ reversion assays to be perfomed), and prophage λ (which allows prophage induction to be scored) (Rossman et al., 1984; Meyer et al., 1984). Exposure to the test agent is carried out in a microtitre plate (volume = 250 µl/well), in twofold serial dilutions. Wells are assayed for toxicity, prophage induction, and Trp$^+$ revertants. The amount of test material needed to perform this assay is approximately 7% of that needed for each tester strain of an Ames test. Our results and those of others (Moreau et al., 1976; Elespuru and Yarmolinsky, 1979) have shown that the induction of λ prophage is an extremely broad genetic end point and responds to chemicals causing a large variety of DNA lesions. No agent has yet been found that is able to cause Trp$^+$ revertants (or His$^+$ revertants in the Ames test) and not λ prophage induction in our assay. For that reason, we report results only on λ prophage induction, and recommend that only it need be used for routine screening purposes. Elespuru has written a recent, thorough review of the theory and practice of using prophage λ induction as an assay for genotoxic agents (Elespuru, in press).

MATERIALS AND METHODS

E. coli Strains and Media

All strains are derivatives of *E. coli* B/r. $WP2_S(\lambda)$ is a λ lysogen of $WP2_S$ (*trpE*, *uvrA*), isolated in our laboratory from $WP2_S$ and wild type λ, both obtained from Dr. Evelyn Witkin, Rutgers University. The indicator strain SR714 (*trpE*, $uvrD_3$) was obtained from Dr. Kendric Smith, Stanford University.

Cultures were grown in Minimal Broth Davis (Difco) containing 0.2% glucose and 20 µg/ml tryptophan (MST). Mid-exponential phase cultures (optical density = ~0.3 at 550 nm) were used for all experiments.

Microsuspension Assay for Growth Inhibition

An adaptation of the method of McCarroll et al. (1981) was used. Eight serial twofold dilutions of the test agent in 150 µl MST are made across a row in a Falcon Micro Test III tissue culture multiwell plate containing 96 wells. The starting concentration is 100 µg/well, or a concentration causing at least one but no more than four wells to show growth inhibition. In cases where no growth inhibition occurs, the starting concentration is the concentration at which precipitation occurs.

Eleven compounds and a row of controls (MST) can be assayed on one plate. The wells are inoculated with 75 µl of a mid-exponential culture of $WP2_S(\lambda)$, diluted so that each well receives approximately 2×10^7 bacteria. At this cell density, no turbidity is evident. The test plate is incubated overnight at 37°C and scored after 20 h for turbidity (see Figure 1). Control (+) wells contain approximately 3×10^8 bacteria per well (approximately 1.3×10^9/ml). Partial growth inhibitions (+/− wells), which are visually less turbid than control wells, were found to contain $1\text{-}5 \times 10^8$ cells/ml.

Assay for Plaque-forming Units

After scoring for growth, an aliquot of 5 µl from each well is diluted in 1 ml MST. Diluted samples (100 µl), which contain approximately 6.5×10^5 lysogens from turbid wells, are added to tubes containing 2.5 ml soft agar (0.65% Bacto agar, 10 mM $MgSO_4$), held at 47°C. A mid-exponential culture of indicator strain (100 µl) is added, and the tubes are mixed and poured onto nutrient broth agar plates (Difco). After overnight incubation at 37°C, the plates are scored for plaques. All assays are run in duplicate, and all experiments have been repeated at least once.

Environmental Samples

High-volume samples of airborne particulate matter were collected during winter or early spring sampling periods in four cities. The sampling and extraction protocols are described in detail elsewhere (Daisey et al., 1980; Daisey et al., 1983; Butler et al., 1985).

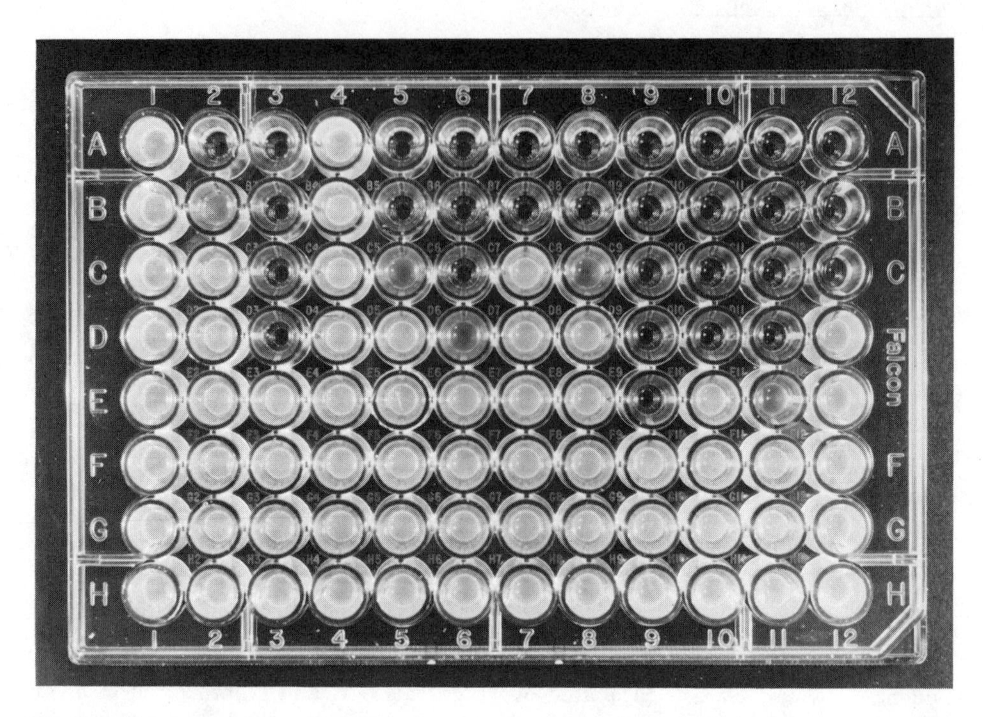

Figure 1. Appearance of the Microscreen assay after the initial overnight incubation. Row 1 (A-H) are all controls (positive and negative) and would be scored + for growth. Rows 2-12 are different test agents titrated in 1/2 dilutions. Scoring of Row 2 for growth would be (from A to H): −, +/−, +, +, +, +, +, +. Row 3 would be: −, −, −, −, +, +, +, +. Samples for λ prophage induction would be taken from 2B, 2C, 2D, 2E, 3D, 3E, 3F, and 3G. Row 4 represents an agent that is not toxic at the concentrations tested. Samples from 4A and 4B would be taken and might give positive results. If results were negative, the compound would be retested at higher concentrations.

In brief, samples were sequentially extracted with cyclohexane (CX), dichloromethane (DCM), and acetone (ACE), to separate nonpolar, moderately polar, and polar fractions, respectively. The samples were reduced to dryness under nitrogen and then redissolved in acetone for bioassay. The highest concentration of acetone was 10% in the wells.

Chemicals

N-methyl-N'-nitro-N-nitrosoguanidine (MNNG) was purchased from Aldrich Chemical Company. Mitomycin C and 2-aminoanthracine (2AA) were obtained from Sigma.

Metabolic Activation

Rat liver S9 from Aroclor 1254 induced animals is obtained from Litton Bionetics. The cofactor mix is a modification of the method of Maron and Ames (1983), and contains:

	Per 10 ml mix
$MgCl_2$ (0.4 M) + KCl (1.65 M)	0.25 ml
NADP	6.1 mg
Glucose 6 phosphate	7.3 mg
Rat liver S9 + Na phosphate buffer (0.2 M, pH 7.4)	6.75 ml

The mix is used at a concentration of 10% in the wells (25 µl/well). The amount of rat liver S9 was varied in preliminary experiments (see Table 1). For testing purposes, a final concentration of 0.25% (0.25 ml per 10 ml mix) was used.

RESULTS

Microscreen Assay of Pure Compounds

In the Microscreen assay, no preliminary toxicity testing is needed, since toxicity is scored in the assay after the initial overnight incubation in the microtitre plate (see Figure 1). Normally, only four wells need to be assayed further, starting with the transition between toxic and nontoxic concentrations. However, for purposes of illustration, six wells were assayed for MNNG and mitomycin C (Figure 2).

MNNG is a simple alkylating agent that methylates DNA predominantly at reactive oxygens (Singer and Kuśmierek, 1982). There is much evidence (summarized by Pegg, 1983) that the O^6 methyl guanine formed is largely responsible for the ability of MNNG to cause base-pair substitutions by causing mispairing during DNA replication. Figure 2A shows that MNNG causes both λ prophage induction and Trp^+ reversion (indicative of base-pair substitution) in the Microscreen assay Prophage induction is the more sensitive end point, and will detect less than 0.4 µg MNNG. The maximum effect is seen in the well containing 1.56 µg. The next well (3 µg) is growth inhibitory (+/−) and results in a downward slope of the dose-response curve for prophage induction, whereas the mutation frequency decreases when growth is completely inhibited at 6 µg (data not shown). As a positive control, 1.5 µg MNNG is routinely used.

Figure 2B shows the assay of mitomycin C, a bifunctional alkylating agent that forms DNA cross-links. Since excision repair is required in order to demonstrate

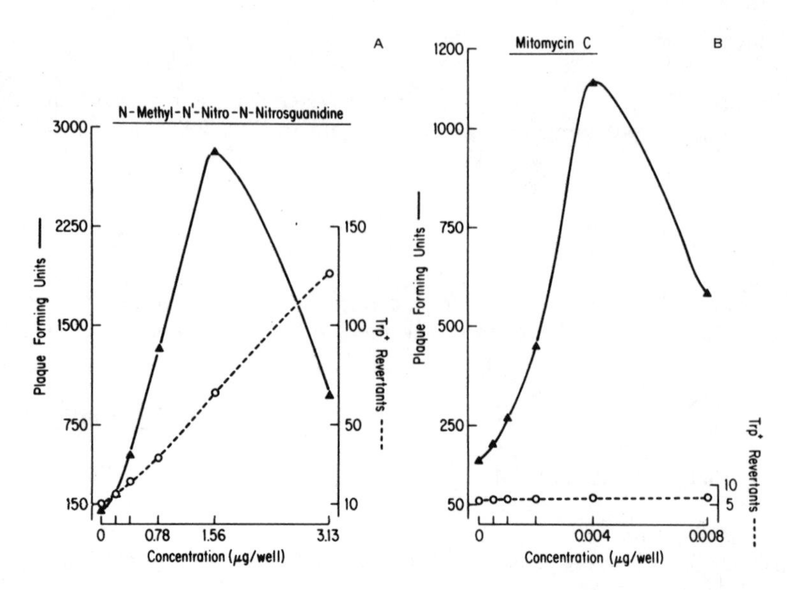

Figure 2. Assay of six wells for λ prophage induction and Trp+ revertants. A, MNNG; B, Mitomycin C. Results are expressed as raw data from the assay performed as described in Materials and Methods.

mutagenesis by cross-linking agents, mitomycin C cannot be detected with the standard Ames strains, which are *uvrB* (and thus lack the ability to excise "bulky lesions") (McCann et al., 1975). The same limitation exists for the Trp+ reversion end point in Microscreen, since the strain carries the *uvrA* mutation (functionally like *uvrB*). However, no such limitation exists for the induction of λ prophage (Figure 2B), which is induced at extremely low concentrations of mitomycin C. Unlike mutagenesis, the induction of λ prophage apparently does not require "processing" of the DNA cross-link by the UVR system. This feature allows one to take advantage of the increased sensitivity of the *uvrA*-carrying strain to many agents without the disadvantage of loss of genetic end point for detecting cross-linking agents.

Metabolic Activation

Using 0.5 µg 2AA per well, the concentration of rat liver S9 from Aroclor-induced rats was titrated in the Microscreen assay, starting at 1% (v/v). The concentration of cofactors remained constant (see Materials and Methods).

Table 1 shows that even in the absence of S9, 2AA has some prophage-inducing activity. However, a clear dose response relationship demonstrates the ability of rat liver S9 to metabolize 2AA to a more active genotoxic agent. In the absence of 2AA, S9 has no effect on prophage induction except at very high (>1%) concentrations. In all future experiments, 0.25% rat liver S9 will be used.

Table 1. Effect of Rat Liver S9 Concentration on Microscreen Assay of 2-Amino-anthracene (2AA)

	PFU/Plate[a]	
Percent S9	−2AA	+2AA (0.5 µg)
0	23	247
0.063	25	663
0.125	10	845
0.25	16	1872
0.5	17	2340
1.0	49	2912

[a]PFU/plate: plaque-forming units per nutrient agar plate in assay performed as described in Materials and Methods. This represents the equivalent of 0.5 µl sample from each well.

Significance

The variability in the Microscreen assay was studied by assaying eight wells each (two samples/well) that were incubated with 10% acetone (the maximum solvent concentration) and with 10% acetone in the presence of the metabolizing system (Table 2). These wells would constitute the negative control wells for assays of complex mixtures dissolved in acetone.

There is no significant difference between wells incubated with acetone alone and those incubated with acetone and S9 mix. At the 99% confidence interval, a result could be considered positive if it were 1.9 times background for acetone alone and 2.1 times background for acetone plus S9 mix. We have chosen to be more stringent in calling a result positive, and in the assay of the air particle extracts described below, a value of 3 times background is used as the minimum significant value.

The amount of spontaneous prophage induction can vary somewhat from experiment to experiment, from a low of about 20 PFU (e.g., Table 1) to a high of about 200. However, the extent of prophage induction in wells containing the test agent also varies in the same direction (i.e., when the background is high, the induced values are also high).

Table 2. Variation in Assay of Control Wells (2 Samples/Well)

Well	Acetone Control (10%)		Acetone Control + S9[a]	
1	118	192	151	186
2	201	147	132	205
3	132	155	186	151
4	108	105	89	170
5	140	170	79	192
6	136	194	128	167
7	147	105	97	136
8	140	187	93	183
Mean ± SD	148.6 ± 32.3		146.6 ± 40.7	

[a]S9: 0.25% (v/v) S9 plus cofactors.

Economy

The amount of test material required in the Microscreen assay is less than 7.7% of that required for *each* tester strain in an Ames test (Table 3). This calculation assumes that five doses are used in the Ames test (the highest being 0.5 mg/plate), and that each dose is run with and without S9 mix and in duplicate (four plates). Material used up in preliminary toxicity testing in the Ames test was not taken into account in the calculation. The Microscreen assay is also run with and without S9 mix (two wells), and duplicate samples are removed from each well.

The dose range in the Ames test can be adjusted higher or lower, depending on the preliminary toxicity test (the top dose should show toxicity). In the Microscreen assay, more than two orders of magnitude are covered by the eight twofold dilutions. Most organic mutagens and carcinogens can be detected within this range. However, the initial 24-h incubation period can also be considered a preliminary toxicity test. No samples need to be taken for prophage induction if there is too much toxicity (eight negative wells); the assay should be rerun starting at a much lower dose (Figure 2B). If no toxicity is seen (eight positive wells), the top two wells should be assayed, since a positive result can often be detected in the absence of toxicity. If the result is negative, the assay should be rerun at higher doses.

Besides the savings in test material, the Microscreen assay also conserves rat liver S9. We calculate that 300 μl (25 μl/plate) of S9 would be consumed using one strain in the Ames test, whereas 5.6 μl would be consumed in the Microscreen assay. Since the Microscreen assay uses only 1.87% as much S9 as does each strain of the Ames test, at a cost of $13.45/5 ml for S9 (Litton), a significant saving should result.

Table 3. Comparison of Material Needed to Perform Ames Test (One Strain) and Microscreen Assay

Ames Test (One Strain)			Microscreen Assay		
Dose/ Plate (mg)		Material Needed (mg)	Dose/ Well (µg)		Material Needed (µg)
0.5	x 4 plates	2.0	100	x 2 wells	200
0.375		1.5	50		100
0.25		1.0	25		50
0.125		0.5	12.5		25
0.05		0.2	6.25		12.5
			3.125		6.25
			1.5625		3.125
			0.78125		1.5625
	Total needed	5.2 mg = 5200 µg		Total needed	398.4375 µg = ~400 µg

Microscreen/Ames = 400/5200 = 7.7%

Microscreen Assay of Air Particle Extracts

The Microscreen assay was performed on the CX, DCM, and ACE extracts of air particle samples (see Materials and Methods). In some cases, too much material had been consumed in the Ames test (Butler et al., 1985) and the sample was not available for the Microscreen assay.

Table 4 shows the results of these assays. The upper limit for the assay is the concentration at which either toxicity or precipitation occurs. For most samples, a starting (highest) concentration of 100 µg/well would suffice.

The nonpolar fractions are limited by precipitation, which tends to occur at 75 µg/well (300 µg/ml). No toxicity is seen up to this concentration. Only two moderately polar fractions were available. The Mexico City fraction showed precipitation at 75 µg/well, whereas the New York City fraction was toxic at 125 µg/well without S9 and precipitated in its presence. At lower concentrations, some

Table 4. Microscreen Assay of Air Particle Extracts

Extract		S9	Upper Limit (µg): Toxicity or Precipitate (P)	Minimum Amount Detectable (µg)[a]	PFU/ng (maximum)
NYC	CX	–	75(P)	37.5	8.7
		+	75(P)	<9.4	89.25
	DCM	–	125	-	-
		+	125(P), +/– tox. to 7.8	62.5	2.1
	ACE	–	50	25	6.75
		+	50	25	9.5
MEX	CX	–	75(P)	18.75	10.8
		+	75(P)	37.5	35.6
	DCM	–	75(P)	-	-
		+	75(P)	-	-
	ACE	–	31.25	-	-
		+	62.5	31.25	9.5
BEIJ	CX	–	75(P)	-	-
		+	75(P)	<9.4	>133
	ACE	–	>100	-	-
		+	>100	<25	14.7
ELIZ	CX	–	75(P)	75	4.2
		+	150(P)	<9.4	>101.5
	ACE	–	>125	125	1.5
		+	>125	62.5	5.35

[a]Amount of extract causing an increase of 3 times background; -: no significant effect.

toxicity was seen in the presence of S9. The most toxic fractions were the polar fractions from New York City and Mexico City.

In most cases, metabolic activation enhanced the genotoxic effect. This is especially true for the nonpolar fractions. The nonpolar fraction from the Beijing sample showed the largest increase, as was true also for the Ames test (Butler et al., 1985). The amount of activation by the S9 mix appears to be much larger for this assay than for the Ames test, where the activities for the polar fractions (except for Beijing) were unchanged by activation (Butler et al., 1985).

Two methods for calculating the genotoxic activities of different samples were compared: (1) the amount of material causing a significant (>3 times) increase over the background; and (2) the maximum activity, expressed as plaque-forming units (PFU) per nanogram of material. The first function can be expressed as a reciprocal x 10, so that larger numbers reflect greater activity and units are in the same numerical range as the second function (Table 5). These two values are in general agreement in ranking the activities of the extracts. The one sample that does not fit is the CX fraction from Mexico City, which gave an extremely sharp dose-response curve in a narrow dose range, causing the minimal detectable amount to be higher (and the reciprocal function to be lower) than expected.

A comparison of activities (with activation) in the Ames strain TA98 and in the Microscreen assay is shown in Figure 3. (Data for the Ames test are taken from Butler et al., 1985.) It is apparent that these two assays detect different classes of compounds with different efficiencies. The Microscreen assay is much more efficient at detecting activity in the nonpolar extracts. However, the two moderately polar fractions tested were not well detected in the Microscreen assay. The polar fractions, while detectable in both systems, represent only a small fraction of the activity seen with the nonpolar fractions in the Microscreen assay (except for Mexico City, where it had approximately 25% of the nonpolar activity).

Table 5. Comparison of Two Methods to Express Genotoxic Activity of Air Particle Extracts (+S9)

Extract		PFU/ng (maximum)	10/Minimum Detectable (μg)
BEIJ	CX	>133.0	>106.0
ELIZ	CX	>101.5	>106.0
NYC	CX	89.3	>106.0
MEX	CX	35.6	26.7
BEIJ	ACE	14.7	>40.0
MEX	ACE	9.5	32.0
NYC	ACE	9.5	40.0
ELIZ	ACE	5.4	16.0
NYC	DCM	2.1	16.0
MEX	DCM	-	-

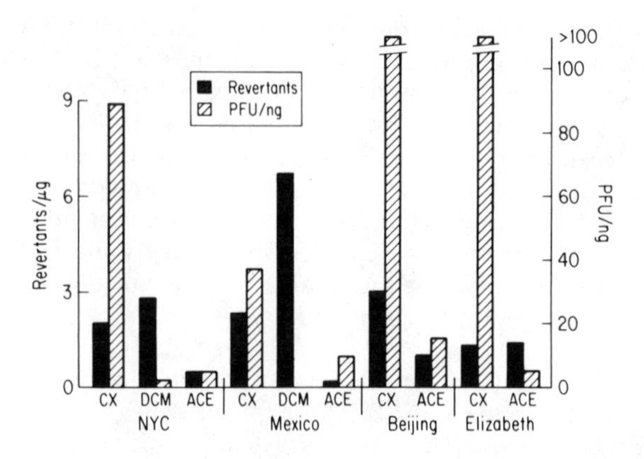

Figure 3. Assay of air particle extracts in the Ames test using TA98 (solid bars), and in the Microscreen assay (hatched bars). All assays were performed with metabolic activation.

DISCUSSION

The Microscreen assay is a short-term assay for DNA-damaging agents that utilizes one strain of *E. coli* containing λ prophage and carrying the *uvrA* mutation (which renders the strain incapable of excising "bulky lesions") and the *trpE* mutation (which allows one to score Trp+ revertants, if desired). The assay is capable of detecting agents that cause a wide variety of DNA lesions, including small group alkylations, bulky group alkylations, intercalations, cross-links, oxidative deaminations, and strand scissions (Meyer et al., 1984; Rossman et al., in preparation).

This assay has a number of features that would make it useful for workers studying complex mixtures:

1. Very small amounts of test agent are required, since the exposure is carried out in microtitre wells containing volumes of only 250 µl.

2. The maximum effect is almost always seen at subtoxic concentrations. Thus, only a few wells need to be assayed in a rapid screening procedure.

3. The assay is fast, easy to perform, and inexpensive.

Only one strain of bacteria is required for exposure to the test agent. The amount of rat liver S9 and plasticware required is greatly reduced.

Uncertainty in some of the calculations for genotoxic activity reflect two features: (1) In some cases, the number of PFU per plate was very large and could not be counted accurately. Results were therefore expressed as a minimum (Table 4, BEIJ-CX and ELIZ-CX). (2) In some cases, the wells that were assayed all contained significant increases in PFU, and no minimum concentration causing a significant increase was determined (Table 4, NYC-CX, BEIJ-CX, BEIJ-ACE, and ELIZ-CX).

A more accurate scoring of PFU per plate can be accomplished simply by making a further 10-fold dilution of the aliquot removed for the plaque assay. A more accurate determination of the minimum amount detectable can be accomplished by assaying the contents of the wells containing less test agent. These two alterations in the assay procedure can easily be incorporated into the repeat assay.

The fact that the CX extracts are particularly active in this assay (Table 5, Figure 3) suggests that perhaps the polycyclic aromatic compounds contained in this fraction are good inducers of λ prophage. We have assayed a number of pure compounds of this class and found that λ prophage induction as a genetic end point is a sensitive indicator of the polycyclic compounds. For example, less than 0.04 μg of 2AA can be detected (Rossman et al., in preparation). In an analysis of about 50 compounds, we find that the most active agents are those that are able to form cross-links (Rossman et al., in preparation; see Figure 2B). The least active compounds are those that cause small modifications in the DNA molecule, such as the base analogue 2-aminopurine and the deaminating agent sodium nitrite (Rossman et al., in preparation). Some metal compounds are also able to induce λ prophage (Rossman et al., 1984).

The Microscreen assay failed to detect genotoxic material in the Mexico City DCM fraction and gave only a barely significant positive result with the New York City DCM fraction (Table 5, Figure 3). Yet the Mexico City DCM fraction gave the largest response with the Ames tester strain TA98 (Figure 3). Two possible interpretations come to mind: (1) Some compound(s) present in the DCM fractions are frameshift mutagens in the Ames test, but fail to induce λ prophage; or (2) some compound(s) present in the DCM fractions are selectively toxic or inhibitory to bacteriophage λ, thereby masking a positive result. We tend to favor the second hypothesis, since we have not yet found any agent capable of causing reversion in *any* Ames tester strain that cannot also induce λ prophage. The second hypothesis can be easily tested by performing a mixing experiment (DCM + CX) and looking for inhibition. This experiment is planned. If the second hypothesis is confirmed, then further fractionation of the DCM fractions should separate the inhibitory agent(s) from other compounds.

The large reduction in the amount of test agent needed has some obvious advantages in using the Microscreen assay for complex mixtures. It would enable workers to screen day-to-day variations in air samples, rather than depending upon seasonal variations on composites. It would also enable assays to be conducted at former dump sites, where the amount of material may be as low as 1 μg/g soil. Since toxicity is measured directly in this assay, variations in toxicity as well as mutagenicity can be measured. Finally, this assay should be of great value in bioassay-directed fractionation procedures. Small samples from chromatography fractions can be assayed without consuming the entire fraction. A variation of a prophage induction assay has also been performed directly on top of a thin-layer chromatography plate (Elespuru and White, 1983).

ACKNOWLEDGMENTS

This study was funded by Grant No. 808482 from the U.S. Environmental Protection Agency and PHS Grant No. CA-29258 awarded by the National Cancer

Institute, DHHS, and is part of Center Programs supported by Grant ES-00260 from the National Institute of Environmental Health Sciences and Grant CA-13343 from the National Cancer Institute.

REFERENCES

Ames, B.R., J. McCann, and E. Yamasaki. 1975. Methods for detecting carcinogens and mutagens with the *Salmonella*/mammalian microsome mutagenicity test. Mutat. Res. 31:347-364.

Butler, J.P., T.J. Kneip, F. Mukai, and J.M. Daisey. 1985. Interurban variations in the mutagenic activity of the ambient aerosol and their relations to fuel use patterns. In: Short-Term Genetic Bioassays in the Analysis of Complex Environmental Mixtures IV. M.D. Waters, S.S. Sandhu, J. Lewtas, L. Claxton, G. Strauss, and S. Nesnow, eds. Plenum Press: New York.

Claxton, L.D. 1983. Characterization of automotive emissions by bacterial mutagenesis bioassay: A review. Environ. Mutagen. 5:609-631.

Daisey, J.M., T.J. Kneip, H. Hawryluk, and F. Mukai. 1980. Seasonal variations in the bacterial mutagenicity of airborne particulate organic matter in New York City. Environ. Sci. Technol. 14:1487-1490.

Daisey, J.M., T.J. Kneip, M.-X. Wang, L.-X. Ren, and W.-X. Lei. 1983. Organic and elemental composition of airborne particulate matter in Beijing, Spring 1981. Aerosol Sci. Technol. 2:407-415.

Elespuru, R.K. (in press). Induction of bacteriophage lambda by DNA-interacting chemicals. In: Chemical Mutagens. F.J. de Serres, ed. Plenum Press: New York.

Elespuru, R.K., and R.J. White. 1983. Biochemical prophage induction assay: A rapid test for antitumor agents that interact with DNA. Cancer Res. 43:2819-2830.

Elespuru, R.K., and M.B. Yarmolinsky. 1979. A colorimetric assay of lysogenic induction designed for screening potential carcinogenic and carcinostatic agents. Environ. Mutagen. 1:65-78.

Kier, L.D., E. Yamasaki, and B.N. Ames. 1974. Detection of mutagenic activity in cigarette smoke condensates. Proc. Natl. Acad. Sci. U.S.A. 71:4159-4163.

Levin, D.E., M. Hollstein, M.F. Christman, E.A. Schwiers, and B.N. Ames. 1982. A new *Salmonella* tester strain (TA102) with A·T base pairs at the site of mutation detects oxidative mutagens. Proc. Natl. Acad. Sci. U.S.A. 79:7445-7449.

Maron, D.M., and B.N. Ames. 1983. Revised methods for the *Salmonella* mutagenicity test. Mutat. Res. 113:173-215.

McCann, J., N.E. Springorn, J. Kobori, and B.N. Ames. 1975. Detection of carcinogens as mutagens: Bacterial tester strains with R factor plasmids. Proc. Natl. Acad. Sci. U.S.A. 72:979-983.

McCarroll, N.E., C.E. Piper, and B.H. Keech. 1981. An *E. coli* microsuspension assay for the detection of DNA damage induced by direct-acting agents and promutagens. Environ. Mutagen. 3:429-444.

Meyer, L.W., M. Molina, and T.G. Rossman. 1984. Microscreen: A rapid small scale system to detect multiple genetic endpoints. Abstract Cb-29, 15th Annual Meeting, Environmental Mutagen Society.

Mohn, G.R. 1981. Bacterial systems for carcinogenicity testing. Mutat. Res. 87:191-210.

Moreau, P., A. Bailone, and R. Devoret. 1976. Prophage λ induction in *Escherichia coli* K12 *envA uvrB*: A highly sensitive test for potential carcinogens. Proc. Natl. Acad. Sci. U.S.A. 73:3700-3704.

Pegg, A.E. 1983. Alkylation and subsequent repair of DNA after exposure to dimethylnitrosamine and related carcinogens. In: Reviews in Biochemical Toxicology, Vol. 5. E. Hodgson, J.R. Bend, and R.M. Philpot, eds. Elsevier: New York. pp. 83-133.

Purchase, I.F.H. 1982. An appraisal of predictive tests for carcinogenicity. Mutat. Res. 99:53-71.

Rossman, T.G., M. Molina, and L.W. Meyer. 1984. The genetic toxicology of metal compounds: I. Induction of λ prophage in *E. coli* $WP2_s(\lambda)$. Environ. Mutagen. 6:59-69

Rossman, T.G., L.W. Meyer, and M. Molina. (in preparation). The microscreen assay detects genotoxic agents of all classes tested.

de Serres, F.J., and M.D. Shelby. 1979. The *Salmonella* mutagenicity assay: Recommendations. Science 203:563-565.

Singer, B., and J.T. Kuśmierek. 1982. Chemical mutagenesis. Ann. Rev. Biochem. 52:655-693.

Skopek, T.R., H.L. Liber, J.L. Krolewski, and W.G. Thilly. 1978. Quantitative forward mutation assay in *Salmonella typhimurium* using 8-azaguanine resistance as a genetic marker. Proc. Natl. Acad. Sci. U.S.A. 76:410-414.

Talcott, R., and E. Wei. 1977. Brief communication: Airborne mutagens bioassayed in *Salmonella typhimurium*. J. Natl. Cancer Inst. 58:449-451.

USDOE. 1982. Workshop on Air Pollutants: Setting Priorities for Long Term Research Needs, Gettysburg, PA, April 19-23 (J.S. Gaffney, ed.).

DEVELOPMENT OF AN *IN SITU* TEST SYSTEM FOR DETECTION OF MUTAGENS IN THE WORKPLACE

Tong-man Ong, John D. Stewart, James D. Tucker, and Wen-Zong Whong

Division of Respiratory Disease Studies, National Institute for Occupational Safety and Health, Morgantown, West Virginia 26505

INTRODUCTION

Air pollutants in the form of particles, gases, or vapors can exist in the occupational setting or in a defined environment. Such pollutants can be caused by the end products, by-products, or materials used in industrial processes. If any of the pollutants are mutagenic, they may pose carcinogenic and genotoxic hazards to the exposed population. Mutagenic monitoring of the workplace environment can be used to determine whether workers are being exposed to mutagenic and potentially carcinogenic compounds. The results of mutagenic monitoring studies can be used to guide human genetic monitoring, epidemiology, and industrial hygiene studies. Efforts can be made immediately to reduce the exposure of workers to mutagenic compounds on a generic control basis.

For the detection of mutagenic pollutants, laboratory analysis of collected and extracted samples has been the primary method used for mutagenesis testing. Filter, impactor, precipitator, sorbent, or cryogenic trapping is conventionally used to collect airborne particles, vapors, or gases. During the processes of collection, transport, and extraction, short-lived mutagenic substances and intermediates may be lost, or artifacts may be generated. Therefore, a desirable test system for detecting airborne mutagens in the workplace would be an *in situ* (on-site) mutagenicity assay that allows for direct exposure of test organisms to conditions that more closely approximate the real environment. To date, very few systems have been developed along this line for the open air or defined environments. Among those systems, the *Tradescantia* and the corn plant assay systems show promise as field monitors for gaseous mutagens (Lower, 1981; Ma, 1982; Plewa, 1982; Schairer et al., 1978).

In this report, we describe an *in situ* assay system in which both mutagenic airborne particles and vapors or gases can be trapped simultaneously into trapping medium containing tester cells. Bacterial cells were employed for most of these studies. In a preliminary effort, human peripheral lymphocytes were used for the sister chromatid exchange (SCE) studies. The results of laboratory validation of this system with chemicals, volatile compounds, gases, and air pollutants are presented.

MATERIALS AND METHODS

Test Chemicals

Dimethylnitrosamine (DMN) and ethylene dibromide (EDB) were obtained from Sigma Chemical Co. (St. Louis, MO). Methyl methanesulfonate (MMS), ethyl methanesulfonate (EMS), 2-aminoanthracene (2AA), and 2,4,7-trinitro-9-fluorenone (TNF) were purchased from Aldrich Chemical Co. (Milwaukee, WI). 2-Methoxy-6-chloro-9-[3(2-chloroethyl)aminopylamino] acridine·2HCl (ICR-191) was donated by Dr. H.J. Creech of the Institute for Cancer Research (Fox Chase, PA). Ethylene oxide (EO) was from Matheson Gas Products (East Rutherford, NJ). Cigarettes used were common American brands obtained from a local store. Diesel exhaust was generated by a diesel engine truck (White Motor Co. Tractor 4864 with NTC 290 Detroit engine).

Bacterial Tester Strains

The two new testers TA98W and TA100W of *Salmonella typhimurium* derived from TA98 and TA100, respectively, were used. These testers were obtained by selecting strains that are resistant to streptomycin (Str^r) and 8-azaguanine (AG^r). Both parental strains carried an ampicillin resistant (Amp^r) plasmid. TA98W and TA100W were used for the Ames histidine (His) reversion assay (Ames et al., 1975). The genetic characteristics of the parental strains have been described previously (Ames et al., 1975).

Sampling Device

The apparatus designed for the *in situ* mutagenesis test system is shown in Figures 1 and 2. It consists of experimental and control sets. Each set has a pump (Model P-4000, du Pont), an impinger, and a cyclone (10-mm Dorr-Oliver cyclone, Mine Safety Appliance Co., Murraysville, PA). The impinger is a 100-ml glass flask (pear-shaped) with a fritted air bubbler containing many small holes (>80 µm in diameter). In the control set, the pump is connected to the cyclone. The filter behind the impinger was used to prevent the release of aerosols containing tester cells into the environment.

Mutagenicity Assay

Cultures of test cells were grown overnight in liquid synthetic medium. The liquid culture medium consisted of 2% glucose in water containing 25 µg histidine/ml, 0.1 µg biotin/ml, and Vogel-Bonner Medium E (Vogel and Bonner, 1956). Ten milliliters of the overnight culture diluted to 20 ml with the same medium was used as

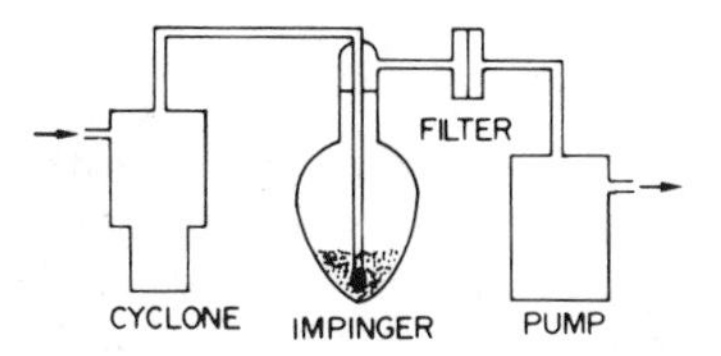

Figure 1. Device used for the *in situ* mutagenicity assay system. From Ong et al. (1984).

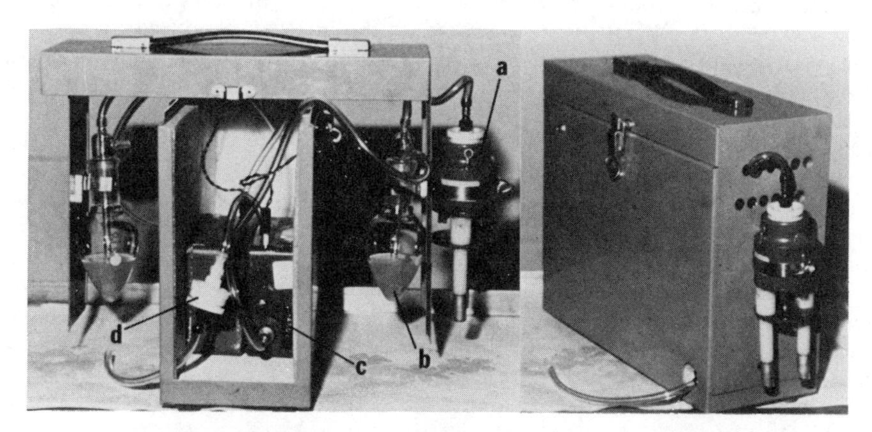

Figure 2. The trapping device for the system can be packed and operated in a small 5.25- x 11- x 14-in. carry box. The components of the device are: a, cyclone; b, impinger; c, pump; d, filter. For the control set, clean air is recirculated in the system. The pump for pulling air through the system can be operated by batteries. From Ong et al. (1984).

trapping medium in the impinging flask. The trapping medium was supplemented with 3% dimethyl sulfoxide (DMSO), 100 μg 8-azaguanine (AG)/ml, 10 μg cycloheximide (Cyc)/ml, 100 μg streptomycin (Str)/ml, and 25 μg ampicillin (Amp)/ml.

During the experiment, air was drawn into the sampling device by the pump at a flow rate of 3 liters/min. The air flow passed through the cyclone, which removed nonrespirable particles. The respirable particles ($\leq$ 10 μm) and vapors or gases along with air were impinged into the trapping medium. Samples were taken from the trapping medium periodically to determine survival and mutagenic response. For survival, 100, 500, and 2500 cells were plated separately onto three culture medium agar plates. For mutagenic response, samples (0.1 ml) were plated onto Ames minimal agar plates. All four drugs (Amp, Str, Cyc, and AG) were also incorporated into the agar plates. The addition of the drugs into the trapping and plating media was used to control microbial contamination. Survivors and His^+ revertants were scored after 2 d incubation at 37°C.

For the study with promutagens, 5 ml S9 mix and 2.5 ml of tester cells were added to dialysis tubing (0.25 in. in diameter, 6 in. in length) that was immersed into the trapping medium. Liver homogenates (male Wistar rats or Syrian golden hamsters pretreated with Aroclor-1254) and S9 mix were prepared according to Ames et al. (1975). After designated impingement times, cell samples from the dialysis tubing were plated to determine mutations. The activation ability of S9 mix at 32°C and 25°C inside the dialysis tubing with bubbling was also measured.

In the vapor trapping study, MMS, EMS, DMN, or EDB was placed in a 125-ml glass bottle that was connected to the front opening of the sampling device. The mutagen vapor was continuously generated as the air flow (produced by the pump) passed over the surface of the mutagen, and was impinged into the trapping medium containing tester cells. For testing the particle trapping, TNF was dissolved in acetone and then added to silica particles (≤ 10 µm). After the acetone was evaporated, the silica particles coated with the mutagen were used for the study. The suspended particles coated with mutagen were generated in a bottle by air flow and were trapped into the trapping medium by impingement. Tester cells in the trapping medium were used to detect mutations.

Side-stream cigarette smoke and diesel exhaust were also tested for mutagenic activity with the *in situ* system. Smoke or diesel exhaust was pulled by the pump into the trapping medium. For the *in vitro* microsomal activation, S9 mix was incorporated into the cigarette smoke experiment. The control set for cigarette smoke or diesel exhaust experiment was run at the same time as the experimental set. In the control set, clean air was recirculated inside the system.

Sister Chromatid Exchange Assay

Heparinized human peripheral lymphocytes from healthy unrelated nonsmokers were used in this study. Whole blood (0.6 ml per 10 ml media) was cultured and treated in the dark at 37°C in RPMI-1640 culture medium supplemented with 15% fetal bovine serum (FBS), L-glutamine (final concentration, 2 mM), 1% penicillin-streptomycin (GIBCO), 0.1 ml phytohemagglutinin (PHA, GIBCO), and 25 µM bromodeoxyuridine (BrdU). Treatment began 22-26 h after culture initiation by transferring the culture contents into sterilized dialysis tubing. This was then placed in a pear-shaped flask containing 20 ml of RPMI-1640 culture medium supplemented as described. Because the CO_2 buffer in RPMI-1640 is rapidly removed by air bubbling, 0.4 ml of 7.5 M phosphate buffer was also added to each flask, and the pH was adjusted to 7.4. Water loss due to evaporation was controlled by careful addition of distilled water as needed. FBS could not be added to the exposure flask because of frothing. The air flow rate was maintained at 1.2 liters/min. After treatment, the cells were washed twice in Hank's balanced salt solution, resuspended in media supplemented as described, and grown for the remainder of the 75-h culture period. Colcemid (0.1 µg/ml) was added for the final 3 h of incubation. The cells were swollen with 0.075 M KCl, washed three times in fresh fixative (methanol:acetic acid, 3:1), dropped onto slides, and air dried. Differential staining was achieved by a modified fluorescent plus Giemsa technique (Perry and Wolff, 1974; Goto et al., 1978). Twenty-five cells were scored per person per dose.

RESULTS

Mutagenic sensitivities to 2AA and ICR-191 between the new tester strains and their parent strains were compared as shown in Table 1. The mutagenic response of the new testers was similar to their corresponding parental strains, indicating that the new testers possess mutagenic sensitivity comparable to their parental strains.

The developed *in situ* microbial system was evaluated with the new tester strains using the volatile mutagens MMS, EMS, and EDB, and the gaseous mutagen EO. Tables 2 and 3 show the mutagenic response of TA100W to the vapor of MMS, EMS, EDB, and EO. A clear mutagenic activity of all four compounds was demonstrated with this system. The mutagenicity increased as the impingement time increased. For the study of trapping efficiency with the impingement, the percent trapping of MMS vapor in trapping medium was determined. By comparing the mutagenic activities obtained from trapping medium to a standard mutagenicity curve of known MMS concentration, recovery of MMS by impingement was found to be between 62 to 77% and was reproducible in three independent experiments (data not shown).

The respirable silica particles coated with TNF were used for the particle trapping experiments. After 1 h trapping, the number of revertants per plate from the

Table 1. Comparison of Mutagenic Response Between Multi-Drug Resistant Strains and Their Parental Strains to 2-Aminoanthracene and ICR-191[a]

		His+ Revertants/Plate			
Mutagen	Concentration (μg/plate)	TA98	TA98W	TA100	TA100W
2AA	0	37	42	103	139
	0.312	194	204	327	299
	0.625	376	351	568	675
	1.25	1094	843	1849	1736
	2.5	2374	2413	2614	2397
ICR-191	0	23	29	144	165
	1.25	186	204	251	218
	2.5	423	387	519	473
	5	607	586	842	813

[a]The plate incorporation test was employed for this study. S9 from rat liver was used for the 2AA test. Data from Whong et al. (1984).

Table 2. Mutagenic Response of TA100W to the Vapor of Methyl Methanesulfonate and Ethyl Methanesulfonate with Impingement[a]

		Control		Experimental	
Mutagen	Impingement Time (h)	Percent Survival	His^+ Revertants/10^7 Survivors	Percent Survival	His^+ Revertants/10^7 Survivors
MMS	0.5	100	11.0	96.0	22.4
	1	100	16.8	92.0	64.7
	2	100	12.2	95.5	134.5
	4	100	17.0	51.0	1056.7
EMS	3	100	10.0	100.0	10.1
	6	100	11.4	78.8	80.3
	9	100	10.6	92.6	145.8
	12	100	9.3	98.4	218.9

[a]Mutagen vapor was generated from a bottle containing 0.3 ml MMS or 0.5 ml EMS by pulling air with a pump at flow rate of 3 liters/min and was impinged into the trapping medium containing TA100W. Data from Whong et al. (1984).

experimental set was 20 times higher than that from the control set (Table 4). Increase in impingement time resulted in increasing mutagenic activity. Previous experiments have indicated that silica alone is not mutagenic (data not shown).

Since S9 mix in dialysis tubing was used in this system to detect promutagens, the stability of S9 mix inside the dialysis tubing was determined with the Ames assay. The results show that the activation ability of S9 remained unaffected even after 6 h impingement (Table 5). Changing the temperature from 32°C to 25°C does not seem to affect the S9 activation (data not shown). With the activation system, the mutagenicity of DMN vapors, which were trapped by impingement, was detected in TA100W after 0.5 h impingement (Table 6). The mutagenicity of DMN in TA100W depended on the impingement time.

In the side-stream smoke experiment, the *in situ* system appears to detect the mutagenic activity of cigarette smoke at a low concentration in ambient air. After 4 h of sampling, the number of revertants increased from 5 in the control sets to 19 in the experimental sets (Table 7). Based on the room size (534.5 ft^3), number of cigarettes (5) burned, the air flow rate (3 liters/min), and the sampling time (4 h), the concentration

Table 3. Mutagenicity of Ethylene Oxide and Ethylene Dibromide in *S. typhimurium* TA100W[a]

Chemical	Impingement Time	Revertants/Plate	
		Control	Experimental
EDB	1 (h)	71	106
	2	62	163
	4	70	264
	6	64	360
EO	5 (min)	67	184
	10	66	346
	20	75	451
	40	75	532

[a]Tester cells were exposed to EO (30 times diluted with air) or EDB (0.15%) vapors at a flow rate of 1.5 liters/min. Results are from two independent experiments.

Table 4. Mutagenicity of 2,4,7-Trinitro-9-fluorenone Coated on Silica Particles Trapped by Impingement[a]

Impingement Time (h)	TA98W Revertants/Plate	
	Control	Experimental
1	13	277
2	19	1809
4	23	3070

[a]1 ml TNF (2 mg/ml) was spiked to 1 g silica particles (<10 μm). 200 mg silica particles was used for trapping. Data from Whong et al. (1984).

Table 5. Mutagenicity of 2-Aminoanthracene Activated by S9 Mix from Dialysis Tubing After Impingement[a]

	TA98W Revertants/Plate	
Impingement Time (h)	Control	Experimental
0	23	1898
1	23	1922
2	22	2050
4	17	1850
6	18	1707

[a]0.5 ml S9 (rat liver) mix from dialysis tubing was sampled during impingement and was used to activate 2AA (2.5 µg/plate) with the plate incorporation test. Data from Whong et al. (1984).

of cigarette smoke that could be detected for the mutagenic activity was determined to be 0.0065 cigarette per milliliter. Increasing the impinging time increased the number of revertants in a time-dependent manner.

The results of diesel emission experiments are shown in Table 8. Since diesel exhaust emission was very toxic and weakly mutagenic to *S. typhimurium*, mutagenicity was determined as reversion frequency related to cell survival. An increase in reversions induced by diesel emission was observed after 0.5 h exposure. The number of reversions increased as impingement time increased. A more than 6-fold increase over control was observed after 4 h impingement.

Table 9 summarizes the results of the SCE assay with human lymphocytes. A clear dose response was seen in both individuals after lymphocytes were exposed to EDB vapor by the impingement. The SCE frequencies were almost double after the cells were exposed for 160 min.

DISCUSSION

Since mutagenic air pollutants may exist in the environment at relatively low quantities, a suitable *in situ* mutagenicity system must fulfill two important criteria: (1) highly sensitive indicator cells that can detect mutagens at very low concentrations, and (2) efficient systems that collect representative whole air samples and promote efficient contact with the tester cells.

Table 6. Mutagenicity of Dimethylnitrosamine Activated by S9 Mix Inside Dialysis Tubing Using *S. typhimurium* TA100W[a]

Impingement Time (h)	His+ Revertants/Plate		DMN Evaporated (mg)
	Control	Experimental	
0.5	103	468	35.0
1.0	176	1743	76.2
2.0	151	2396	139.2
4.0	224	2509	243.6

[a]DMN vapor was generated and impinged at 25°C by drawing air with a pump. Mutagenic activity was determined by sampling tester cells from the dialysis tubing containing S9 (hamster liver) and tester cells. Data from Whong et al. (1984).

Table 7. Detection of the Mutagenic Activity of Cigarette Smoke in a Controlled Environment

Impingement Time (h)	His+ Revertants/Plate[a]			
	Control Set		Experimental Set	
	Exp. 1	Exp. 2	Exp. 1	Exp. 2
2	7	7	10	10
4	6	3	16	22
6	4	4	30	22
8	3	3	44	28

[a]Each experiment was conducted in duplicate. TA98W was tested with rat S9. Data from Ong et al. (1984).

Table 8. Mutagenicity of Diesel Emission with the *In Situ* Assay System[a]

Impingement Time (h)	Control		Experimental	
	Percent Survival	Revertants/ 10^8 Survival	Percent Survival	Revertants/ 10^8 Survival
0.5	100	14.6	96.8	24.6
1.0	100	14.8	65.6	34.5
2.0	100	13.5	37.1	84.7

[a]Results are averages of four independent experiments. TA98W was tested without S9 activation. The trapping apparatus was set 6 ft apart from the rear of a diesel engine truck exhaust pipe.

A comparison of mutagenic sensitivities between the new bacterial testers and their corresponding parental strains indicated that the new testers possess high sensitivity and, therefore, are suitable for use in the *in situ* system. Results of the mutagenic activity of MMS, EMS, DMN, and EDB vapors and EO gas trapped by the system demonstrate that the trapping efficiency by impingement is satisfactory. The fritted air bubbler used in impingement generates many small bubbles in the trapping medium, which may help increase the trapping efficiency. The results for the TNF-coated silica particles indicate that respirable particles can be efficiently trapped by impingement and mutagenic material can be released readily into the trapping medium from the particle. The uniqueness of this system is that both mutagenic airborne particles and vapors can be trapped simultaneously into one device. Furthermore, by the use of multi-drug resistant bacterial testers and presence of Str, Amp, Cyc, and AG in the trapping and plating media, the entrapment of airborne mutagens from unfiltered ambient air without microbial contamination becomes possible.

It appears that during the impingement the activation ability of S9 can be destroyed rapidly, and a flask full of foam can be generated from the trapping medium if S9 is added directly into the trapping medium. In our study, the use of dialysis tubing to confine S9 mix from trapping medium effectively avoided these technical problems. Prolonged stability of S9 activation inside the dialysis tubing at room temperature suggests that this system can detect promutagens. The dialysis tubing with a 12,000 molecular weight cutoff may sufficiently retain microsomal enzymes inside the tubing and allow mutagenic molecules to diffuse in or out of the tubing. Satisfactory activation of promutagens to mutagenic reactants by S9 in dialysis tubing was demonstrated with 2AA (data not shown), DMN, and cigarette smoke.

Table 9. Sister Chromatid Exchanges Induced by Ethylene Dibromide

Treatment Time (min)	Donor	Number of SCE's	SCE's per Chromosome	SCE's/Cell ± S.E.
0 (control)	D	230	0.201	9.20 ± 0.56
	B	235	0.205	9.40 ± 0.73
5	D	265	0.232	10.60 ± 0.81
	B	261	0.228	10.44 ± 0.60
16	D	310	0.271	12.40[b] ± 1.11
	B	272	0.237	10.88 ± 0.88
48	D	379	0.333	15.16[c] ± 1.00
	B	368	0.320	14.72[c] ± 0.79
160	D	449	0.391	17.96[c] ± 1.08
	B	407	0.356	16.28[c] ± 1.04

[a]Cells from the two donors were treated on separate occasions. The concentration of EDB was 245 and 418 ppm for donors D and B, respectively.
[b]$p < 0.01$, t-test performed on square root transformed data.
[c]$p < 0.0001$.

Addition of DMSO (3%) in the trapping medium does not affect the viability of tester cells, whereas it may actually help to retain volatile vapors and to dissolve nonpolar mutagens from trapped particles. In field studies, however, it is conceivable that the system described here may encounter problems such as insufficient entrapment of mutagenic gases and possible inability to detect mutagens that are strongly bound to particles. This would necessitate improvement of the system. Nevertheless, the results from laboratory studies and the studies with side-stream cigarette smoke and diesel exhaust suggest that this *in situ* microbial mutagenicity test system involving impingement is potentially useful for the on-site detection of airborne mutagens in the workplace. The results of preliminary SCE studies with human lymphocytes indicate that this system may be used for cultured mammalian cells and for other genetic end points.

ACKNOWLEDGMENT

The authors would like to thank Professor B.N. Ames for generously providing *S. typhimurium* TA98 and TA100.

REFERENCES

Ames, B.N., J. McCann, and E. Yamasaki. 1975. Method for detecting carcinogens and mutagens with the Salmonella/mammalian-microsome mutagenicity test. Mutat. Res. 31:347-364.

Goto, K., S. Maeda, Y. Kano, and T. Sugiyama. 1978. Factors involved in differential Giemsa-staining of sister chromatids. Chromosoma 66:351-359.

Lower, W.R. 1981. Mutagenic effects of petro chemical complexes on *Zea mays*, Tradescantia, and Salmonella. Environ. Mutagen. 3:400.

Ma, T-H. 1982. Tradescantia cytogenetic tests (root-tip mitosis, pollen mitosis, pollen mother-cell meiosis): A report of the U.S. Environmental Protection Agency Gene-Tox Program. Mutat. Res. 99:293-302.

Ong, T., J. Stewart, and W-Z. Whong. 1984. A simple *in situ* mutagenicity test system for detection of mutagenic air pollutants. Mutat. Res. 139:177-181.

Perry, P., and S. Wolff. 1974. New Giemsa method for the differential staining of sister chromatids. Nature 251:156-158.

Plewa, M.J. 1982. Specific-locus mutation assay in *Zea mays*: A report of the U.S. Environmental Protection Agency Gene-Tox Program. Mutat. Res. 99:317-337.

Schairer, L.A., J. Van't Hof, C.G. Hayes, R.M. Burton, and F.J. de Serres. 1978. Exploratory monitoring of air pollutants for mutagenicity activity with the Tradescantia stamen hair system. Environ. Health Perspect. 27:51-60.

Vogel, H.J., and D.M. Bonner. 1956. Acetylornithinase of *E. coli*, partial purification and some properties. J. Biol. Chem. 218:97-106.

Whong, W-Z., J.D. Stewart, and T. Ong. 1984. Development of an *in situ* microbial mutagenicity test system for airborne workplace mutagens: Laboratory evaluation. Mutat. Res. 130:45-51.

PERTURBATIONS OF ENZYMIC INITIATION OF EXCISION REPAIR DUE TO GUANINE MODIFICATION IN DNA

Nahum J. Duker

Department of Pathology and Fels Research Institute, Temple University School of Medicine, Philadelphia, Pennsylvania 19140

INTRODUCTION

Complex mixtures of genotoxic agents exist in the environment, and therefore it is important to analyze the consequences of such exposures. Excision repair of DNA base alterations is initiated by DNA glycosylases, each of which releases a specific damaged base by cleaving the glycosylic bond linking it to the sugar. The DNA is nicked at the resultant base loss site by an apurinic/apyrimidinic site endonuclease, followed by further endonucleolytic or exonucleolytic degradation of the DNA, resynthesis of the excised portion by DNA polymerase, and closure of the last phosphodiester bond by DNA ligase (Lindahl, 1979).

The use of assays of DNA repair synthesis by mammalian cells as a means of detection of putative mutagens entails measurement of DNA polymerase activity. This system detects primary DNA damage as incorporation of (^{3}H)thymidine and is evaluated as unscheduled DNA synthesis. This system has been used with hepatocytes capable of activating chemical carcinogens to their active forms and therefore can detect environmental precursors of mutagens (Casciano, 1979, Williams, 1977). Uptake of (^{3}H)thymidine is measured by liquid scintillation counting or by autoradiography (Williams, 1980). The latter assay can measure both the amount of repair synthesis per cell and the proportion of cells performing DNA repair, enhancing its usefulness in the detection of weak carcinogens (Casciano and Gaylor, 1983). It has been used to detect *in vivo* DNA damage to mouse germ cells (Sega and Sotomayor, 1982). However, because this test, in all its variations, is in essence the assay of DNA polymerase activity, its utility is dependent on proper initiation of the excision repair pathway by DNA glycosylases.

The problem of interaction between the DNA glycosylases that initiate excision repair and their substrates is therefore a vital one. Recognition of DNA modifications by repair enzymes is the *sine qua non* in prevention of permanent genomic alterations. It is therefore imperative to determine which types of DNA damage are recognized and repaired and which ones are unrecognized, remain uncorrected, and whose persistence may result in mutagenesis or carcinogenesis. Even when the enzymes are well characterized, the relationship to *in vivo* DNA repair is not elucidated thereby. The environment often contains several agents that may damage DNA independently, and an organism may be exposed to several of them simultaneously. Determination of the modes by which different genotoxins interact to influence DNA repair is central to evaluation of proper environmental levels of such chemicals.

In order to eliminate sources of uncertainty that necessarily exist in the study of a complex pathway, the activities of two purified repair enzymes against well-characterized substrates containing more than one type of DNA modification were investigated. DNA was reacted with the ultimate carcinogen N-acetoxy-N-2-acetylaminofluorene (AAAF). This introduced the covalent adduct N-(deoxyguanosin-8-yl)acetylaminofluorene (dGuo-AAF) into substrate molecules. The effects of this moiety on two enzymes that initiate DNA base excision repair, uracil-DNA glycosylase and pyrimidine dimer-DNA glycosylase, were assayed against substrates containing either uracil or pyrimidine dimers, respectively. Both enzyme activities were reduced by the presence of dGuo-AAF in their substrates. Therefore, interference with the enzymic initiation of excision repair of other damaged moieties may be an indirect mechanism of mutagenesis by carcinogen-DNA adducts. These results raise the possibility of false negative results when DNA repair synthesis is used to detect a mixture of diverse environmental mutagens.

MATERIALS AND METHODS

DNA Preparations and Modifications

Phage PBS2 was grown in *B. subtilis* and its DNA was purified and stored according to Lindahl et al. (1977). PBS2 DNA, which contains uracil in place of thymine and is substrate for uracil-DNA glycosylase, was radiolabeled by addition of (^{3}H)uridine or (^{14}C)guanosine as previously described (Duker et al., 1982). The DNA was reacted for 20 min with AAAF (obtained from the National Cancer Institute Repository, Bethesda, MD) and the carcinogen was extracted as previously described (Duker and Hart, 1984). The substrate for pyrimidine dimer-DNA glycosylase was (^{3}H)thymidine-labeled *E. coli* DNA prepared according to Frenkel et al (1981) It was reacted with AAAF according to Santella et al. (1981) and then irradiated by 254-nm light in the presence of the photosensitizer silver nitrate (Radany and Friedberg, 1981) using the apparatus previously described (Duker et al., 1981). This treatment yielded a substrate DNA with 15% of the thymine radioactivity as dimers.

Analysis of DNA Modifications

The extent of dGuo-AAF in (^{14}C)guanosine-labeled PBS2 DNA was determined as previously described (Duker and Hart, 1984). The dGuo-AAF content of *E. coli* DNA was estimated before irradiation according to Fuchs and Daune (1972). Thymine dimers were assayed according to Carrier and Setlow (1971).

Enzymology

Uracil-DNA glycosylase was purified from *B. subtilis* according to Cone et al. (1977). The enzyme assays, including reaction conditions, identification of the released product, and analysis of the Lineweaver-Burk plot, were performed as previously described (Duker et al., 1981; Duker et al., 1982). Pyrimidine dimer-DNA glycosylase was purified according to Friedberg et al. (1980). The enzyme reaction conditions were those of Radany and Friedberg (1980). Activity was assayed by release of free thymine following photoreversal of substrate DNA (Radany and Friedberg, 1980).

RESULTS

The activity of *B. subtilis* uracil-DNA glycosylase is shown in Figure 1. The damaged DNA contained 1.4% of the guanines as dGuo-AAF. This alteration decreased the rate of enzymic uracil release by 35% as compared with the control undamaged PBS2 DNA. None of the modified purines were released from the DNA. The V_{max} was reduced from 1.07 ± 0.06 nmol of uracil released/min/ml of enzyme for the control DNA to 0.70 ± 0.04 for the arylamidated substrate. There was no significant alteration of the apparent K_m, which was 3.47 ± 0.89 x 10^{-7} M for dUMP in DNA for the control and 3.83 ± 0.91 x 10^{-7} M for the AAAF-reacted substrate.

The effects of dGuo-AAF in DNA on pyrimidine dimer-DNA glycosylase were assayed. Figure 2 shows a marked decrease in enzyme activity when guanines in the substrate were modified by AAAF. This reduction was not due to a decreased DNA thymine dimer content. Assays using paper chromatography showed no significant

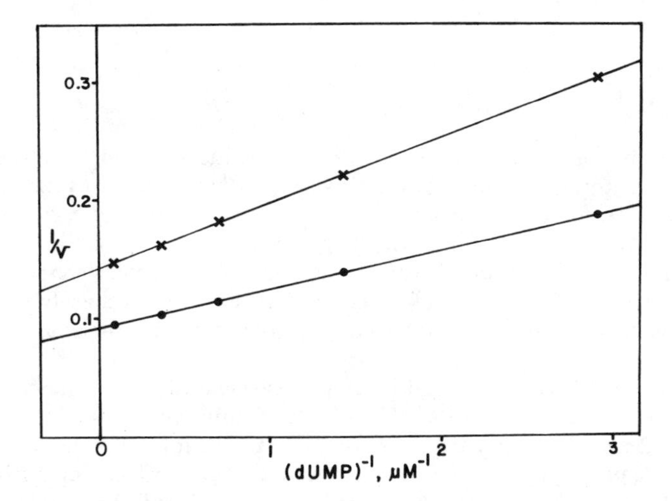

Figure 1. Double-reciprocal plot of *B. subtilis* uracil-DNA glycosylase activity with AAAF-reacted (x) and control (•) PBS2 DNA. The reaction time was 10 min. Each point represents the average of four determinations.

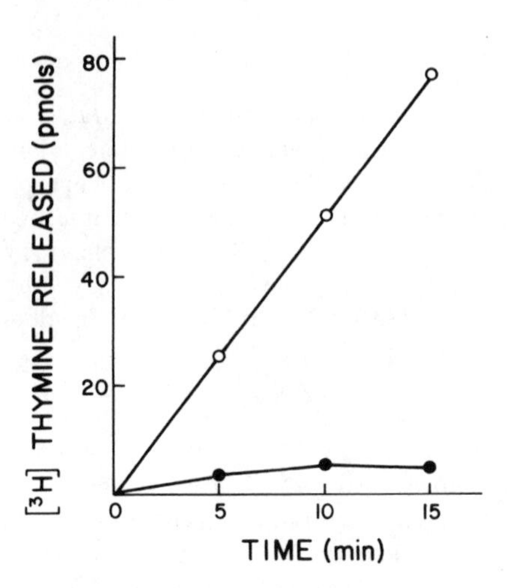

Figure 2. Release of free thymine from DNA containing thymine dimers by pyrimidine dimer-DNA glycosylase. 5 µl of purified enzyme was reacted with 2.4 µg of substrate containing dGuo-AAF (●) and control DNA (o). Each point represents the average of two determinations.

reduction of dimer yield when arylamidated DNA was photosensitized and irradiated by 254-nm light (data not shown). The dGuo-AAF content of the modified DNA was 16% of the guanines. Incision of thymine dimers was linear with time of enzyme reaction for the control DNA, but was barely detectable with the AAAF-reacted substrate.

DISCUSSION

The presence of uracil in DNA may result from incorporation in place of thymine during DNA synthesis or from hydrolytic deamination of DNA cytosines (Lindahl, 1979). Transition mutations result if such deaminated cytosines remain unrepaired (Duncan and Miller, 1980). Such uracils are removed by uracil-DNA glycosylase. Deamination of cytosine to uracil can occur under physiological conditions (Lindahl and Nyberg, 1974). The activity of uracil-DNA glycosylase is therefore necessary to preserve genomic integrity, and its reduction can result in mutagenesis.

The activity of *B. subtilis* uracil-DNA glycosylase is altered by modification of the other, nonsubstrate bases in DNA. The effects of different types of modifications on rate of enzymic uracil excision are shown in Table 1. The inhibition index is the ratio of the V_{max} of the glycosylase with control DNA to its V_{max} with the modified substrate. Therefore, the index increases as the latter V_{max} is decreased. The agents causing the greatest increases in the inhibition index are apurinic sites and 8-(2-hydroxy-2-propyl)purines (Duker et al., 1982). The presence of 7-methylguanine or 5-azacytosine, which cause few deformities or local denaturations in DNA, do not

Table 1. Inhibition of Uracil-DNA Glycosylase Activity by DNA Damage

Damaged Adducts		
Number per PBS2 DNA Molecule	Type	Inhibition Index
600	8-(2-Hydroxy-2-propyl)purine	1.2[a]
1540	8-(2-Hydroxy-2-propyl)purine	3.2[a]
3200	7-Methylguanine	1.1[a]
1600/1600	7-Methylguanine/apurinic site	2.5[a]
1000	5-Azacytosine	1.0[b]
1200	dGuo-AAF	1.5
2200	dGuo-AAF	1.5[c]
9000	Uracil dimer	1.9[d]

[a]Duker et al., 1982.
[b]Chao and Duker, in press.
[c]Duker and Hart, 1984.
[d]Duker et al., 1981.

affect uracil-DNA glycosylase activity (Duker et al., 1982; Chao and Duker, in press). Uracil dimers and dGuo-AAF are intermediate in their inhibition of uracil-DNA glycosylase.

Since a number of different types of DNA alterations do interfere with uracil-DNA glycosylase activity, it becomes important to determine if such inhibition is restricted to enzymic uracil excision or if other glycosylases are similarly affected. Therefore, the effects of dGuo-AAF on pyrimidine dimer-DNA glycosylase were examined. This enzyme cleaves the glycosylic bond between the 5' pyrimidine of the photodimer and the corresponding deoxyribose (Gordon and Haseltine, 1980). An endonuclease that hydrolyzes the phosphodiester bond at the resultant apyrimidinic site is also present on the same molecule (Nakabeppu et al., 1982). Because the glycosylase hydrolyzes only one of the glycosylic bonds linking the dimer to the sugars, the two adjacent thymines remain linked to the DNA only by the cyclobutane ring between them. Monomerization of the dimer by subsequent photoreversal at 254 nm then results in release of the 5' thymine from the DNA (Radany and Friedberg, 1980). This assay is specific for the glycosylase activity and avoids the uncertainties obtained with other measurements of this enzyme. It is apparent that this activity is reduced by DNA dGuo-AAF, although quantitative comparisons with inhibition of uracil excision cannot be made.

This work shows that covalently bound dGuo-AAF in DNA reduces enzymic excision of two unrelated forms of DNA moieties, uracil and pyrimidine dimers. Such interference with the enzymic initiation of DNA excision repair of unrelated types of DNA damaged sites may be an indirect mechanism of mutagenesis by carcinogen-DNA adducts. Left unrepaired, uracil in DNA is mutagenic (Duncan and Miller. 1980) and the pyrimidine dimer is carcinogenic (Hart et al., 1977). This suggests that complex mixtures of environmental mutagens can exert their biological effects by hindrance of excision repair of chemically and etiogenically unrelated types of modified DNA bases.

These results indicate that difficulties may be expected if DNA repair synthesis is used to detect a mixture of environmental genotoxins. Inhibition of the initiation of DNA excision repair will result in a decreased incorporation of (^{3}H)thymidine during the resultant non-semiconservative synthesis. This would result in a false negative result, and the mutagenic activity of the mixture would remain undetected. This would also result if the modified DNA sites were not recognized by repair glycosylases. Sites of DNA damage can reduce DNA polymerase activity by blockage of the enzyme as well. Sequence studies have demonstrated blockage of a number of DNA polymerases at or adjacent to dGuo-AAF adducts (Moore et al., 1981; Moore and Strauss, 1979). This would also reduce repair synthesis in cultured cells. Finally, repair enzymes themselves might be inactivated by reactions with the mutagens (Park et al., 1981). All these pitfalls might be encountered in the application of measurement of DNA repair synthesis to detect a mixture of environmental mutagens.

The danger of false negative results can be minimized in a number of ways. Careful purification of the putative mutagen will avoid the danger of an unknown agent causing a second set of DNA lesions that might impede repair of the major sites of genomic alterations. Examination of the chemical structure of the agent will often indicate if it can produce the types of DNA distortions associated with interference with repair glycosylases (Table 1). However, because one form of damage can interfere with the repair of another, caution should be exercised in the application of cellular DNA repair synthesis to detect a mixture of mutagens. Therefore, it appears that the chief application for this test is the detection of DNA-damaging activity of purified premutagens and mutagens.

ACKNOWLEDGMENTS

The technical assistance of Ms. Donna M. Hart and Mr. George W. Merkel is gratefully acknowledged. This work was supported by U.S. Public Health Service Grants ES-02935 from the National Institute of Environmental Health Sciences and CA-12923 from the National Cancer Institute. The author is the recipient of Public

Health Service Research Career Development Award CA-00796 from the National Cancer Institute.

REFERENCES

Carrier, W.L., and R.B. Setlow. 1971. The excision of pyrimidine dimers (the detection of dimers in small amounts). Methods Enzymol. 21:230-237.

Casciano, D.A. 1979. Use of isolated rodent hepatocytes to evaluate potential premutagens and precarcinogens. Banbury Rep. 2:125-131.

Casciano, D.A., and D.W. Gaylor. 1983. Statistical criteria for evaluating chemicals as positive or negative in the hepatocyte/DNA repair assay. Mutat. Res. 122:81-86.

Chao, T.L., and N.J. Duker. (in press). Effects of 5-azacytosine in DNA on enzymic uracil excision. Mutat. Res.

Cone, R., J. Duncan, L. Hamilton, and E.C. Friedberg. 1977. Partial purification and characterization of a uracil-DNA glycosylase from *Bacillus subtilis*. Biochemistry 16:3194-3201.

Duker, N.J., and D.M. Hart. 1984. Perturbations of enzymic uracil excision due to guanine modifications in DNA. Cancer Res. 44:602-604.

Duker, N.J., W.A. Davies, and D.M. Hart. 1981. Alteration of uracil-DNA glycosylase activity by pyrimidine dimers in DNA. Photochem. Photobiol. 34:191-195.

Duker, N.J., D.E. Jensen, D.M. Hart, and D.E. Fishbein. 1982. Perturbations of enzymic uracil excision due to purine damage in DNA. Proc. Natl. Acad. Sci. U.S.A. 79:4878-4882.

Duncan, B.K., and J.H. Miller. 1980. Mutagenic deamination of cytosine residues in DNA. Nature 287:560-561.

Frenkel, K., M.S. Goldstein, N.J. Duker, and G.W. Teebor. 1981. Identification of the *cis*-thymine glycol moiety in oxidized deoxyribonucleic acid. Biochemistry 20:750-754.

Friedberg, E.C., A.K. Ganesan, and P.C. Seawell. 1980. Purification and properties of a pyrimidine dimer-specific endonuclease from *E. coli* infected with bacteriophage T4. Methods Enzymol. 65:191-201.

Fuchs, R., and M. Daune. 1972. Physical studies on deoxyribonucleic acid after covalent binding of a carcinogen. Biochemistry 11:2659-2666.

Gordon, L.K., and W.A. Haseltine. 1980. Comparison of the cleavage of pyrimidine dimers by the bacteriophage T4 and *Micrococcus luteus* UV-specific endonucleases. J. Biol. Chem. 255:12047-12050.

Hart, R.W., R.B. Setlow, and A.D. Woodhead. 1977. Evidence that pyrimidine dimers in DNA can give rise to tumors. Proc. Natl. Acad. Sci. U.S.A. 74:5574-5578.

Lindahl, T. 1979. DNA glycosylases, endonucleases for apurinic/apyrimidinic sites, and base excision repair. Prog. Nucleic Acids Res. Mol. Biol. 22:135-192.

Lindahl, T., and B. Nyberg. 1974. Heat-induced deamination of cytosine residues in deoxyribonucleic acid. Biochemistry 13:3405-3410.

Lindahl, T., S. Ljungquist, W. Siegert, B. Nyberg, and B. Sperens. 1977. DNA N-glycosidases: properties of uracil-DNA glycosidase from *Escherichia coli*. J. Biol. Chem. 252:3286-3294.

Moore, P., and B.S. Strauss. 1979. Sites of inhibition of *in vitro* DNA synthesis in carcinogen and UV-treated ΦX174 DNA. Nature 278:664-666.

Moore, P.D., K.K. Bose, S.D. Rabkin, and B.S. Strauss. 1981. Sites of termination of *in vitro* DNA synthesis on ultraviolet- and N-acetylaminofluorene-treated ΦX174 templates by prokaryotic and eukaryotic DNA polymerases. Proc. Natl. Acad. Sci. U.S.A. 78:110-114.

Nakabeppu, Y., K. Yamashita, and M. Sekiguchi. 1982. Purification and characterization of normal and mutant forms of T4 endonuclease V. J. Biol. Chem. 257:2556-2562.

Park, S.D., K.H. Choi, S.W. Hong, and J.E. Cleaver. 1981. Inhibition of excision-repair of ultraviolet damage in human cells by exposure to methylmethanesulfonate. Mutat. Res. 82:365-371.

Radany, E.H., and E.C. Friedberg. 1980. A pyrimidine dimer-DNA glycosylase activity associated with the *V* gene product of bacteriophage T4. Nature 286:182-185.

Santella, R.M., D. Grunberger, I.B. Weinstein, and A. Rich. 1981. Induction of the Z conformation in poly(dG-dC):poly(dG-dC) by binding of N-2-acetylaminofluorene to guanine residues. Proc. Natl. Acad. Sci. U.S.A. 78:1451-1455.

Sega, G.A., and R.E. Sotomayor. 1982. Unscheduled DNA synthesis in mammalian germ cells--its potential use in mutagenicity testing. Chemical Mutagens 7:421-445.

Williams, G.M. 1977. Detection of chemical carcinogens by unscheduled DNA synthesis in rat liver primary cell cultures. Cancer Res. 37:1845-1851.

Williams, G.M. 1980. The detection of chemical mutagens/carcinogens by DNA repair and mutagenesis in liver cultures. Chemical Mutagens 6:61-107.

PLANT GENETIC ASSAYS TO EVALUATE COMPLEX ENVIRONMENTAL MIXTURES

Michael J. Plewa

Institute for Environmental Studies, Department of Agronomy and Department of Genetics and Development, University of Illinois, Urbana, Illinois 61801

INTRODUCTION

Higher plants possess a variety of well defined genetic end points that include the quantitative detection of chromosome damage and point mutations. Plant assays are the only higher eukaryotic system currently employed as *in situ* monitors of polluted air (Lower et al., 1978; Ma, 1981; Schairer and Sautkulis, 1982), polluted water (Heartlein et al., 1981; DeMarini et al., 1982; Klekowski, 1978), agricultural chemicals (Plewa and Wagner, 1981; Gentile et al., 1982; Plewa et al., 1984), municipal sewage sludges (Hopke et al., 1982; Plewa and Hopke, 1983; Hopke et al., in press), and nuclear power plants (Ichikawa, 1981). Review papers on the use of plant systems in genetic toxicology are available (Nilan, 1978; Klekowski, 1982).

An environmental mutagen, or genotoxin, is a physical or chemical agent released into the environment that can alter the genome or its proper functioning. The presence of genotoxic agents in the environment constitutes a serious threat to the public health (Crow and Abrahamson, 1982). The purpose of this paper is to explore the use of two higher plants, *Zea mays* and *Tradescantia paludosa*, in the analysis of mutagenic properties of complex environmental mixtures.

IN SITU ANALYSIS USING MAIZE

Pesticides

A compehensive analysis of the mutagenic properties of pesticides used in commercial corn (maize) production in the United States was conducted under the auspices of the U.S. Environmental Protection Agency (EPA) (Plewa and Gentile, 1983; Gentile et al., 1982; Plewa et al., 1984). Twenty-one pesticides and twelve combinations of herbicides were analyzed. Three genetic assays were employed: reverse mutation in *Salmonella typhimurium*, gene conversion in *Saccharomyces cerevisiae*, and reverse mutation in *Z. mays*. The use of the genetic indicator organisms established a comprehensive data base that resolved a spectrum of genetic damage. The pesticides were evaluated with the microbial assays directly, after *in vitro* mammalian microsomal activation, and after *in vivo* plant activation techniques. All of the pesticides were evaluated for their ability to induce mutation in maize under *in situ* modern agricultural conditions. The analysis of pesticides under field conditions constitutes the study of a complex (soil-pesticide) mixture.

The insecticides that were evaluated included carbofuran (2,3-dihydro-2,2-dimethylbenzofuran-7-yl methylcarbamate), chlordane (1,2,4,5,6,7,8,8-octachloro-2,3,3a,4,7,7a-hexahydro-4,7-methanoindene and related compounds), chlorpyrifos (0,0-diethyl 0-(3,5,6-trichloro-2-pyridyl) phosphorothioate), curacron (0-(4-bromo-2-chloro-phenyl)-0-ethyl-S-propyl phosphorothioate), ethoprop (0-ethyl S,S-dipropyl phosphorodithioate), fonofos (0-ethyl S-phenyl ethyl-phosphonodithioate), heptachlor (1,4,5,6,7,8,8-heptachloro-3a,4,7,7a-tetrahydro-4,7-methanoindene and related compounds), metham (sodium methyldithiocarbamate), phorate (0,0-diethyl-S-[(ethylthio)methyl] phosphorodithioate), and terbufos (S-{[(1,1-dimethylethyl) thio] methyl} 0,0-diethyl phosphorodithioate). The herbicides that were evaluated were alachlor (2-chloro-2',6'-diethyl-N-(methoxymethyl) acetanilide), atrazine (2-chloro-4(ethylamino)-6(isopropylamino)-*s*-triazine), bifenox (methyl 5-(2,4-dichlorophenoxy)-2-nitrobenzoate), cyanazine (2-[4-chloro-6-(ethylamino)-*s*-triazine-2-yl]-2-methyl-propionitrile), dicamba (2-methoxy-3,6-dichlorobenzoic acid 3,5-dichloro-0-anisic acid), eradicane (S-ethyl dipropylthiocarbamate), metolachlor (2-chloro-N-(2-ethyl-6-methylphenyl)-N-(2-methoxy-1-methylethyl) acetamide), procyazine (2-chloro-N-isopropyl-acetanilide), SD50093 (formulation of atrazine plus cyanazine), and simazine (2-chloro-4,6-bis-(ethylamino)-*s*-triazine). The *in situ* test plots for the evaluation of insecticides were constructed at the Illinois Natural History Survey. Five kernels of homozygous *wx-C* inbred W22 maize were planted in each subplot, which measured 2 m x 0.5 m. A field-grade formulation of each insecticide was applied to its assigned subplot prior to seedling emergence. The rates of application are presented in Figure 1. Appropriate control subplots were included within the test plot.

The *in situ* test plots for the evaluation of herbicides were constructed at the South Farms of the Department of Agronomy at the University of Illinois. Each subplot was approximately 10 m x 3 m and consisted of three rows. The outer two rows were planted with commercial corn, while the center row was planted with inbred W22 *wx-C*/*wx-C* kernels. A field-grade formulation of each herbicide was applied to its assigned subplot prior to seedling emergence. The rates of application are presented in Figures 2 and 3. Appropriate control subplots were included within the test plot.

Maize *waxy* Locus Pollen Assay

The *waxy* (*wx*) locus is located on the short arm of chromosome 9. The *wx* locus controls the synthesis of the carbohydrate amylose. The advantage of this assay is that the pollen grain, which is a functional haploid, is the unit of measurement; therefore, large numbers can be analyzed, providing a high degree of genetic resolution. A pollen grain carrying a *Wx* allele can synthesize amylose, while a pollen grain carrying a *wx*

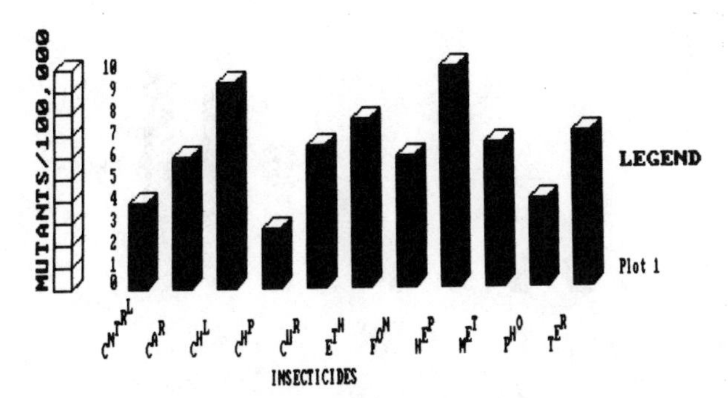

Figure 1. Induction of mutant pollen grains by insecticides. The abbreviation for each insecticide and the application rate are the following: CNTRL (control, 0 kg/ha), CAR (carbofuran, 2.24 kg/ha), CHL (chlordane, 2.24 kg/ha), CHP (chlorpyrifos, 2.24 kg/ha), CUR (curacron, 2.24 kg/ha), ETH (ethoprop, 2.24 kg/ha), FON (fonofos, 2.24 kg/ha), HEP (heptachlor, 1.12 kg/ha), MET (metham, 2.24 kg/ha), PHO (phorate, 2.24 kg/ha), TER (terbufos, 2.24 kg/ha).

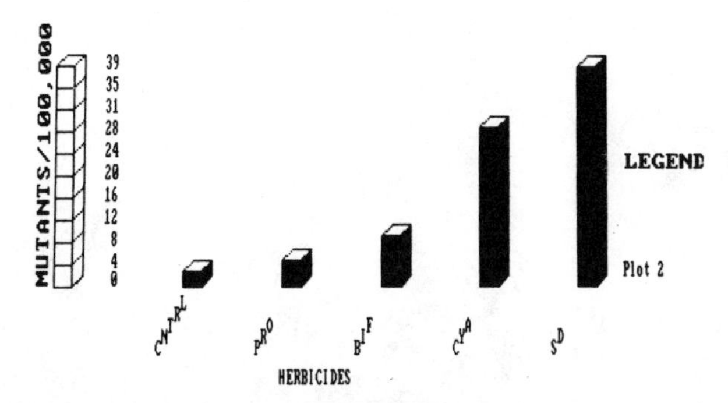

Figure 2. Induction of mutant pollen grains by herbicides. The abbreviation for each herbicide and the application rate are the following: CNTRL (control, 0 kg/ha), PRO (procyazine, 3.58 kg/ha), BIF (bifenox, 2.24 kg/ha), CYA (cyanazine, 3.58 kg/ha), SD (SD50093, 4.48 kg/ha).

allele cannot. When an iodine solution is reacted with amylose, a black complex is formed. Thus, pollen grains that carry a *Wx* allele stain black when exposed to an iodine solution, while pollen grains that carry a *wx* allele stain a tan color. The starch type of a pollen grain is controlled by the genetic constitution of that pollen grain and not by the parental sporophyte. A genetic reversion of *wx* to *Wx* can be detected in a homozygous *wx* plant by scoring for pollen grains that stain a black color after treatment with an iodine solution (Figure 4). Likewise, forward mutation can also be measured by scoring for tan-staining, *wx*-carrying pollen grains obtained from sporophytes that are *Wx*/*Wx* (Figure 5).

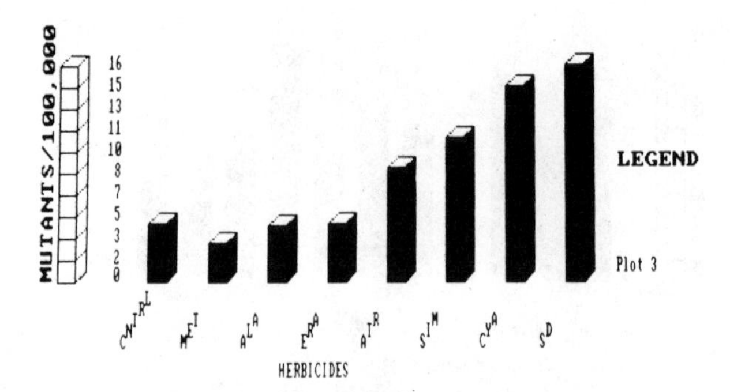

Figure 3. Induction of mutant pollen grains by herbicides. The abbreviation for each herbicide and the application rate are the following: CNTRL (control, 0 kg/ha), MET (metolachlor, 8.40 kg/ha), ALA (alachlor, 6.00 kg/ha), ERA (eradicane, 7.20 kg/ha), ATR (atrazine, 3.84 kg/ha), SIM (simazine, 3.84 kg/ha).

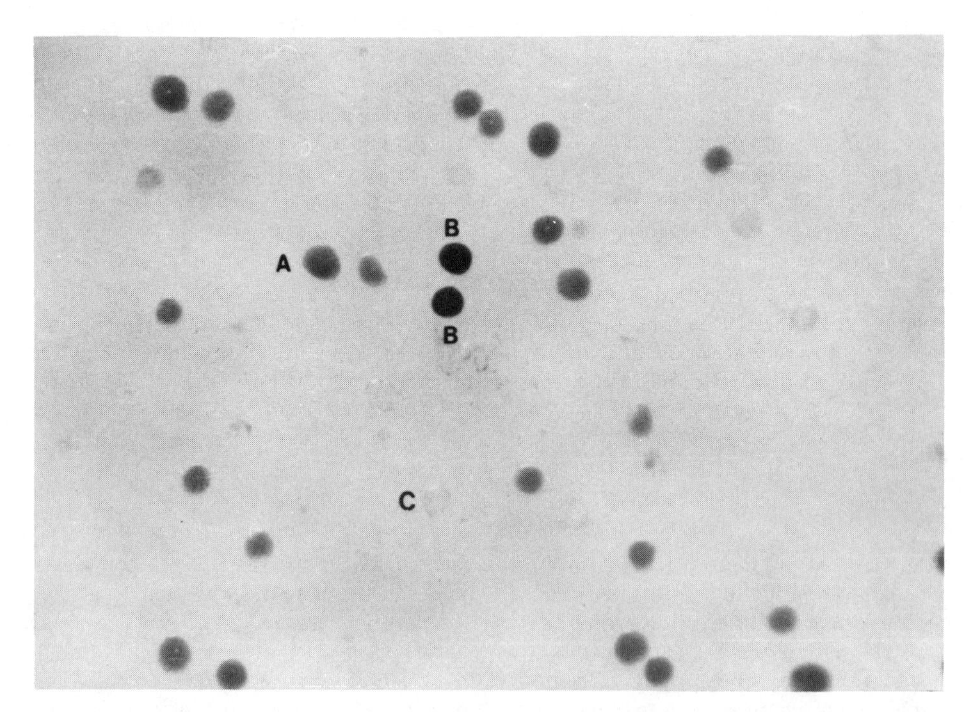

Figure 4. Reverse mutation at the *wx* locus in maize pollen grains. A, a *wx-C* pollen grain; B, a revertant pollen grain; C, an aborted pollen grain. The pollen were from an inbred M14 plant that was *wx-C/wx-C*.

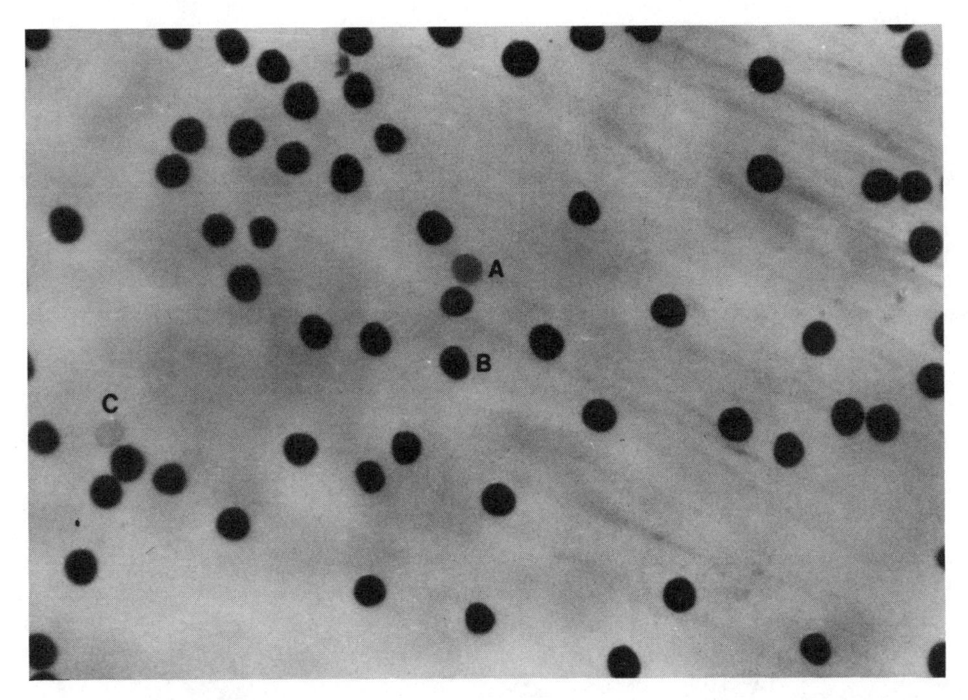

Figure 5. Forward mutation at the *wx* locus in maize pollen grains. A, a forward mutant pollen grain; B, a *Wx* pollen grain; C, an aborted pollen grain. The pollen were from an inbred Early-Early Synthetic plant that was *Wx*/*Wx*.

The plants grew to early anthesis, the tassels were harvested, and the pollen grains were analyzed as outlined by Plewa and Wagner (1981). The frequency of mutant pollen grains from each subplot was compared with the mutant frequency from the appropriate control subplots. The results are presented in Figures 1 to 3. The data were statistically analyzed. (For the final EPA report, see Plewa and Gentile, 1983.) A significant increase in the frequency of mutant pollen grains was induced by chlordane, ethoprop, and heptachlor (Figure 1) (Gentile et al., 1982). The herbicides cyanazine and SD50093 exhibited mutagenic properties in Test Plot 2 (Figure 2), and the herbicides atrazine, simazine, cyanazine, and SD50093 exhibited mutagenic activity in Test Plot 3 (Figure 3) (Plewa et al., 1984).

In Situ Analysis of Municipal Sewage Sludge Applied to Soil

A problem facing the United States is the prudent disposal of the sludge resulting from the treatment of municipal sewage. An attractive disposal method is the application of this organic matter to agricultural lands or its use in the reclamation of land stripped of its organic matter by surface mining.

There may be problems associated with municipal sewage sludge disposal by application on agricultural lands. Research to analyze the mutagenic potential of various municipal sewage sludges was funded by the EPA (Hopke and Plewa, 1983). Part of that study was an *in situ* evaluation of Chicago, IL, municipal sewage sludge that was applied to agricultural soils between May and August 1979 (Plewa and Hopke, 1983). A test plot was constructed on the NW900 plots at the University of Illinois Agronomy Research Center near Elwood, IL. One row each of *Z. mays* kernels of inbred M14 homozygous for either the *wx-C* or *wx-90* allele was planted in each subplot (Figure 6). The maximum-treatment subplot received 17.8 cm of liquid sludge during the growing season. This was equivalent to 21.0 t/ha of dry material. The other subplots received one half and one quarter of this amount, respectively. The control subplot received no sludge, and adequate nutrients were provided by the application of chemical fertilizer. No pesticides were applied to any subplot. The plants grew to early anthesis and the tassels were harvested. The pollen grains were analyzed for reverse mutation to the *Wx* allele and for pollen abortion.

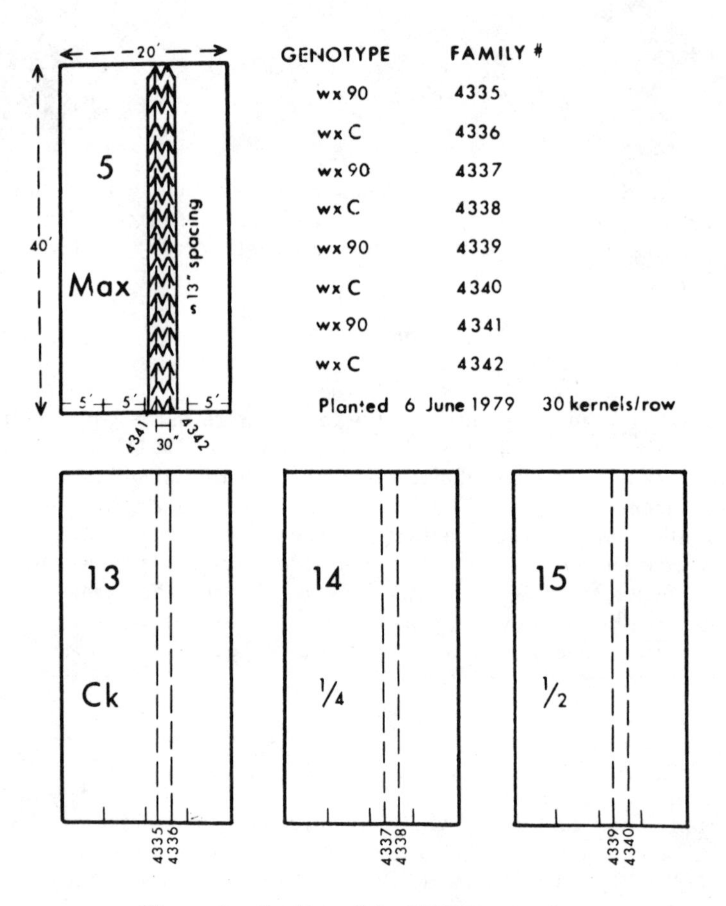

Figure 6. Outline of the NW900 test plot.

For the *wx-C* allele, over 5.1×10^6 pollen grains were analyzed. The frequency of mutant pollen grains did not increase with increasing amounts of sewage sludge. However, a direct relationship was observed between the amount of applied sludge and the frequency of pollen abortion. The range of pollen abortion was from 19.4% for the control to 49.9% for the maximum-sludge subplot.

The *wx-90* allele responded positively to increasing sludge amendment. The control plants had a mean frequency of pollen mutants of 1.6×10^{-6}. Plants grown on the one-fourth-maximum subplot did not have an increased frequency. However, plants grown on the half-maximum and maximum subplots had frequencies of revertant pollen grains of 3.2×10^{-6} and 1.8×10^{-5}, respectively (Figure 7). Thus, the sludge-amended soil exhibited a concentration-dependent increase in the induction of point mutations in maize germinal cells. These data indicate that mutagens present in Chicago sewage sludge were translocated into a crop plant grown on the amended soil. The general conclusion of the *in situ* experiments agreed with the results of experiments on Chicago municipal sewage sludge assayed with *S. typhimurium* and *Tradescantia paludosa* (Hopke et al., 1982).

Recommendations for *In Situ* Analysis

The above experiments demonstrate the utility of the maize *wx* locus pollen assay for the *in situ* evaluation of complex environmental mixtures. The most difficult component of the experimental design is identifying suitable control subplots and a suitable area to be assayed. The history of the NW900 plot used in the sewage sludge study is presented in Table 1.

Although *in situ* environmental monitoring is an attractive approach in the identification of genotoxic components in a complex environmental setting, several requirements must be fulfilled to insure adequate experimental designs. These requirements include (1) a knowledge of the history of contamination in the area to be assayed, (2) the construction of test plots with proper and concurrent controls, (3) the use of well defined assays with specific and calibrated genetic end points, (4) the use of

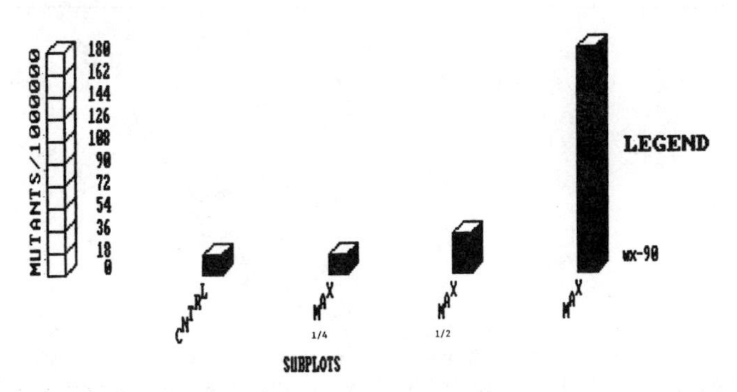

Figure 7. Induction of reverse mutation in *wx-90* pollen grains by Chicago sewage sludge. The rates of sludge application (in equivalent tons per hectare) were: CNTRL = 0, 1/4 MAX = 5.25, 1/2 MAX = 10.5, MAX = 21.

Table 1. History of the NW900 Plot

Year	Treatment
1968	Plot received 61.06 t/ha sludge solids.
1969	Plot was planted to kenaf and received 41.48 t/ha sludge solids.
1970	0.5 g Eptam 6E was applied and disced into the soil, and alfalfa was seeded; 32 g dalapon (Dowpon) in 1.5 gal water was applied to plot; plot received 28.28 t/ha sludge solids.
1971	No sludge was applied; plot was in alfalfa, and Dowpon was sprayed for grass control.
1972	Plot was in alfalfa; plot received 12.16 t/ha sludge solids.
1973	Plot was in alfalfa; plot received 20.77 t/ha sludge solids.
1974	No crop; plot received 46.21 t/ha sludge solids.
1975	No sludge was applied; plot was in alfalfa.
1976	No sludge was applied; plot was in spinach.
1977	No sludge was applied; plot was in spinach.
1978	No sludge was applied; plot was in spinach.
1979	Subplots No. 5, 13, 14, and 15 were used in present study.

appropriate numbers of indicator organisms to provide a sufficiently large data base, (5) trained personnel to maintain the test plots and to label and harvest the plants at correct times, (6) laboratory personnel to conduct the competent analysis and the proper interpretation of the observed genetic events, and (7) the use of appropriate statistical analysis in the evaluation of the data.

LABORATORY ANALYSIS OF COMPLEX ENVIRONMENTAL MIXTURES USING MAIZE

Development and Calibration of Inbred Early-Early Synthetic

A new maize inbred has been developed for use as an assay in genetic toxicology. A description of inbred Early-Early Synthetic maize was published by Plewa and Wagner (1981). This inbred grows from kernel to tassel emergence in less than 4 wk and attains a height of approximately 50 cm. Its size and rapid maturity make it useful for laboratory analysis. The genetic end points that are available include mutation induction at the *wx* locus in pollen grains, induction of forward mutation at the *yg2* locus in leaf embryo (primordial) cells, and the induction of micronuclei in primary root-tip cells. Thus, mutation in germinal and somatic cells and mitotic chromosome aberrations may be assayed simultaneously in a higher eukaryote within a 4-wk time frame.

Inbred Early-Early Synthetic or F_1's derived from Early-Early Synthetic have been calibrated with chemical mutagens and ionizing radiation (Plewa et al., in press). These calibration studies define the levels of sensitivity of the assays and provide data for fundamental studies in mutation research.

Forward mutation at the *wx* locus in pollen grains has been calibrated after chronic exposure to the chemical mutagens ethyl methanesulfonate (EMS) or maleic hydrazide (MH) administered via the soil (Plewa and Wagner, 1981). Individual Early-Early Synthetic kernels were planted in 10-cm plastic pots in standard soil (a mixture of loam, peat moss, and sand [4:2:1, v/v/v]). Each pot was enveloped in a plastic bag; five pots were collected in a plastic tub and designated as a control or treatment group. The pots were placed in a plant growth chamber adjusted to a photoperiod of 14 h at an illumination of 300 $\mu E\ m^{-2}\ s^{-1}$ PRR with day and night temperatures at 25°C and 20°C, respectively. Three times a week, 50 ml of a known molar concentration of either EMS or MH was administered onto the soil of each pot. The control pots received water only. After the tassels emerged, the mutagen treatments were halted and the plants received water. All the plants received adequate water and nutrients. After the plants reached anthesis, tassels were harvested and pollen grains were analyzed for forward *wx* mutants. Since the amounts and concentration of each mutagen were recorded, the total moles of mutagen to which each plant was exposed was calculated. The amount of EMS exposure ranged from 1.1×10^{-6} to 6.8×10^{-6} moles per plant. The data from one EMS experiment are presented in Figure 8. The frequency of mutant pollen grains in every treatment group was significantly different ($p \leq 0.001$) from the control group. Approximately 6.5×10^{5} pollen grains were analyzed for each group, and the frequency of mutant pollen grains ranged from 5.11×10^{-5} for the control group to 2.05×10^{-4} for the highest treatment group. The concentration-response curve was linear ($r^2 = 0.93$). However, there was no correlation between the concentration of EMS and the percentage of pollen abortions ($r = 0.22$). Therefore, EMS is a potent mutagen in this assay when chronically administered through the soil.

The data from one MH experiment are presented in Figure 9. The chronic exposure to MH ranged from 5.0×10^{-9} to 5.0×10^{-7} moles per plant. The frequency of mutant pollen grains ranged from 4.59×10^{-5} for the control to 5.54×10^{-4} for the highest treatment group. Each treatment group had a significantly increased

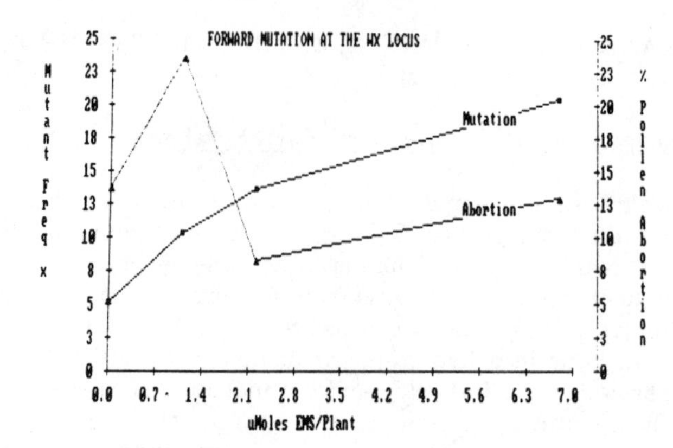

Figure 8. Mutation at the *wx* locus in maize pollen grains induced by chronic exposure to ethyl methanesulfonate. Mutation frequency is presented as x 10^{-5}.

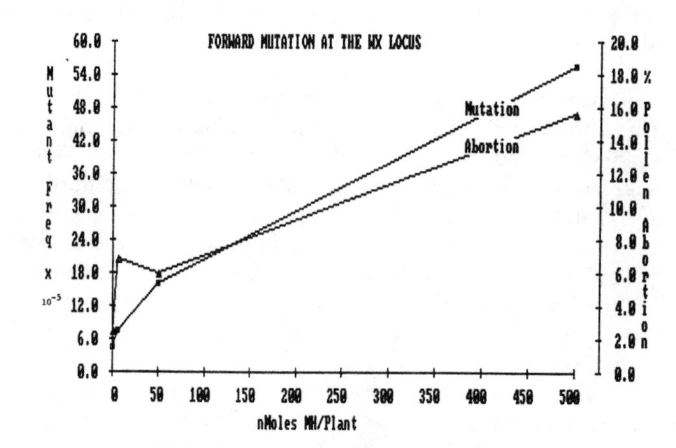

Figure 9. Mutation at the *wx* locus in maize pollen grains induced by chronic exposure to maleic hydrazide. Mutation frequency is presented as x 10^{-5}.

frequency of mutant pollen grains when compared to the control ($p \leq 0.001$). This increase was concentration-dependent and linear ($r^2 = 0.99$). The percentage of pollen abortion increased directly with the concentration of MH ($r = 0.96$). Thus, MH induces gametophytic death at concentrations that are mutagenic.

This demonstrates that maize can detect mutagens that are present in soil. The mutagen-soil combination is a complex mixture. Therefore, it appears that maize would be a suitable genetic indicator organism for the analysis of complex environmental mixtures that are present in soil.

Analysis of Municipal Sewage Sludge

Forward mutation at the *wx* locus was assayed after exposure to extremely high concentrations of municipal sewage sludge from Chicago or Champaign, IL (Hopke et al., 1982). The sludge was from the same lot used in the *in situ* test plots described above. The experiment consisted of three treatment groups of five plants each. The highest treatment group consisted of a mixture of 1 part sludge with 2 parts soil. The plants were watered with a 1/3 dilution of sludge in deionized water. The middle treatment group contained 1 part sludge with 5 parts soil, and the plants were watered with a 1/6 dilution of sludge. The low treatment group contained 1 part sludge and 11 parts soil, and the plants were watered with a 1/12 dilution of sludge. The treatment groups contained concentrations of sludge that far exceeded any environmental use. A concurrent control of sibling kernels was conducted using standard soil and water only. The plants were grown in a plant growth chamber, the tassels were harvested, and the pollen grains were analyzed for the frequency of forward mutants and the percentage of pollen abortion.

The data are presented in Figure 10. The results indicate that the administration of high amounts of Chicago sludge increased the frequency of *wx* mutant pollen grains by two orders of magnitude. The frequency of mutant pollen grains for the control group was 4.31×10^{-5}, which was consistent with the control values in the EMS and MH calibration studies (Figures 8 and 9). The frequency of mutant pollen grains was inversely related to the concentration of sludge, indicating that even the lowest concentration of applied sludge was high enough to reach the toxic region of the mutational concentration-response curve. The percentage of pollen abortions was higher at the higher sludge concentrations, indicating some gametophytic toxicity, and the plants grown at the two higher concentrations exhibited toxic effects. Thus, the sewage sludge contained toxic as well as genotoxic components. These data confirm the results of the *in situ* maize assays.

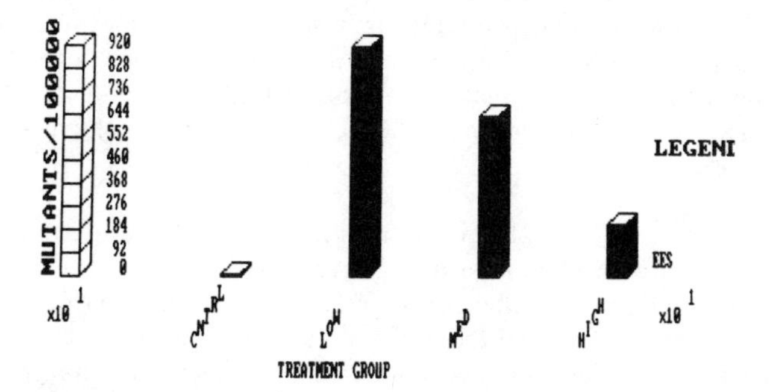

Figure 10. Forward mutation at the *wx* locus in maize pollen grains induced by high concentrations of Chicago sewage sludge. CNTRL = control, LOW = low treatment group (1/12 sludge:soil amendment), MED = medium treatment group (1/6 sludge:soil amendment), HIGH = high treatment group (1/3 sludge:soil amendment).

Analysis of Municipal Water

Reverse mutation at the *wx* locus was one of four genetic assays used in the analysis of the mutagenicity of municipal water obtained from an agricultural area (DeMarini et al., 1982; Heartlein et al., 1981). Water from Lake Bloomington and the tap water for Bloomington, IL (population 44,000), was analyzed for toxic and mutagenic products. Gas chromatographic and mass spectrometric analyses of mutagenic water samples indicated the presence of the carcinogen dimethyl dithiocarbamate, a commonly used agricultural fungicide.

Additional Genetic End Points In Maize

Two additional genetic end points are being introduced and calibrated in Early-Early Synthetic maize. They are forward mutation at the *yellow-green 2* (*yg2*) locus and the induction of micronuclei in primary root-tip cells. Although these assays have not been used in the analysis of complex environmental mixtures, the laboratory experiments to date indicate that they should be sensitive, useful, and rapid genetic assays in a higher eukaryotic organism.

The *yg2* locus is on the short arm of chromosome 9. Homozygous *yg2* plants are a pale yellow-green color, while heterozygous plants exhibit a normal green color. The assay is based on *Yg2/yg2* plants. If the dominant allele is lost due to a gene mutation, a chromosome aberration, or the loss of the chromosome that carries *Yg2*, the recessive allele is expressed and the cell is deficient in the production of chlorophyll. When a mutant cell divides, a sector or clone of mutant progeny cells is produced (Figure 11). If a leaf primordial cell in an embryo suffers the loss of the dominant allele, the resulting leaf will be a chimera and the clone of *yg2* mutant cells will appear as a yellow-green sector within the green leaf.

The *yg2* assay has been standardized and calibrated with gamma radiation and EMS (Plewa et al., in press). The calibration curves for ^{137}Cs gamma rays and for EMS are presented in Figures 12 and 13, respectively. The advantages of this assay are that it detects forward mutation at a specific locus in somatic cells of a higher eukaryote. It is a quantitative assay with high sensitivity. The large number of target cells in the leaf primordium of the analyzed leaves provides a high degree of genetic resolution. Finally, the assay is inexpensive and relatively rapid for a test based on an intact higher eukaryotic organism.

A second genetic end point that is being calibrated is the induction of micronuclei in primary root-tips of maize (Wagner and Plewa, 1983). Micronuclei are formed from acentric fragments, lagging chromosomes, or multicentric chromosomes connected by bridges. Experiments were conducted to determine if gamma rays could induce micronuclei at the same concentrations that were used for the *yg2* calibration experiments. Kernels were treated according to the procedures of Plewa et al. (in press) with gamma rays that ranged from 0 to 500 rads. Following treatment, the kernels were rinsed and planted in vermiculite. After 24 h, the primary root-tips were removed and the cells were cytologically analyzed. The data are presented in Figure 14 (Wagner and Plewa, unpublished). The data indicate that the induction of micronuclei is a feasible system to detect clastogens. The advantages of the micronucleus test in maize

Figure 11. A *yg2* sector on Leaf 4.

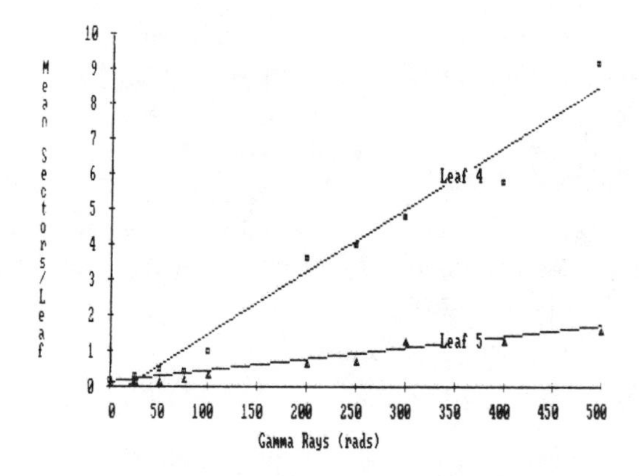

Figure 12. Induction of forward mutation at the *yg2* locus in maize by gamma radiation in Leaves 4 and 5.

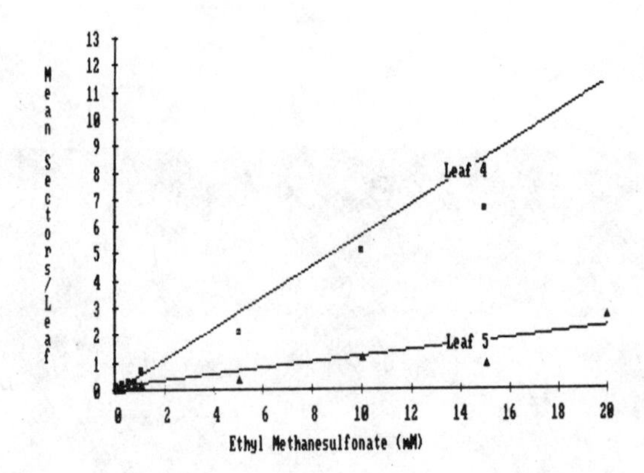

Figure 13. Induction of forward mutation at the *yg2* locus in maize by ethyl methanesulfonate in Leaves 4 and 5.

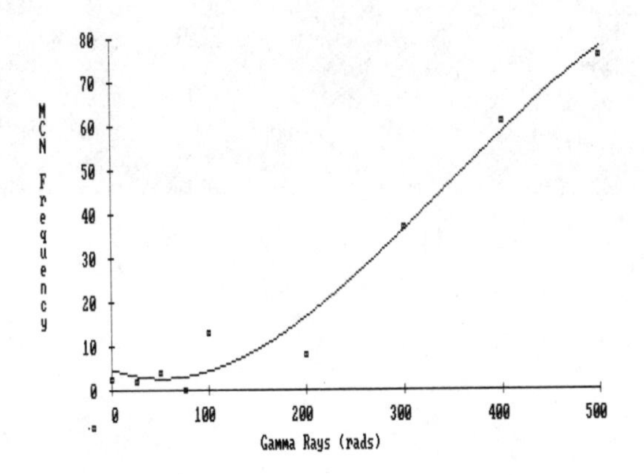

Figure 14. Micronuclei in maize primary root-tip cells induced by gamma radiation. Frequency of micronuclei is presented as x 10^{-4}.

root-tip cells for the analysis of environmental pollutants include its ease and rapidity. It is sensitive to acute treatments, and the end point can be used simultaneously with the *wx* and *yg2* point mutation assays.

INDUCTION OF MICRONUCLEI IN *TRADESCANTIA* BY COMPLEX MIXTURES

Analysis of Municipal Sewage Sludge

The induction of micronuclei in *T. paludosa* is a sensitive assay that detects clastogens that induce chromosome damage in tetrads following meiosis of a pollen grain mother cell (Ma, 1981). A micronucleus may result from the induction of multipolar nuclear divisions or acentric fragments. The fragment of a broken chromosome can be observed as a separate micronucleus within the cells of a tetrad (Figure 15).

Inflorescences of *Tradescantia* clone 03 were used in the analysis of municipal sewage sludges. The use of a single clone ensured that all of the inflorescences were isogenic. Typically, 8 to 10 inflorescences were used for each treatment group. A negative control of Hoagland's solution (a plant nutrient solution), a positive control of 50 mM MH in Hoagland's solution, and various mixtures of sludge and Hoagland's solution were used for each experiment. The inflorescences were treated for 24 h with the test mixture and then placed into fresh Hoagland's solution for an additional 24 h. The solutions were aerated continuously, and the temperature was maintained at 25°C by using a water bath. The inflorescences were then removed from the tubes and fixed in ethanol:glacial acetic acid (3:1, v/v) for 48 h. The inflorescences were stored in 70% ethanol and subsequently analyzed for micronuclei.

The procedure for scoring micronuclei in tetrads has been described (Ma, 1981). The frequency of micronuclei induction was determined by dividing the total number of micronuclei observed by the number of tetrads analyzed. The ratio was multiplied by 100 to yield the number of micronuclei per 100 tetrads. Samples of whole municipal sewage sludge and sludge concentrated by the removal of water were analyzed. Sludge samples were from the Illinois cities of Chicago, Champaign, Hinsdale, Kankakee, and Sauget (Plewa and Hopke, 1983).

Champaign sludge. Neither whole sludge nor a 1/2 dilution of Champaign sludge induced a significant increase in the frequency of micronuclei. The whole sludge was concentrated by lyophilization. The sludge solids were mixed with Hoagland's solution

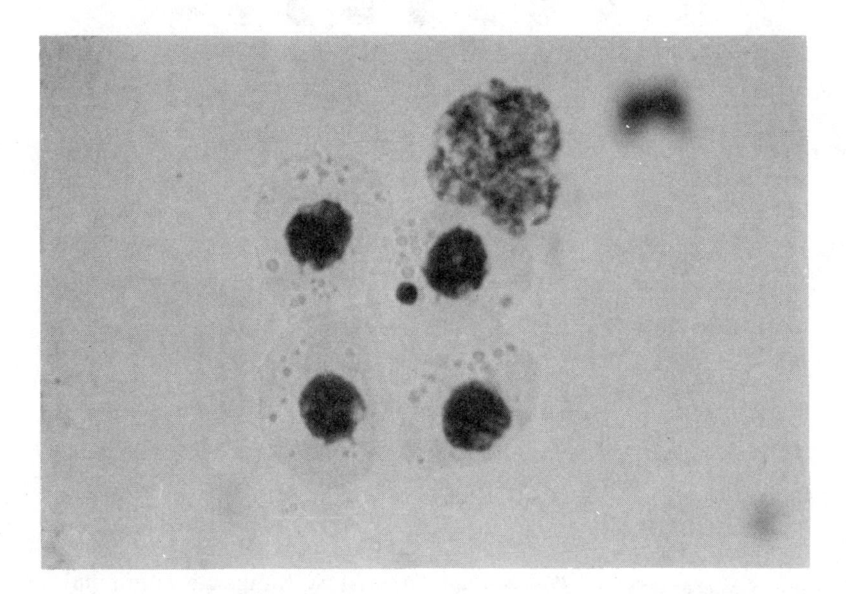

Figure 15. Micronucleus in a tetrad of *Tradescantia paludosa* clone 03.

to give 2X and 4X concentrations. None of the Champaign sludge samples induced a significant increase in the frequency of micronuclei.

Chicago sludge. The data with whole Chicago sludge and various dilutions demonstrate an increased frequency of induced micronuclei with increased sludge concentration (Figure 16). Dilutions above 1/4 caused a significant increase in the frequency of micronuclei.

Hinsdale sludge. A negative response was observed with whole sludge samples. The Hinsdale sludge was concentrated by lyophilization. The sludge solids were mixed with Hoagland's solution to give 6.7X, 13.3X, and 20.0X concentrations. None of these concentrated sludge samples significantly increased the frequency of micronuclei in *Tradescantia*.

Kankakee sludge. Kankakee whole sludge samples were negative when assayed. The whole sludge was concentrated, and 6.7X, 13.3X, and 20.0X concentrations in Hoagland's solution were prepared. The results demonstrate that only the 20.0X concentration induced a positive response (Figure 17). Thus, a low concentration of clastogenic agents was present in Kankakee sludge.

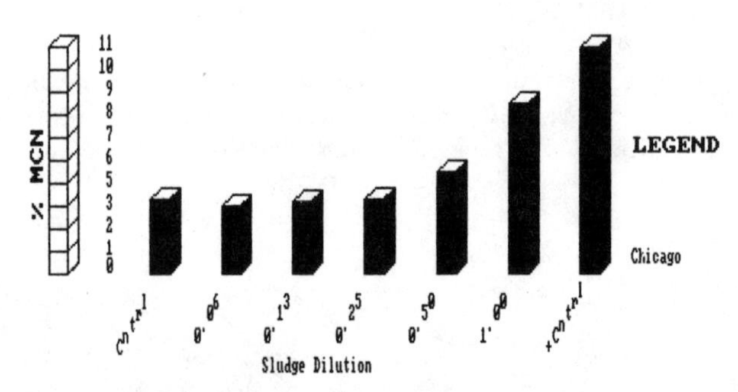

Figure 16. Micronuclei in *Tradescantia* induced by Chicago municipal sewage sludge.

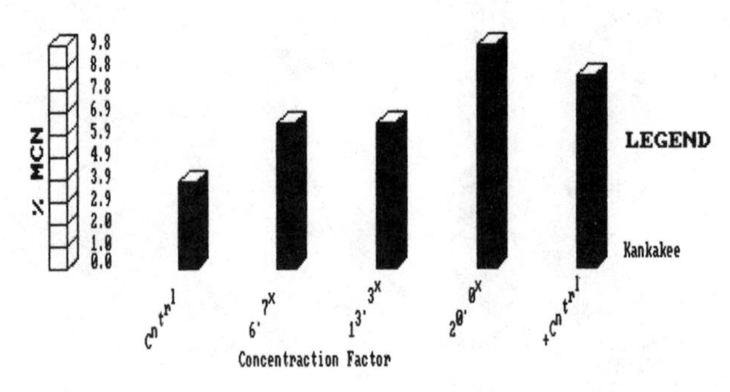

Figure 17. Micronuclei in *Tradescantia* induced by Kankakee municipal sewage sludge.

Sauget sludge. The Sauget sludge was difficult to assay because it was so toxic to *Tradescantia*. The whole Sauget sludge inhibited meiosis and killed the cells in the stem of the inflorescence. The sludge sample was lyophilized, and the dry material was reconstituted with distilled water and 0.3 ml dimethylsulfoxide. Samples of 3.53%, 7.07%, and 10.6% dry weight of solids were prepared (dilution factors of 0.25X, 0.50X, and 0.76X). The original sludge sample contained 13.9% dry weight of solids. The 0.76X dilution was so toxic that meiosis was inhibited in 9 of the 10 inflorescences. Of those tetrads that could be analyzed in the 0.76X dilution sample, a very high frequency of micronuclei was observed (Figure 18). The two lower concentrations were analyzed, and significant high frequencies of micronuclei were observed.

The municipal sewage sludges discussed above were also fractionated into crude chloroform:methanol, neutral, basic, weak acidic, and strong acidic fractions. These fractions were assayed for their mutagenicity with and without mammalian S9 activation with *Salmonella* strains TA98 and TA100 (Plewa and Hopke, 1983; Hopke et al., in press). In general, the Champaign, Hinsdale, and Kankakee sludges demonstrated no direct mutagenicity in the *Salmonella* tests. In strain TA98, only a weak response was observed following S9 activation with samples obtained by the chemical fractionation of chloroform:methanol extracts. Only the Sauget sample demonstrated direct-acting mutagenic activity as well as very strong mutagenicity after S9 activation.

The *Tradescantia* data indicate that whole or diluted Chicago and Sauget sludges induce meiotic chromosome aberrations. When whole sludge samples from each municipality were evaluated, a pattern of response emerged (Figure 19). A ranking in order of increasing mutagenicity was Champaign ≈ Hinsdale < Kankakee < Chicago ≪ Sauget. When concentrated sludge samples were evaluated, the same ranked order was maintained. The Sauget sludge was much more potent than any of the other sludges, and Chicago sludge was more mutagenic than Kankakee. There is an apparent correlation of the cytogenetic potency of the sludge sample and the relative industrialization of the municipality.

The *in situ* and laboratory studies using the *wx* locus assay in maize pollen grains demonstrate substantial uptake of mutagens from Chicago sludge amended soils. These studies indicate that mutagens present in the sludge-amended soils are

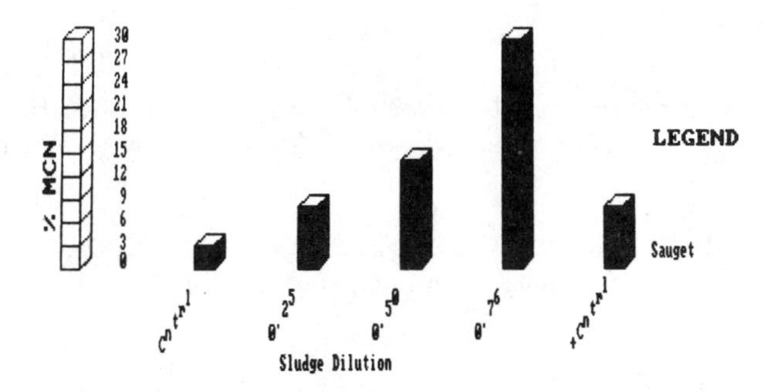

Figure 18. Micronuclei in *Tradescantia* induced by Sauget municipal sewage sludge.

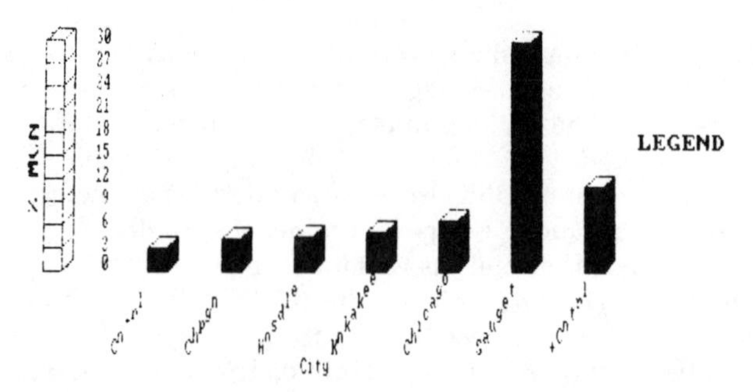

Figure 19. Comparison of the genotoxic potency of whole municipal sewage sludges to induce micronuclei in *Tradescantia*. The Sauget sludge was a 0.76X concentration of whole Sauget sludge.

transported into a crop plant and can induce genetic damage in germinal cells. These studies do not provide any information of the possible mutagenicity of the grain grown on sludge-amended soil or the transport of mutagens from the sludge to ground or surface waters. However, these results indicate the acute need for further study of the possible adverse effects of land application of municipal sewage sludges that exhibit substantial mutagenic potency. It is clear that some sludges, like that from Sauget, are exceedingly genotoxic, and their disposal should be conducted with considerable care. Such sludges should not be land applied except in dedicated facilities. Highly mutagenic sludges should not be amended with agricultural soils until a better understanding of the degradation and transport of the constituent mutagenic agents in soil has been elucidated and the safety of such disposal practices can be thoroughly delineated.

ACKNOWLEDGMENTS

The research discussed in this paper has been supported in part by U.S. Environmental Protection Agency contracts 68-02-2704 and R807009010 and NIEHS grant ES01895. I thank Ms. E.D. Wagner for her critical review of the manuscript.

REFERENCES

Crow, J.F., and S. Abrahamson, eds. 1982. Identifying and Estimating the Genetic Impact of Chemical Mutagens. National Academy Press: Washington, DC. 316 pp.

DeMarini, D.M., M.J. Plewa, and H.E. Brockman. 1982. Use of four short-term tests to evaluate the mutagenicity of municipal water. J. Toxicol. Environ. Health 9:127-140.

Gentile, J.M., G.J. Gentile, J. Bultman, R. Sechriest, E.D. Wagner, and M.J. Plewa. 1982. An evaluation of the genotoxic properties of insecticides following plant and animal activation. Mutat. Res. 101:19-29.

Heartlein, M.W., D.M. DeMarini, A.J. Katz, J.C. Means, M.J. Plewa, and H.E. Brockman. 1981. Mutagenicity of municipal water obtained from an agricultural area. Environ. Mutagen. 3:519-530.

Hopke, P.K., and M.J. Plewa. 1983. The Evaluation of the Mutagenicity of Municipal Sewage Sludge. Final Report EPA-600/1-83-016, U.S. Environmental Protection Agency. 85 pp.

Hopke, P.K., M.J. Plewa, J.B. Johnston, D. Weaver, S.G. Wood, R.A. Larson, and T. Hinesly. 1982. Multitechnique screening of Chicago municipal sewage sludge for mutagenic activity. Environ. Sci. Technol. 16:140-147.

Hopke, P.K., M.J. Plewa, P.A. Stapleton, and D.L. Weaver. (in press). Comparison of the mutagenicity of sewage sludges. Environ. Sci. Technol.

Ichikawa, S. 1981. In situ monitoring with Tradescantia around nuclear power plants. Environ. Health Perspect. 37:145-164.

Klekowski, E.J., Jr. 1978. Screening aquatic ecosystems for mutagens with fern bioassays. Environ. Health Perspect. 27:99-102.

Klekowski, E.J., Jr., ed. 1982. Environmental Mutagenesis, Carcinogenesis and Plant Biology, Vols. 1 and 2. Praeger Publishers: New York.

Lower, W.R., P.S. Rose, and V.K. Drobney. 1978. In situ mutagenic and other effects associated with lead smelting. Mutat. Res. 54:83-93.

Ma, T.H. 1981. Tradescantia micronucleus bioassay and pollen tube chromatid aberration test for in situ monitoring and mutagen screening. Environ. Health Perspect. 37:85-90.

Nilan, R.A. 1978. Potential of plant genetic systems for monitoring and screening mutagens. Environ. Health Perspect. 27:181-196.

Plewa, M.J., and J.M. Gentile. 1983. The Detection of Mutagenic Properties of Pesticides Used in Commercial Corn Production. Final Report EPA-600/S1-83-0006, U.S. Environmental Protection Agency: Research Triangle Park, NC. 191 pp.

Plewa, M.J., and P.K. Hopke. 1983. Mutagenicity of municipal sewage. In: Carcinogens and Mutagens in the Environment, Vol. 3. H.F. Stich, ed. CRC Press: Boca Raton, FL. pp. 155-175.

Plewa, M.J., and E.D. Wagner. 1981. Germinal cell mutagenesis in specially designed maize genotypes. Environ. Health Perspect. 37:61-73.

Plewa, M.J., E.D. Wagner, G.J. Gentile, and J.M. Gentile. 1984. An evaluation of the genotoxic properties of herbicides following plant and animal activation. Mutat. Res. 136:233-245.

Plewa, M.J., P.A. Dowd, and E.D. Wagner. (in press). Calibration of the maize yg2 assay using gamma radiation and ethyl methanesulfonate. Environ. Mutagen.

Schairer, L.A., and R.C. Sautkulis. 1982. Detection of ambient levels of mutagenic atmospheric pollutants with the higher plant Tradescantia. In: Environmental Mutagenesis, Carcinogenesis, and Plant Biology, Vol. 2. E.J. Klekowski, Jr., ed. Praeger Publishers: New York. pp. 155-194.

Wagner, E.D., and M.J. Plewa. 1983. The induction of micronuclei in root tip cells. Maize Genet. Coop. News Letter 57:141-143.

STUDIES WITH GASEOUS MUTAGENS IN *DROSOPHILA MELANOGASTER*

P.G.N. Kramers, B. Bissumbhar, and H.C.A. Mout

Laboratory for Carcinogenesis and Mutagenesis, National Institute of Public Health and Environmental Hygiene, Bilthoven, The Netherlands

INTRODUCTION

In the great majority of mutagenicity studies using *Drosophila*, feeding or injection has been utilized as the route of administration of the test chemical. The exposure of *Drosophila* to potential mutagens in the gas phase has been used by a limited number of investigators, including Verburgt and Vogel (1977), Nomura (1979), Kramers and Burm (1979), and Kale and Baum (1981). One of the advantages of using inhalation exposure is the organization of the respiratory system in insects that allows for the transport of gases directly into the organs of the fly, including the gonads. Furthermore, this method of exposure provides the possibility to study the effect of variations of exposure duration with less technical difficulties than with other routes of administration. This paper reports on an investigation into the relations between exposure concentration, exposure time, and mutation induction by two gaseous mutagens, 1,2-dichloroethane (DCE) and methylbromide (MeBr). The question is discussed whether *Drosophila* is sensitive enough to be used as an *in situ* monitor for gaseous air pollutants.

MATERIALS AND METHODS

Several techniques have been used for the exposure of *Drosophila* to gases using either a static (Verburgt and Vogel, 1977; Nomura, 1979) or a dynamic (Kale and Baum, 1981) system. We have devised a setup consisting of glass vessels large enough (18 liter) to allow for a static exposure of several days without running into the problem of atmospheric changes caused by the metabolism of the flies. At the same time, the apparatus provides the possibility of a dynamic (continuous flow) exposure (Figure 1). During the exposure, the flies are kept in glass containers with perforated Teflon screw caps and exchangeable medium cups (Figure 2). Most of the experiments described

Figure 1. Mobile apparatus with four 18-liter exposure chambers.

below were carried out using static exposures. When treatment times exceeded 48 h, the vessels were flushed every 2-3 d, fresh medium was given, and a new amount of the test compound was introduced. Concentrations of the test compound were regularly checked by gas chromatography.

Mutation induction was assayed using the standard sex-linked recessive lethal test, using wild type males (Berlin K) for treatment and *Basc* virgins as their mates. After treatment, the males were mated every 2-3 d to fresh virgins in order to obtain subsequent samples (broods) representing germ cells exposed at different stages of development. For details on the procedures, see Würgler et al. (1977).

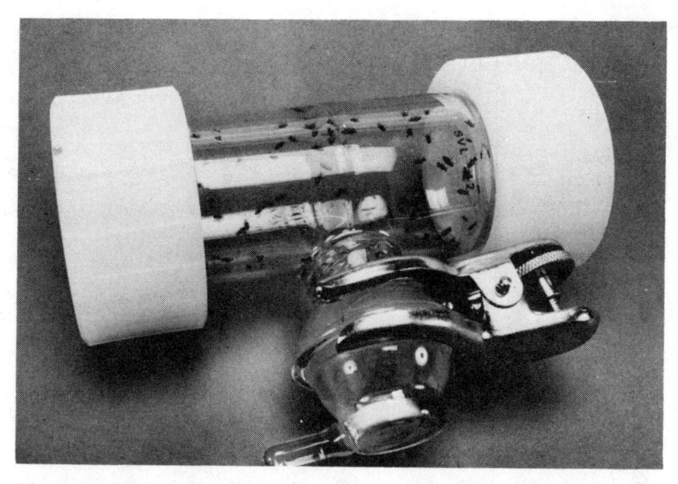

Figure 2. Fly container with perforated screw caps and exchangeable medium cup.

RESULTS AND DISCUSSION

Exposures to DCE were performed for periods of from 6 h to 3 wk at a range of concentrations. The results are shown in Tables 1 and 2 and Figure 3.

The ability of DCE to produce mutations in *Drosophila* had been shown earlier (Rapoport, Shakarnis, reviewed by Rannug, 1980; King et al., 1979). The present work clearly demonstrates that the mutagenic effect of gaseous DCE is still apparent at rather low exposure concentrations (1% sex-linked recessive lethals at 8 mg/m^3 [about 2 ppm]) for 96 h. Using the adult feeding technique, considerable concentrations are needed in comparison with many known mutagens to produce the effect: 3.5% lethals at 3 d feeding on 50 mM (King et al., 1979). From the 6-h and 96-h series it is clear that spermatocytes and spermatogonia and perhaps early spermatids (broods C-E) are far more sensitive than the more advanced stages (broods A-B). This observation is of interest since one of the possible ways of action of DCE involves conjugation with glutathione, resulting in the formation of S(2-chloroethyl)cysteine (Rannug, 1980). The fact that the latter compound when tested in *Drosophila* (Fahmy and Fahmy, 1960) showed a germ cell stage-specific response very similar to the one observed for DCE supports the view that in *Drosophila* glutathione conjugation also plays a role in the activation of DCE.

In the graphical representation of the results (Figure 3), the different concentration-response curves and points for each of the selected exposure periods (6 h, 96 h, 1 wk, 2 wk, and 3 wk) have been plotted on an integrated exposure scale expressed in milligrams per meter cubed times hours (times 10^{-3} to avoid large numbers). The data used in the graph are the pooled frequencies for the responsive matings (broods C-E in the 6-h and 96-h series and broods B-D, A-C, and A-B in the 1-, 2-, and 3-wk series, respectively). For 6 h as well as for 96 h, the concentration-response relation

Table 1. Induction of Sex-Linked Recessive Lethals by Gaseous 1,2-Dichloroethane

Exposure time (h)	Concentration (mg/m3)	Exposure (h·mg/m3·10-3)	Brood A		Brood B		Brood C		Brood D		Brood E		Brood F	
			nchr.[a]	% l.[b]	nchr.	% l.	nchr.	% l.	nchr.	% l.	nchr.	% l.	nchr.	% l.
6	0	0	392	0.5	396	0	391	0.3	193	0	194	0		
6	150	0.9	197	0	199	0	198	0.5	198	0	198	1.0		
6	400	2.4	213	0	219	0	212	2.4						
					389	1.3	390	3.3	387	1.6	382	0.5	352	0.8
6	800	4.8	215	1.4	218	0.5	216	8.3						
					382	2.1	385	4.7	379	2.4	383	1.0	382	2.9
6	1200	7.2	216	0	219	0.5	165	9.1						
					377	2.4	383	6.0	385	8.6	388	3.6	385	3.1
6	2300	13.8	199	0.5	198	0.5	195	0.5	194	5.2	54	5.6		
96	0	0			594	0.3	588	0.3	585	0	588	0		
96	8	0.8			394	0	390	1.0	387	0.8	391	1.3		
96	20	1.9			394	0.8	393	3.1	395	1.0	395	0.8		
96	50	4.8			380	1.8	181	0.6	392	3.8	396	3.5		
96	125	12			384	1.6	248	2.4	387	9.8	380	7.4		
96	330	32	188	0.5	189	0.5	198	1.0	25	24.0	61	21.0		
96	1000	96	189	0	206	1.5	178	0	185	0	60	1.7		

[a]nchr.: number of chromosomes tested. [b]%l.: percentage of lethals.

Table 2. Induction of Sex-Linked Recessive Lethals by Gaseous 1,2-Dichloroethane: Chronic Exposure at 7 mg/m^3

Exposure time (wk)	Exposure ($h \cdot mg/m^3 \cdot 10^{-3}$)	Brood A		Brood B		Brood C		Brood D	
		nchr.[a]	% l.[b]	nchr.	% l.	nchr.	% l.	nchr.	% l.
1	1.2	583	0.5	586	0.7	585	1.9	588	0.3
2	2.4	594	1.2	483	2.7	334	3.0		
3	3.5	303	1.3	185	2.2				

[a]nchr.: number of chromosomes tested.
[b]%l.: percentage of lethals.

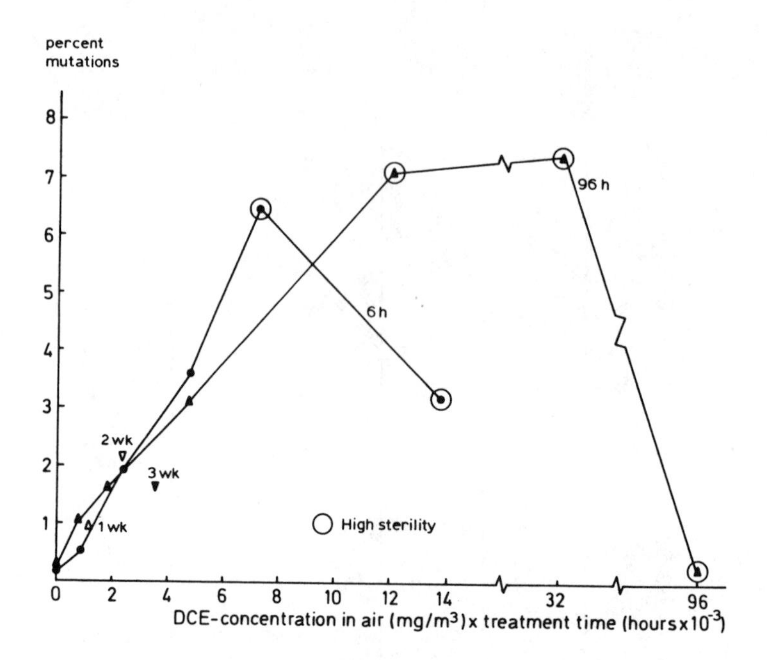

Figure 3. Sex-linked recessive lethal induction after exposure of *Drosophila* males to gaseous 1,2-dichloroethane, in matings representing mainly immature germ cell stages (see text).

appears approximately linear over one order of magnitude. At high exposure levels, the mutation rates are decreasing, in association with a high degree of sterility, which indicates a toxic action of DCE toward the same cell type that is sensitive to mutation induction. Since exposure is expressed in milligrams per meter cubed times hours, the close overlapping of the 6-h and 96-h concentration-response curves and of the 1- and 2-wk points indicates that mutation induction is not only proportional to exposure concentration at a given exposure period, but it also appears to be approximately proportional to exposure time up to exposure levels of about 5000 $mg/m^3 \cdot h$. In other words, increasing the exposure time seems to proportionally lower the concentration needed to obtain a certain mutation frequency. The implication is that the use of longer exposure times permits the detection of lower concentrations of the mutagen and thus increases the sensitivity of the test in a proportional manner.

When comparing the DCE data with results obtained on its bromine analogue, 1,2-dibromoethane (DBE), also applied to *Drosophila* by inhalation (Kale and Baum, 1979, 1981), several similarities and differences are apparent: (1) the germ cell stage-specific response for both compounds is similar, (2) the exposure level needed to produce a given frequency of recessive lethals is about 20 times higher for DCE than for DBE (compared on a parts per million times hours basis), (3) acute toxicity to the flies seems somewhat more restrictive with DBE than DCE with respect to the range of mutation frequencies that could be recovered: up to about 5% for DBE and up to 10% or even more for DCE, and (4) similar to DCE, DBE showed a proportionality between exposure time and mutation frequency (or lack of a dose-rate effect), except for the highest exposure levels.

Treatments with MeBr were carried out on a schedule of 6 h/d, 5 d/wk because specific exposure facilities were available at that time. It appeared that the mutagenic effect of MeBr was most pronounced in postmeiotic germ cell stages. The pooled values for these stages are represented in Figure 4. As was the case with DCE, prolongation of exposure time permits lower concentrations to be detected as mutagenic: 375 and 487 mg/m^3 for 5 x 6 h and 200 mg/m^3 for 15 x 6 h were effective exposures whereas treatment up to 750 mg/m^3 for 6 h was not sufficient to produce significantly increased mutation frequencies. The results are being published elsewhere in full detail (Kramers et al., 1984).

It must be noted that direct comparisons of short- versus long-term exposures are subject to bias when the sensitivities of the various germ cell stages differ; this is the case with both DCE and MeBr. After short-term exposures, the repeated mating of treated males will properly fractionate the various treated germ cell stages. Upon chronic exposure, however, a batch of sperm sampled in a particular mating will carry a mutation frequency that is the result of the accumulation of mutations according to the sensitivities of the various stages through which the cells have proceeded during treatment. Thus, when for each separate germ cell stage the relation between exposure time and mutation induction would be linear, prolongation of exposure time should result in a less than proportional increase of mutation frequencies, observed in the most responsive matings. Such a less than proportionality does appear in the

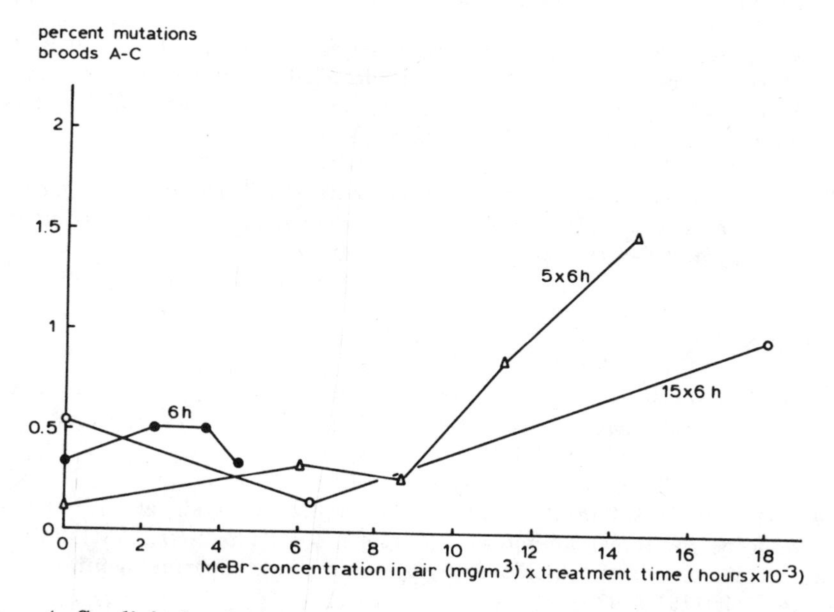

Figure 4. Sex-linked recessive lethal induction after exposure of *Drosophila* males to gaseous methyl bromide, in matings representing predominantly postmeiotic germ cell stages (see text).

comparison of the 5 x 6 h and 15 x 6 h data with MeBr, but is not apparent in the DCE data, except perhaps for the highest exposure levels and the 3-wk exposure. It must be stated, however, that time-effect relations can be affected by other factors such as saturable metabolism (cf. Verburgt and Vogel, 1977) or repair.

Having established that application of prolonged exposure times can lead to the detection of lower concentrations of mutagens, the question arises as to the applicability of *Drosophila* as an *in situ* monitor for the presence of mutagens in ambient air. Donner et al. (1983) exposed flies for several weeks at different locations in a rubber factory where elevated concentrations of pollutants were known to occur and observed a slight but significant increase in the lethal frequency at several locations. In preliminary experiments carried out by ourselves, flies were placed in an automobile traffic tunnel for 2 or 3 wk, resulting in a sex-linked recessive lethal frequency of 14/7079 = 0.20%, which is about our laboratory control value. Clearly, the approach has to be extended to a larger variety of polluted situations to be able to evaluate the possibilities and the restrictions of *in situ* applications of mutagenicity tests with *Drosophila*.

CONCLUSIONS

The findings presented in this paper can be summarized as follows:

1. Inhalation exposure can be a very efficient way of administering volatile substances to *Drosophila* in mutagenesis testing; this is in confirmation of work published by other investigators.

2. Prolongation of the exposure period increases the sensitivity of the test in the sense that lower concentrations can be detected as mutagenic; in the case of DCE this increase in sensitivity is approximately proportional within 2 wk at nontoxic and nonsterilizing exposure levels.

3. The high effectiveness of some gaseous mutagens in *Drosophila* suggests its use as an *in situ* monitor; the question of the possibilities and restrictions of this approach deserves further investigation.

REFERENCES

Donner, M., S. Hytönen, and M. Sorsa. 1983. Application of the sex-linked recessive lethal test in *Drosophila melanogaster* for monitoring the work environment of a rubber factory. Hereditas 99:7-10.

Fahmy, O.G., and M.J. Fahmy. 1960. Cytogenetic analysis of the action of carcinogens and tumor inhibitors in *Drosophila melanogaster*. VIII. Selective mutagenic activity of S-2-chloroethylcysteine on the spermatogonial stages. Genetics 45:1191-1203.

Kale, P.G., and J.W. Baum. 1979. Sensitivity of *Drosophila melanogaster* to low concentrations of gaseous mutagens. II. Chronic exposures. Mutat. Res. 68:59-68.

Kale, P.G., and J.W. Baum. 1981. Sensitivity of *Drosophila melanogaster* to low concentrations of gaseous mutagens. III. Dose-rate effects. Environ. Mutagen. 3:65-70.

King, M.T., H. Beikirch, K. Eckhardt, E. Gocke, and D. Wild. 1979. Mutagenicity studies with X-ray-contrast media, analgesics, antipyretics, antirheumatics and some other pharmaceutical drugs in bacterial, Drosophila and mammalian test systems. Mutat. Res. 66:33-43.

Kramers, P.G.N., and A.G.L. Burm. 1979. Mutagenicity studies with halothane in *Drosophila melanogaster*. Anesthesiology 50:510-513.

Kramers, P.G.N., C.E. Voogd, A.G.A.C. Knaap, and C.A. van der Heijden. 1984. Mutagenicity of methyl bromide in a series of short-term tests. Mutat. Res.

Nomura, T. 1979. Potent mutagenicity of urethan (ethyl carbamate) gas in *Drosophila melanogaster*. Cancer Res. 39:4224-4227.

Rannug, U. 1980. Genotoxic effects of 1,2-dibromoethane and 1,2-dichloroethane. Mutat. Res. 76:269-295.

Verburgt, F.G., and E. Vogel. 1977. Vinyl chloride mutagenesis in *Drosophila melanogaster*. Mutat. Res. 48:327-336.

Würgler, F.E., F.H. Sobels, and E. Vogel. 1977. Drosophila as assay system for detecting genetic changes. In: Handbook of Mutagenicity Test Procedures. B.J. Kilbey, M.S. Legator, W.W. Nichols, and C. Ramel, eds. Elsevier: Amsterdam. pp. 355-373.

ANAPHASE ABERRATIONS: AN *IN VITRO* TEST FOR ASSESSING THE GENOTOXICITY OF INDIVIDUAL CHEMICALS AND COMPLEX MIXTURES

Richard M. Kocan and David B. Powell

School of Fisheries, University of Washington, Seattle, Washington 98195

INTRODUCTION

Anaphase aberrations (aa) were proposed as a genotoxicity test several years ago by the Ad Hoc Committee of the Environmental Mutagen Society (Nichols et al., 1972) and have been described in procedural texts as a recommended technique (Nichols et al., 1977). Since that time, however, a large number of *in vitro* and *in vivo* test systems have been proposed, while the aa test has gone unexploited. On the other hand, the micronucleus test (MNT) has gained wide acceptance, even though it measures only the final events of anaphase damage (Schmid, 1982) and is considerably less sensitive than the aa test for quantitating *in vitro* chromosome damage.

The aa test is a relatively simple procedure that measures any abnormal events in chromosome segregation that might occur during anaphase. When the test is performed on cells grown on a solid surface, all of the cells are in a monolayer and all anaphase chromosomes in cells undergoing mitosis are in a plane parallel to the surface of the culture vessel (Nichols et al., 1977; Kocan et al., 1982). Because of this feature, it is quite easy to see acentric chromosome fragments, bridges, and, in the case of spindle damage, entire displaced chromosomes recognized as nondisjunctions. Since anaphase cells are being examined, there is no need to add spindle poisons (e.g., colcimide) to the cultures; likewise, no spreading or "squashing" of the mitotic cells is required.

Since our laboratory's major emphasis is aquatic toxicology, we have adapted an established fish cell line for use with the aa test. Using this system, we have demonstrated that a wide range of known genotoxic organic compounds is capable of producing heritable chromosomal damage that is visible in cells undergoing mitosis (Kocan et al., 1982). This same genotoxic response can be observed in whole animals of the same species, giving even greater significance to the *in vitro* response (Liguori and Landolt, 1985).

MATERIALS AND METHODS

Cells and Culture Conditions

The RTG-2 continuous cell line developed by Wolf and Quimby (1962) from rainbow trout gonadal tissues was selected because it has been well characterized over the years, is commercially available (American Type Culture Collection, Rockville, MD), and contains high levels of mixed function oxygenase (MFO) and aryl hydrocarbon hydroxylase (AHH) activity (Diamond and Clark, 1970). The presence of an active MFO system eliminates the need for adding exogenous enzymes to the test system, which is necessary in most *in vitro* genotoxicity test systems (Ames et al., 1973; Stetka and Wolff, 1976).

Cells were maintained and exposures were carried out at 18°C in Leibovitz L-15 medium (pH 7.1-7.3) supplemented with 10% fetal calf serum and antibiotics. Stock cultures were routinely transferred at confluency. Plating densities were 20,000 to 25,000 cells per cubic centimeter of growth surface. The cells were determined to be mycoplasma-free by the method of Russell et al. (1975).

Experimental Procedures

Cells were removed from the stock cultures with versene/trypsin (Wolf and Quimby, 1962), diluted with fresh L-15 medium, and placed onto 1- x 5-cm^2 coverslips in Leighton tubes at a density of 20,000 cells/cm^2. The cells were allowed to settle and attach for 18 h; then the medium was removed and fresh medium containing the test chemicals was added to the tubes.

For aa determinations, cells were treated with the test substance for 48 h, washed with phosphate-buffered saline (pH 6.8), and fixed in a mixture of methanol and acetic acid (3:1, v/v) for 1 h. After fixing, coverslips were air-dried and then placed into 3% Gurr's R66 Giesmsa stain in Sorrensen's buffer (pH 6.8) for 10 min. After staining, the coverslips were washed in distilled water, air-dried, and mounted cell side down on clean glass slides. All slides were coded and examined blind on a standard compound microscope (470X). Scoring of cells followed the description of Nichols et al. (1977). For survey purposes, 100 anaphase cells per slide were examined (10 slides/control and 2 slides/test substance). The test cultures were considered positive if the mean abnormal anaphase count of the replicates exceeded the 95% confidence limit determined from the 10 control cultures by a one-sample t test.

Individual Chemicals

Three classes of organic chemicals were used to demonstrate the range of sensitivity of the aa test. These were benzo(a)pyrene (B(a)P; Aldrich Chemical Company, Milwaukee, WI), 9-aminoacridine (9-AA; Sigma Chemical Company, St. Louis, MO), and N-methyl-N'-nitro-N-nitrosoguanidine (MNNG; Sigma). B(a)P is a polycyclic aromatic hydrocarbon, 9-AA is an aromatic amine, and MNNG is a nitrosamide. Each compound was dissolved in sterile spectrophotometric-grade DMSO (Schwartz-Mann, Spring Valley, NY) to make a 1-mg/ml stock solution. Each stock solution was added to culture medium in appropriate amounts to make the desired final concentration. In each case, the DMSO was adjusted to 0.5% final concentration in the culture medium.

Flow Cytometry

Cells that were to be used for flow cytometry were plated onto 35-mm culture dishes at 20,000 cells/cm^2 and allowed to settle and attach overnight. These were then exposed to B(a)P at the same concentrations as those cultures being used for anaphase examination. Following the 48-h exposure period, the cultures were replenished with new medium and allowed to grow in the absence of the test substance for 7-14 d. During this period, the cultures were split 1:2 on Days 5 and 12 and replated. This insured the removal of the test material and allowed continued growth of the cultures. On Days 7 and 14, triplicate cultures were washed with PBS and treated with the fluorescent dye DAPI (4',6-diamidino-2-phenylindole-2HCl) (Accurate Scientific Company, Westbury, NY) in NP-40 detergent solution to selectively stain and osmotically separate the nuclei from the cytoplasm. The nuclei were then examined with a nonsorting flow cytometer (ICP-22) and the coefficient of variation (CV) of nuclear DNA fluorescence was determined for G1 cells. The fluorescent intensity measurement of each nucleus is directly proportional to its total DNA content (Dean et al., 1982). The measurements from the treated cultures were compared to those of the untreated and solvent controls, to determine whether there was an uneven distribution of chromosomes or chromosome fragments to daughter cells as a result of exposure to the test compounds (Deaven and Petersen, 1973; Callis and Hoehn, 1976; Cram and Lehman, 1977; Otto and Oldiges, 1980).

Because these procedures require the presence of dividing cells, it was necessary to measure the cytotoxicity of each compound prior to doing the test. This enabled us to determine the maximum chemical concentration that would not inhibit mitosis yet would produce sufficient anaphase cells for scoring. This maximum concentration became the high dose from which dilutions were made for constructing dose-response curves for both the aa test and DNA content.

Sediment Extract Preparation

Multiple sediment samples were collected at each of six sites around Puget Sound using a 0.1-m^2 van Veen grab. These samples were returned to the laboratory and frozen at −20°C until analyzed. Each 150-g (wet weight) sediment sample was serially extracted with pesticide-grade (nanograde) methanol and dichloromethane as solvents. Details of the field collection protocol and extraction procedures are described in Chapman et al. (1982) and Malins et al. (1980), respectively.

Following extraction, the combined solvents were evaporated with a stream of nitrogen, and the total weight of the organic extract was determined gravimetrically. These residues were back-extracted with DMSO and reweighed, and the amount soluble in DMSO was determined. The amount soluble in DMSO was used as a known concentration (mg/ml) stock solution from which dilutions into culture medium were made in a manner similar to that described under Individual Chemicals.

The selection of sample sites was based on their known human use and history of sediment contamination. Polycyclic aromatic hydrocarbons (PAH), polychlorinated biphenyls (PCB), and metals were present at elevated concentrations in Elliott Bay, the Duwamish River, Blair Waterway, City Waterway, and Sinclair Inlet. Blair and City Waterways also had high levels of other types of chlorinated hydrocarbons (Malins et al., 1980; Dexter et al., 1981; Riley et al., 1981). The Port Madison site is remote from any industry and is only sparsely populated relative to the other sites. However, it is surrounded by residential areas and has several small sewage discharge sites. Malins et al. (1980) reported that compared to the other sites, this area had relatively low yet readily detectable levels of pollutants.

RESULTS

Anaphase aberrations produced by both individual compounds and marine sediment extracts resembled those previously described (Nichols et al., 1977; Kocan et al., 1982). Figure 1 shows examples of the various types of anaphase disruption that were observed and used for scoring the genotoxic potential of the substances under study.

Individual Chemicals

Each of the three classes of organic chemicals tested produced an increase in the number of damaged anaphase cells in proportion to the concentration of the compound present in the culture medium. Figure 2 summarizes the results of this experiment and shows that the lower limits of detection range from 0.05 μg/ml for B(a)P to 0.5 μg/ml for 9-AA. Concentrations beyond the end of each curve shown in Figure 2 were cytotoxic or inhibited mitosis, and resulted in too few anaphase cells for scoring and analysis. Consequently, the need for preliminary titration of the test material in order to find the maximum concentration that still permits mitotic activity is obvious.

Flow Cytometry

There was a significant difference in the CV of DNA content in treated cells when compared to controls 7 and 14 d post treatment (ANOVA, $p < 0.05$). From Table 1 it can be seen that the DNA variation is proportional to the increase in the number of damaged anaphase cells, indicating that the damage that occurred at the time of exposure was incorporated into the daughter cells and duplicated in subsequent generations. It is also evident from the table that cell replication was also reduced in proportion to the concentration of chemical present in the initial exposure medium. Such results indicate the presence of some residual cytotoxic effect due either to the effects of the parent compound or its metabolites.

Puget Sound Sediment Extracts

Organic extracts of Puget Sound sediments were found to be genotoxic as measured by the aa test. Among the positive sites, the levels of anaphase damage

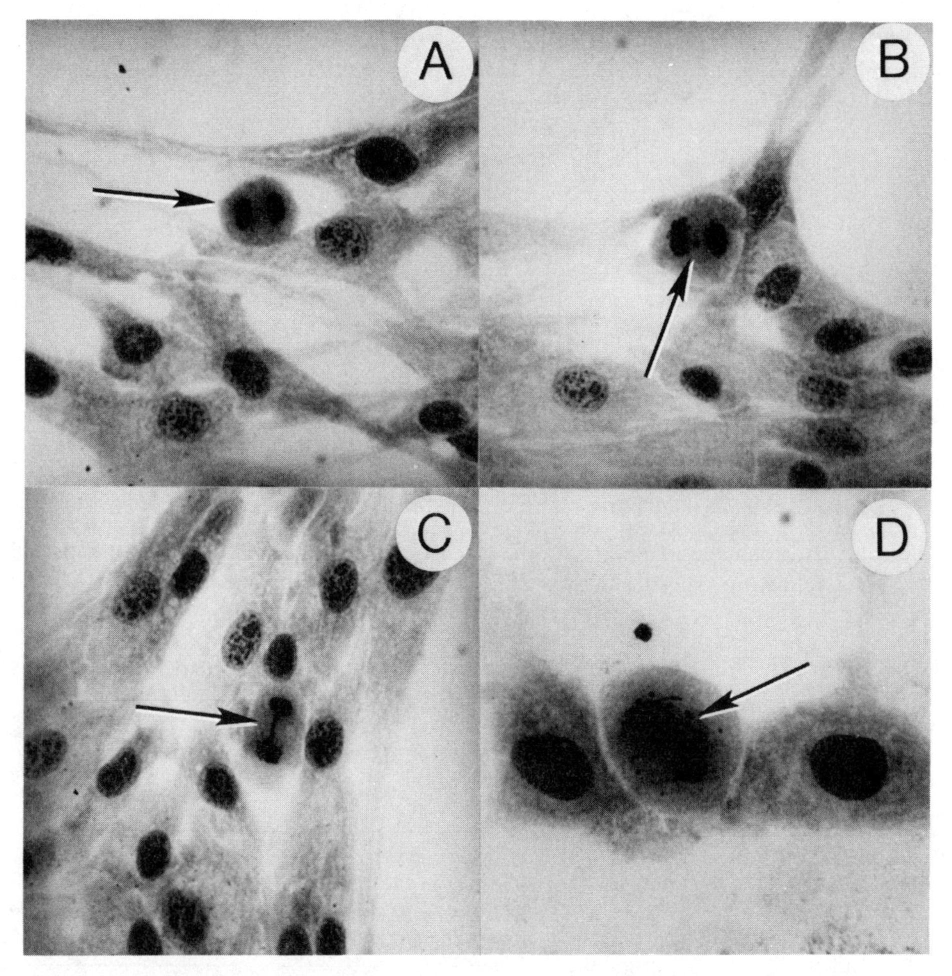

Figure 1. Anaphase cells showing various types of aberrations. A, normal cell; B, lagging chromosome; C, chromosome bridge; D, bridges, fragments, and displaced chromosomes.

varied from just above background to >80%. Because the area from which bottom samples were collected at each site was so large, the individual samples from each site could not be treated as replicates. The data are therefore presented as the mean response for all samples from a single site; the range of responses is also given for comparison. Table 2 summarizes the results of aa tests conducted on sediments from six sites around Puget Sound and compares the data to those obtained from untreated control cultures.

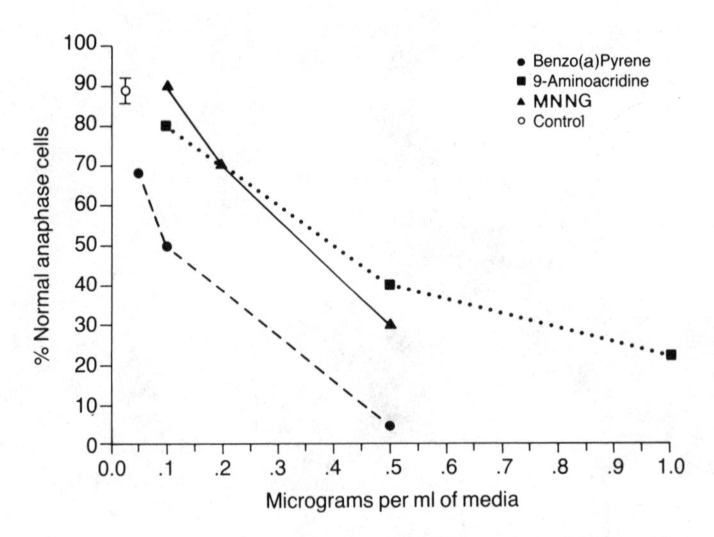

Figure 2. Decrease in normal anaphase cells observed in RTG-2 cell cultures following exposure to three classes of genotoxic chemicals.

Table 1. Relationship of Nuclear DNA Content (CV) and Anaphase Aberrations in Cells Treated with Benzo(a)pyrene

B(a)P Concentration (μg/ml)	Anaphase Aberrations (%)	CV (%)		Population Doublings	
		Day 7	Day 14	Day 7	Day 14
0	10.5	4.24	4.55	2.28	4.64
DMSO	10.0	4.31	4.85	2.55	4.64
0.10	12.0	4.95	4.69	2.54	4.34
0.20	21.5[a]	5.04[a]	5.09[a]	1.23	4.05
0.50	59.5[a]	6.06[a]	6.63[a]	0.95	1.78

[a]Significantly higher than control cultures (ANOVA, $p < 0.05$).

Table 2. Genotoxic Response of RTG-2 Cells Following 48-hour Exposure to Organic Extracts of Various Puget Sound Sediment Samples

Site Location	Samples per Site	Percent Positive Samples per Site[a]	Mean Positive Extract Concentration (μg/ml)[b]	Mean Percent Abnormal Anaphase Cells
Elliott Bay	17	71	7.3 (1.6-13)	25 (18-35)
Duwamish River	23	52	10.4 (0.9-13.5)	23 (18-58)
City Waterway	4	100	20.2 (4.9-25)	27 (21-34)
Blair Waterway	9	78	6.4 (0.4-15.5)	29 (18-48)
Sinclair Inlet	12	67	14.4 (4-25)	52 (23-82)
Port Madison	6	33	4.3 (2.8-5.8)	19 (18-19)
Control cultures	10	0	N/A	12 (8-14)

[a]A sample was rated as positive when the number of abnormal anaphase cells exceeded the 95% confidence limits determined for the control cultures (i.e., 18 or more abnormal cells per 100 cells).
[b]Numbers in parentheses represent the range of values obtained from the various samples taken from a single site.

The number of positive samples obtained from the five known contaminated sites ranged from 52% (12/23) to 100% (4/4). In contrast, only 33% (2/6) of the samples from the remote Port Madison site were positive (Table 2). It can also be seen from the table that the positive sediment samples from the contaminated sites produced higher levels of aa than did those from Port Madison or the untreated controls.

The amount of extract required to produce a genotoxic effect ranged from <1 μg/ml to >20 μg/ml in culture medium (Table 2). Five sediment extracts were selected specifically to demonstrate the wide range of responses that can result from exposure of cells to these extracts, and to demonstrate the quantitative nature of the aa test (Figure 3). These five sediments were chosen because of their differences and are not meant to represent any particular site within Puget Sound.

DISCUSSION

Chromosomal variation is not unusual within a species or cell population, but generally remains relatively low and constant over time. If, however, the cells are exposed to physical or chemical mutagens, the frequency of chromosomal variation increases proportionately. This increase in variation is of concern in the context of risk assessment because of its potential to cause long-term genetic effects. It has been estimated that 1/3 of all human abnormalities can be attributed to structural rearrangement of chromosomes, with the other 2/3 being attributed to the presence or absence of extra chromosomes (Evans, 1983). This unequal distribution of chromosomes and chromosome fragments is the basis of the aa test.

The aa test, when performed with the RTG-2 cell line, has been shown to be sensitive to a wide range of chemical mutagen/carcinogens as well as unknown mixtures. It is likely that other cell types will perform equally well and for some purposes even better.

The aa test is relatively easy to perform, requiring no specialized facilities or equipment. It is capable of detecting lesions that are secondary consequences of

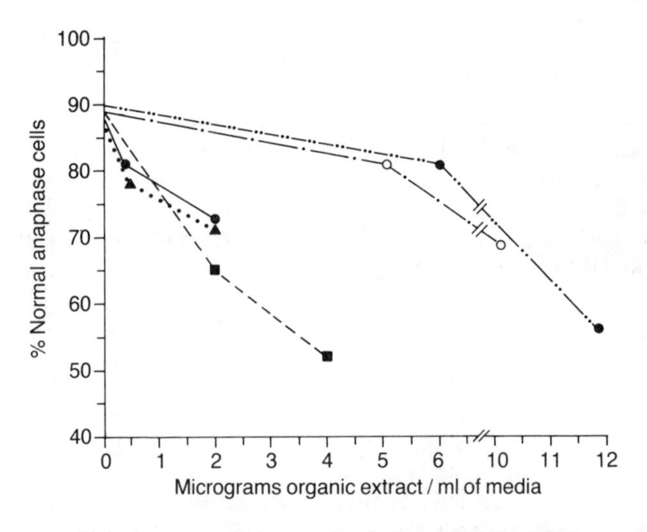

Figure 3. Dose-response curves for anaphase damage in RTG-2 cells following exposure to five different Puget Sound sediment extracts.

structural changes in chromosomes or chromatids as well as microtubule (spindle) damage, each of which could result in missegregation of chromosomes or chromatids, which can lead to aneuploidy (Evans, 1983). Aneuploidy has further been demonstrated to be perpetuated into subsequent generations of cells by the use of the relatively new technique of flow cytometry. Both techniques can measure chromosomal damage that has occurred and can detect the damage that is passed on to subsequent generations of cells. Therefore, it may now even be possible to eliminate visual scoring of cells and rely only on flow cytometry to evaluate the degree of cytogenetic damage in terms of increased aneuploidy resulting from exposure to genotoxic agents.

Scoring and interpretation of aa test data is relatively uncomplicated. Each anaphase cell is scored as either normal or damaged (+ or −) as determined by microscopic examination. The final count is compared to that obtained from a set of control cultures that have been processed concurrently with the experimental cultures. Control cultures should consist of untreated (negative) controls, solvent/diluent controls, and positive controls. The positive controls serve to verify the sensitivity and proper functioning of the aa test during the particular set of experiments. As pointed out earlier, the RTG-2 cell line contains a relatively high level of MFO activity, thereby eliminating the need to add exogenous enzymes (microsomes). We have found that over a period of four years, the baseline level of abnormal anaphases has remained between 8 and 12%. The use of historical controls is recommended for procedures that may require comparisons over a long period of time, such as the environmental assessments before and after a "cleanup" activity or prior to new human uses (Brusick, 1980).

The aa test has several features that make it especially relevant for assessing the genotoxicity of unknown evironmental samples. One is that it is not restricted to a particular species or cell type. The RTG-2 cell line served our purposes; other investigators might choose a species peculiar to the area they are studying or select a cell type with a specific set of characteristics (e.g., presence or absence of MFO system), thus tailoring the test system to the study area or some set of specific requirements. Secondly, the aa test has an *in vivo* counterpart. Liguori and Landolt (1985) have shown that the same genotoxic response we observed also occurs in trout embryos, and that the increase in chromosome damage is related to other types of abnormalities observed later in the embryo's development. This same type of anaphase lesion was observed by Longwell and Hughes (1980) in marine fish embryos collected from a contaminated area, but not in those collected from a clean area. Our data support their conclusion that the lesions in marine fish embryos were due to previous exposure to waterborne contaminants.

Since a dose-response curve can be generated by using increasing concentrations of test material, the test can be used to quantitate toxic levels of unknown substances and thus make it possible to compare the toxic potential of samples collected from different geographic areas. Such a comparison can be made on the basis of total organic content, as demonstrated in Figure 3, or can be extrapolated back to the original sample weight.

Although metaphase cells have been considered the more desirable cell type for genotoxicity testing, we feel that the aa test has good potential as an *in vitro* cytogenetic assay for complex mixtures if the appropriate cell type is chosen, and that the use of flow cytometry may be a logical extension of the aa test that would eliminate the need to microscopically examine each culture.

ACKNOWLEDGMENTS

This work was supported in part by grants from the U.S. Environmental Protection Agency (R810057); the National Institute of Environmental Health Sciences (ES 02190); and the Office of Marine Pollution Assessment, National Oceanographic and Atmospheric Administration (NA80RAD00053). The authors wish to thank K.M. Sabo and R.N. Dexter (E.V.S. Company) for technical assistance and R. Stahl for manuscript review.

REFERENCES

Ames, B.N., W.E. Durston, E. Yamasaki, and R.D. Lee. 1973. Carcinogens are mutagens: a simple test system combining liver homogenates for activation and bacteria for detection. Proc. Natl. Acad. Sci. U.S.A. 70:2281-2285.

Brusick, D. 1980. Principles of Genetic Toxicology. Plenum Press: New York and London. 279 pp.

Callis, J., and H. Hoehn. 1976. Flow-fluorometric diagnosis of euploid and aneuploid human lymphocytes. Am. J. Hum. Genet. 28:577-584.

Chapman, P.M., G.A. Vigers, M.A. Farrell, R.N. Dexter, E.A. Quinlan, R.M. Kocan, and M.L. Landolt. 1982. Survey of Biological Effects of Toxicants Upon Puget Sound Biota. I. Broad-scale Toxicity Survey. Technical Memorandum OMPA-25, National Oceanographic and Atmospheric Administration. 98 pp.

Cram, L.S., and J.M. Lehman. 1977. Flow microfluorometric DNA content measurements of tissue culture cells and peripheral lymphocytes. Hum. Genet. 37:201-206.

Dean, P.N., J.W. Gray, and F.A. Dolbeare. 1982. The analysis and interpretation of DNA distributions measured by flow cytometry. Cytometry 3:188-195.

Deaven, L.L., and D.F. Petersen. 1973. The chromosomes of CHO, an aneuploid Chinese hamster cell line: G-band, C-band and autoradiographic analysis. Chromosoma 41:129-144.

Dexter, R.N., D.E. Anderson, E.A. Quilan, L.S. Goldstein, R.M. Strickland, S.P. Pavlou, J.R. Clyton, Jr., R.M. Kocan, and M.L. Landolt. 1981. A Summary of Knowledge of Puget Sound Related to Chemical Contaminants. Technical Memorandum OMPA-13, National Oceanographic and Atmospheric Administration. 435 pp.

Diamond, L., and H.F. Clark. 1970. Comparative studies on the interaction of benzo(a)pyrene with cells derived from poikilothermic and homeothermic vertebrates. I. Metabolism of benzo(a)pyrene. J. Natl. Cancer Inst. 45:1005-1011.

Evans, H.J. 1983. Cytogenetic methods for detecting effects of chemical mutagens. In: Cellular Systems for Toxicity Testing. G.M. Williams, V.C. Dunkel, and V.A. Ray, eds. Ann. N.Y. Acad. Sci. 407:131-141.

Kocan, R.M., M.L. Landolt, and K.M. Sabo. 1982. Anaphase aberrations: A measure of genotoxicity in mutagen-treated fish cells. Environ. Mutagen. 4:181-189.

Longwell, A.C., and J.B. Hughes. 1980. Cytologic, cytogenetic and developmental state of Atlantic mackeral eggs from sea surface waters of the New York Bight, and prospects for biological effects monitoring with ichthyoplankton. Rapp. P-v. Reun. Cons. Int. Explor. Mer. 179:275-291.

Liguori, V.M., and M.L. Landolt. 1985. Anaphase aberrations: An *in vivo* measure of genotoxicity. In: Short-Term Genetic Bioassays in the Analysis of Complex Environmental Mixtures IV. M.D. Waters, S.S. Sandhu, J. Lewtas, L. Claxton, G. Strauss, and S. Nesnow, eds. Plenum Press: New York.

Malins, D.C., B.B. McCain, D.W. Brown, A.K. Sparks, and H.O. Hodgins. 1980. Chemical Contaminants and Biological Abnormalities in Central and Southern Puget Sound. Technical Memorandum OMPA-2, National Oceanographic and Atmospheric Administration. 295 pp.

Nichols, W.W., P. Moorhead, and G. Brewen. 1972. Chromosome methodologies in mutation testing--Report of the Ad Hoc Committee of the Environmental Mutagen Society and the Institute for Medical Research. Toxicol. Appl. Pharmacol. 22:269-275.

Nichols, W.W., R.C. Miller, and C. Brandt. 1977. In vitro anaphase and metaphase preparations in mutation testing. In: Handbook of Mutagenicity Test Procedures. B.J. Kilbey, M. Legator, W.W. Nichols, and C. Ramel, eds. Elsevier/North Holland: New York. p. 485.

Otto, J.J., and H. Oldiges. 1980. Flow cytogenetic studies in chromosomes and whole cells for the detection of clastogenetic effects. Cytometry 1:13-17.

Riley, R.G., E.A. Crecelius, M.L. O'Malley, K.H. Abel, and D.C. Mann. 1981. Organic Pollutants in Waterways Adjacent to Commencement Bay (Puget Sound). Technical Memorandum OMPA-12, National Oceanographic and Atmospheric Administration. 81 pp.

Russell, W.C., C. Newman, and D.H. Williamson. 1975. A simple cytochemical technique for demonstration of DNA in cells infected with mycoplasmas and viruses. Nature 253:461-462.

Schmid, W. 1982. The micronucleus test: An in vivo bone marrow method. In: Cytogenetic Assays of Environmental Mutagens. T.C. Hsu, ed. Allanheld, Osmun and Co.: New Jersey. pp. 221-229.

Stetka, D.G., and S. Wolff. 1976. Sister chromatid exchange as an assay for genetic damage induced by mutagen-carcinogens. II. In vitro test for compounds requiring metabolic activation. Mutat. Res. 41:343-350.

Wolf, K., and M.C. Quimby. 1962. Established eurythermic line of fish cells in vitro. Science 135:1065-1066.

ANAPHASE ABERRATIONS: AN *IN VIVO* MEASURE OF GENOTOXICITY

Vincent M. Liguori and Marsha L. Landolt

School of Fisheries, University of Washington, Seattle, Washington 98195

INTRODUCTION

A large number of short-term genotoxicity tests, ranging from microbial to eukaryotic cell assays, have been developed for use in toxicological investigations. Most of these have proven to be very useful for determining the genotoxic potential of pure compounds and complex mixtures (Landolt and Kocan, 1983; Schmid, 1982; Stetka and Wolff, 1976; Ames et al., 1973). One of the major controversies that surrounds the use of these tests is their significance as predictors of risk to whole-animal systems. The purpose of the investigation reported here was to address that question by comparing the results of one short-term *in vivo* assay with long-term biological effects.

The assay selected for use was the anaphase aberration test, a relatively simple procedure that allows one to visualize chromosomal macrolesions in mitotic cells during anaphase. The test has been used *in vivo* (Hose et al., 1984; Longwell and Hughes, 1980; Nichols et al., 1977) and *in vitro* (Kocan et al., 1982), and has been recommended for use as a measure of toxicity and environmental contamination by the Ad Hoc Committee of the Environmental Mutagen Society and the Institute for Medical Research (1972). The sensitivity and utility of the test have been verified *in vitro* using both pure compounds (Kocan et al., 1982) and complex mixtures (Chapman et al., 1982). Work in our laboratory has shown that visible chromosomal damage can be induced in cultured cells at very low exposure levels and that the damage is perpetuated into subsequent cell generations (Kocan and Powell, 1985). Biological effects measurable as cell death or mitotic inhibition (Kocan et al., 1979), mutation (Kocan et al., 1981), or cellular inclusions (Kocan et al., 1983) have also been demonstrated. In the current study, we attempted to translate our *in vitro* techniques to an *in vivo* system, to determine whether anaphase aberrations similar to those observed *in vitro* could be produced *in vivo* with the same species, and to determine whether mutagen-exposed animals demonstrated other biological changes.

Because our laboratory is interested in aquatic toxicology, we chose an aquatic animal species as our test system. Rainbow trout (*Salmo gairdneri*) were used as a model because (1) all life stages are commercially available on a year-round basis, (2) their husbandry requirements are well established, (3) there is a broad literature base describing their microanatomy and pathological conditions, (4) an established, well-characterized rainbow trout cell line is available and has been used in our *in vitro* studies, and (5) primary cell cultures can be readily established. Embryos were selected as the life stage of choice for this study because their cells are rapidly dividing and provide large numbers of anaphase figures for quantification.

By using such a model system, we hoped to establish relationships among results obtained from whole-animal tests and those obtained from subcellular/cellular tests (Kocan and Powell, 1985; Kocan et al., 1982). The objectives of the present study were (1) to determine whether chromosomal damage seen *in vitro* was reproducible *in vivo* using the same genotoxic agent, (2) to determine whether a correlation existed between *in vivo* chromosomal damage and developmental or pathological defects in the whole animal, and (3) to provide further information on the sensitivity and standardization of the *in vivo* assay by using a range of concentrations and by examining the persistence of the damage following exposure.

MATERIALS AND METHODS

Test Organism

Rainbow trout (*Salmo gairdneri*) eggs and sperm were obtained from the University of Washington Hatchery (Seattle, WA). The sperm were thoroughly mixed with the eggs and allowed to sit for 15 min. Then the eggs were washed free of the sperm. The fertilized eggs were water-hardened for 1 h, separated into subgroups, and placed in flowing dechlorinated city water (10°C) until the blastodisc stage of development (2-3 d post fertilization). To minimize genetic variability, all of the eggs and sperm used were derived from a single female/male.

Test Compound

N-Methyl-N'-nitro-N-nitrosoguanidine (MNNG) was purchased from Sigma Chemical Company (St. Louis, MO). MNNG is a direct-acting, alkylating agent that is genotoxic to rainbow trout cells *in vivo* (Hendricks et al., 1980a, b) and *in vitro* (Kocan et al., 1982).

MNNG was dissolved in spectrophotometric-grade dimethylsulfoxide (DMSO; Schwartz-Mann, Spring Valley, NY) to form a stock solution (1 mg MNNG/ml DMSO). Aqueous working solutions (0.0, 0.25, 2.5, 5.0 μg/ml) were immediately prepared and used for embryo exposure. The DMSO concentration was held constant (5 μg/ml) in all of the working solutions.

At the completion of the exposure period, all MNNG solutions and contaminated glassware were treated with full-strength hypochlorite prior to disposal.

Exposure

Twenty-four hours prior to exposure, blastodisc-stage embryos were gently transferred to 100-ml plastic specimen cups containing 1.5 ml water (30 eggs/cup) and placed in an incubator (10°C). After 24 h, the cups were removed from the incubator, and injured eggs were culled and replaced with viable eggs. The water was then withdrawn, and 1.5 ml of test solution was placed into each cup. Five replicates (30 eggs/replicate) were used for each of five exposure groups (0.0, 0.25, 2.5, 5.0 µg/ml MNNG; 5.0 µg/ml DMSO). The eggs were returned to the incubator (10°C) and left for 24 h. At the end of the exposure period, the eggs were rinsed twice with clean water and transferred to flowing dechlorinated city water (10°C). For 10 d following exposure, three eggs per replicate were removed daily, placed in 10% neutral buffered formalin, and examined for the presence of anaphase aberrations.

A second group of eggs was exposed to MNNG and examined for the presence of hatching or developmental defects. The exposure procedures were similar to those described above except that replicates consisted of 50 eggs per cup (in 2.5 ml test solution) and only two MNNG concentrations (0.0, 2.5 µg/ml) were tested. After exposure, the eggs were placed in flowing dechlorinated city water (10°C) and monitored for 6 wk post hatching.

Anaphase Aberrations

Formalin-fixed embryos were dissected from the yolk and placed in acetic acid (50-70%) for 2-3 h. The embryos were then placed for 2-3 h in 2% aceto orcein stain (Humason, 1967) that contained proprionic acid in a ratio of 19 parts stain to 1 part acid (Longwell and Hughes, 1980). The stained embryo was placed in a drop of acetic acid on a glass microscope slide and gently squashed between the slide and a glass coverslip. The squash preparations were sealed with clear fingernail polish and examined by light microscopy (1000X) for the presence of aberrant anaphase figures. The number of aberrant figures was determined either by counting the first 100 anaphase cells encountered or, in very young embryos, by counting all of the anaphase cells present on the slide (Figure 1).

Hatching and Developmental Defects

The number of eggs that completed embryonic development and hatched successfully was determined for each treatment group. After hatching, the emergent eleutheroembryos ("sac fry") were examined for the presence of grossly visible defects, and their survival and development were monitored for six additional weeks. Some fish from each treatment group were maintained for a full year after hatching. At the end of that period, they were sacrificed, necropsied, and examined grossly and microscopically.

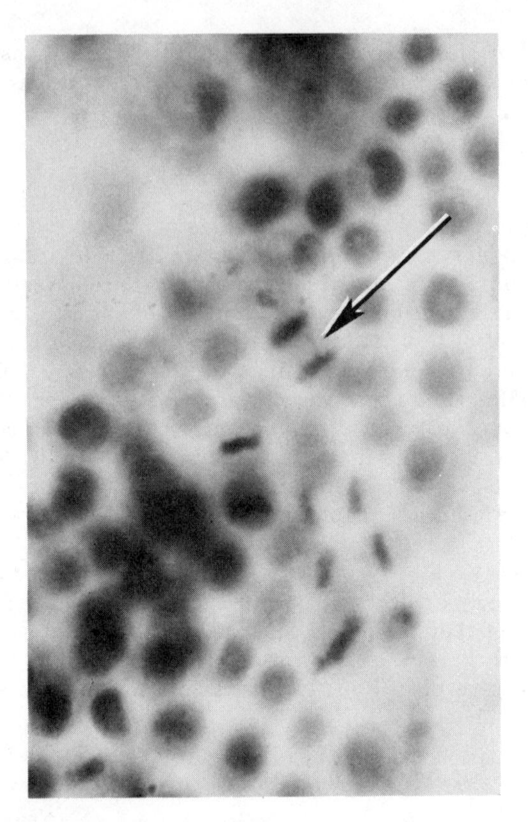

Figure 1. Squash preparation of blastodisc-stage rainbow trout (*Salmo gairdneri*) embryo showing several anaphases, including one with an attached fragment (arrow). 400X.

Statistical Analysis

One-way analysis of variance (ANOVA) was used to determine whether differences existed among the treatment groups. Differences were considered to be significant if $p < 0.05$.

RESULTS

Anaphase Aberrations

Anaphase aberrations were observed in the embryo squash preparations. The aberrations induced *in vivo* were similar to those produced *in vitro* in rainbow trout cells in our laboratory (Kocan et al., 1982) and consisted of bridges, chromosomal fragments, and attached fragments (Figure 2).

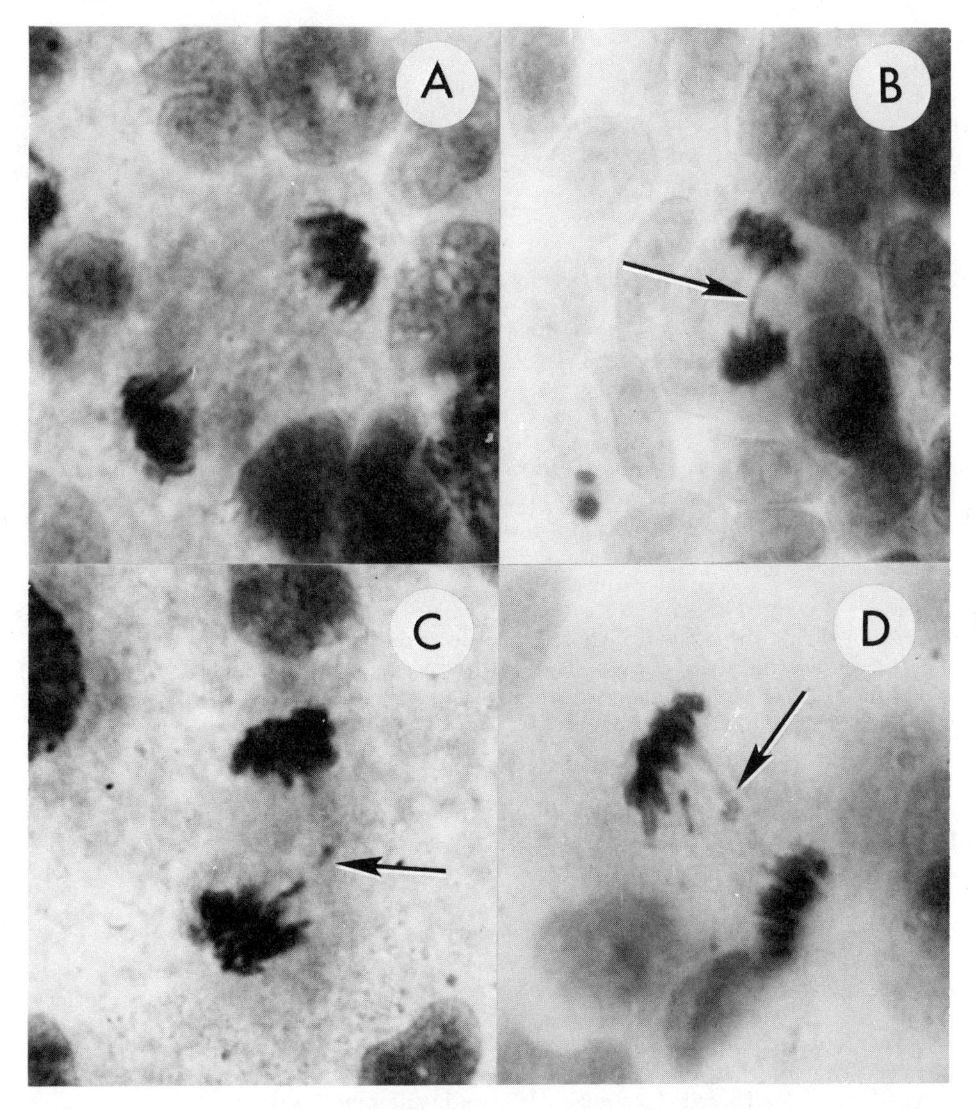

Figure 2. Several anaphase figures from rainbow trout embryos. A, normal anaphase. Aberrations appear as bridges (B), trailing fragments (C), and attached fragments (D). 1000X.

The incidence of aberrations was dose responsive, with increasing numbers of aberrant figures occurring at increasing MNNG dosage levels (Figure 3). For untreated and solvent-treated embryos, the incidence of aberrations averaged from 3.4% to 8.7% over the first 7 d following exposure. Embryos exposed to 0.25 μg/ml MNNG had 17.0% abnormal anaphase figures on Day 1, 18.3% on Day 3, and 22.6% on Day 4. Embryos exposed to 2.5 and 5.0 μg/ml MNNG had initial (Day 1) incidences of

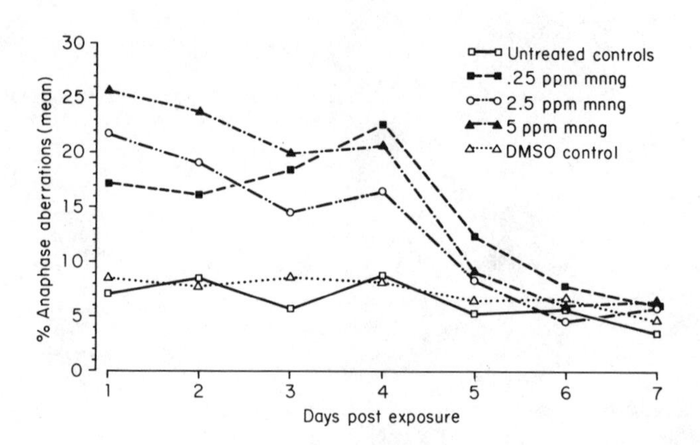

Figure 3. Dose response to MNNG in rainbow trout embryos exposed at blastodisc stage. Data shown are mean percent anaphase aberrations for 7 d following exposure to several concentrations of MNNG.

21.6% and 25.5%, respectively. Those levels decreased somewhat on Days 2 and 3 and then increased slightly on Day 4. By Day 5 post exposure, the anaphase aberration levels in all three treatment groups dropped to values that were not significantly different from those of the control groups.

Analysis of the data showed that the anaphase aberration levels for the three exposure groups were statistically different from those of the controls on Days 1-4 ($p < 0.05$). During the same period, the incidence levels of embryos exposed to 2.5 µg/ml MNNG could not be distinguished from those of either the higher or lower dosages.

Hatching and Developmental Defects

In the untreated and solvent-treated controls, approximately 88% of the eggs hatched. This contrasted with the 2.5 µg/ml MNNG exposure group, in which only 75% hatched (Table 1). These values were statistically different ($p < 0.05$).

At hatching, differences in pigmentation were noted. In the control groups, all of the fry were light tan, while all of the exposed fry were nonpigmented. Within 1 wk, the control fish darkened to brown and the exposed fish turned black. The black coloration persisted through the third week post hatching. By Week 4, all of the exposed fish turned brown and were thereafter indistinguishable in color from the controls (Table 2).

In the control groups, 3% of the fish died during the first 6 wk of life. Mortality levels in the exposure groups at 6 wk post hatching averaged 29.1% (Table 1).

Six months post hatching, moribund fish began to appear in the exposure group. These fish and representative individuals from each of the control groups were

Table 1. Hatching Success and Post Hatching (6 wk) Mortality of Rainbow Trout Embryos Exposed to MNNG (24 h) During the Blastodisc Stage of Development

MNNG (μg/ml)	Hatch (%)	Post Hatching Mortality (%)
0.0 (untreated)	87.4	2.7
0.0 (DMSO)	88.3	3.1
2.5	74.6	29.1

Table 2. Pigment Development in Rainbow Trout Alevins Exposed to MNNG (24 h) During the Blastodisc Stage of Development

MNNG (μg/ml)	Number of Weeks Post Hatching	Color
0.0 (untreated)	0	Tan
0.0 (DMSO)		Tan
2.5		No pigment
0.0 (untreated)	1-3	Brown
0.0 (DMSO)		Brown
2.5		Black
0.0 (untreated)	4	Brown
0.0 (DMSO)		Brown
2.5		Brown

sacrificed. At necropsy, the MNNG-exposed fish exhibited abnormal renal development. The right anterior kidney was hypoplastic and, in some fish, the tissue was barely visible (Figure 4). By contrast, the kidneys of control fish were fully developed and were bilaterally symmetrical. Abnormal renal development occurred in 16.9% of the exposed fish versus 0% of the control fish. Exposed fish that survived for 1 yr post hatching did not exhibit renal hypoplasia.

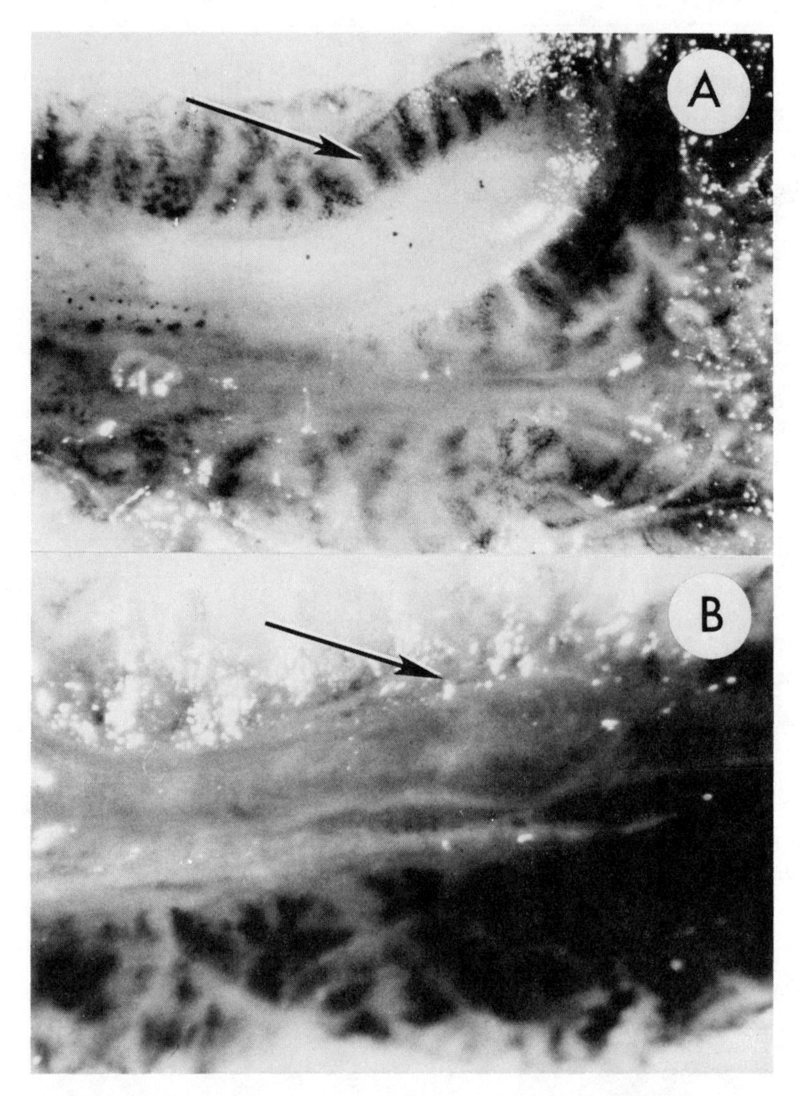

Figure 4. Abnormal renal development in trout fingerlings exposed to MNNG during blastodisc stage of development. A, normal anterior kidney from control rainbow trout fingerling. B, hypoplasia of right anterior kidney in fingerling exposed to MNNG. 6.4X.

DISCUSSION

Anaphase Aberrations

In this study, anaphase aberrations were observed in embryos exposed to MNNG during the blastodisc stage of development. The aberrations were comparable to those that have been observed *in vivo* and *in vitro* by other investigators (Kocan et al., 1982; Hose et al., 1984; Longwell and Hughes, 1980). Anaphase aberrations occurred at significantly elevated levels during the first 4 d following exposure and returned to baseline levels by Day 5-7 post exposure. A dose response was initially observed; increased anaphase aberrations occurred at increased MNNG concentrations.

Squash preparations of blastodisc-stage embryos provided a monolayer of cells that afforded a degree of resolution similar to that obtained in cell cultures. Significant increases in anaphase aberrations were detected *in vivo* at exposure levels as low as 0.25 µg/ml MNNG; thus, the sensitivity of the *in vivo* assay proved comparable to that of the *in vitro* system (Kocan et al., 1982). Clearly, a correlation exists between chromosomal damage seen *in vivo* and *in vitro* when cells of the same species are exposed to the same genotoxic agent.

On Days 1-4, anaphase aberration levels were significantly higher in exposed embryos than in untreated and solvent-treated controls. On Day 4, an increase in anaphase aberrations (from Day 3) was observed in all treatment groups but was most pronounced at the lowest (0.25 µg/ml) exposure level. This increase on Day 4 probably represented expression of damage occurring during a second wave of mitotic division. It is possible that the higher exposure concentrations were toxic, and that the lower relative incidence of aberrations noted in these groups reflected inhibition of mitosis or actual cell death.

By Day 5, anaphase aberrations in all treatment groups returned to baseline levels. There are several possible explanations for this observation: (1) The decrease in anaphase aberrations may have resulted from the action of DNA repair mechanisms to reverse some of the damage caused by MNNG. Since MNNG is an unstable compound that does not persist for long periods of time, damage due to MNNG probably occurs very soon following exposure. (2) It is possible that aberrations can be detected visually only during the first or second mitotic cycles following exposure, since damage occurring during that period would rapidly be incorporated into the genome of subsequent cell generations and would no longer be visible by light microscopic examination. (3) The aberrations may be diluted by the rapid proliferation of embryonic cells. (4) Some of the damage may result in cell death, thereby eliminating those cells from the older embryo.

Hatching and Developmental Defects

Immediately following hatching, the MNNG-exposed fish showed dramatically different pigmentation from that of the unexposed fish. This difference in pigmentation persisted for up to 4 wk, at which time all treatment groups developed normal color. This phenomenon could be attributed to genetic alterations, but may well have been the result of alterations in protein synthesis or homonal production. MNNG is known to bind both to DNA and RNA; thus, in addition to causing cytogenetic damage, MNNG has the potential to alter protein synthesis.

Unilateral renal hypoplasia was noted in 16.9% of the MNNG-exposed fish. This defect was highly site specific and was restricted to the right anterior kidney. The regularity of the lesion and its occurrence only in MNNG-exposed fish was strong evidence that it resulted from treatment of the early embryo with a genotoxic agent. Lesions of this type are very likely to vary with the species or strain of experimental animal; nonetheless, they can provide valuable insight into the mechanisms of injury exerted by a given compound.

This study has demonstrated that it is possible to correlate short-term cytogenetic responses with pathological conditions in whole organisms. The correlation provides support for the use of short-term assays as measures of environmental contamination and as indicators of potential hazard. In addition, the relationship between cytogenetic responses observed *in vivo* and those observed *in vitro* enables one to predict risk to whole organisms based on the results of *in vitro* tests.

ACKNOWLEDGMENTS

This work was supported in part by grants from the U.S. Environmental Protection Agency (R-810057-01-0), the National Institute of Environmental Health Sciences (5-P30-ES-02190-02), and the Office of Marine Pollution Assessment, National Oceanographic and Atmospheric Administration (NA80RAD-00053), and a Grant-in-Aid of Research from Sigma Xi, The Scientific Research Society.

REFERENCES

Ad Hoc Committee of the Environmental Mutagen Society and the Institute for Medical Research. 1972. Chromosome methodologies in mutagen testing. Toxicol. Appl. Pharmacol. 22:269-275.

Ames, B.N., W.E. Durston, E. Yamasaki, and F.D. Lee. 1973. Carcinogens are mutagens: a simple test system combining liver homogenates for activation and bacteria for detection. Proc. Natl. Acad. Sci. U.S.A. 70:2281.

Chapman, P.M., G.A. Vigers, M.A. Farrell, R.N. Dexter, E.A. Quinlan, R.M. Kocan, and M.L. Landolt. 1982. Survey of Biological Effects of Toxicants Upon Puget Sound Biota. I. Broad-scale Toxicity Survey. Technical Memorandum OMPA-25, National Oceanographic and Atmospheric Administration. 98 pp.

Hendricks, J.D., J.H. Wales, R.O. Sinnhuber, J.E. Nixon, P.M. Loveland, and R.A. Scanlan. 1980a. Rainbow trout (*Salmo gairdneri*) embryos: a sensitive model for animal carcinogenesis. Fed. Proc. 39:3222-3229.

Hendricks, J.D., R.A. Scanlan, J. Williams, and R.O. Sinnhuber. 1980b. The carcinogenicity of N-methyl-N'-nitro-N-nitrosoguanidine to the livers and kidneys of rainbow trout (*Salmo gairdneri*) exposed as embryos. J. Natl. Cancer Inst. 64:1511-1519.

Hose, J.E., V.M. Liguori, and H.W. Puffer. 1984. A method for evaluating pollutant teratogenicity and genotoxicity using freshwater or marine embryos. Presented at the Aquatic Toxicology/Seventh Symposium (American Society for Testing and Materials), Milwaukee, WI.

Humason, G.L. 1967. Animal Tissue Techniques, 2nd Ed. W.H. Freeman and Co.: San Francisco. 559 pp.

Kocan, R.M., and D.B. Powell. 1985. Anaphase aberrations: An *in vitro* test for assessing the genotoxicity of individual compounds and complex mixtures. In: Short-Term Genetic Bioassays in the Analysis of Complex Environmental Mixtures IV. M.D. Waters, S.S. Sandhu, J. Lewtas, L. Claxton, G. Strauss, and S. Nesnow, eds. Plenum Press: New York.

Kocan, R.M., M.L. Landolt, and K.M. Sabo. 1979. *In vitro* toxicity of eight mutagens/carcinogens for three fish cell lines. Bull. Environ. Contam. Toxicol. 23:269-274.

Kocan, R.M., M.L. Landolt, J. Bond, and E.P. Benditt. 1981. *In vitro* effect of some mutagens/carcinogens on cultured fish cells. Arch. Environ. Contam. Toxicol. 10:663-671.

Kocan, R.M., M.L. Landolt, and K.M. Sabo. 1982. Anaphase aberrations: a measure of genotoxicity in mutagen treated fish cells. Environ. Mutagen. 4:181-189.

Kocan, R.M., E.Y. Chi, N. Eriksen, E.P. Benditt, and M.L. Landolt. 1983. Sequestration and release of polycyclic aromatic hydrocarbons and vertebrate cells *in vitro*. Environ. Mutagen. 5:643-656.

Landolt, M.L., and R.M. Kocan. 1983. Fish cell cytogenetics: a measure of the genotoxic effects of environmental pollutants. In: Aquatic Toxicology. J. Nriagu, ed. John Wiley and Sons, Inc.: New York. pp. 335-353.

Longwell, A.C., and J.B. Hughes. 1980. Cytologic, cytogenic and developmental state of the Atlantic mackeral eggs from sea surface waters of the New York Bight, and prospects for biological effects monitoring with ichthyoplankton. Rapp. P-v. Reun. Cons. Int. Explor. Mer. 179:275-291.

Nichols, W.W., R.C. Miller, and C. Brandt. 1977. *In vitro* anaphase and metaphase preparations in mutagen testing. In: Handbook of Mutagenicity Test Procedures. B.J. Kilbey, W.W. Nichols, M. Legator, and C. Ramel, eds. Elsevier Science Publishing Company: New York. 485 pp.

Schmid, W. 1982. The micronucleus test: an *in vivo* bone marrow method. In: Cytogenetic Assays of Environmental Mutagens. T.C. Hsu, ed. Allanheld, Osmun and Company: New Jersey. pp. 221-229.

Stetka, D.G., and S. Wolff. 1976. Sister chromatid exchange as an assay for genetic damage induced by mutagen-carcinogens. II. *In vitro* test for compounds requiring metabolic activation. Mutat. Res. 41:343-350.

MEASUREMENT OF CHEMICALLY INDUCED MITOTIC NONDISJUNCTION USING A MONOCHROMOSOMAL HUMAN/MOUSE HYBRID CELL LINE

Raghbir S. Athwal[1] and Shahbeg S. Sandhu[2]

[1]Department of Microbiology, New Jersey Medical School, Newark, New Jersey 07103, and [2]Health Effects Research Laboratory, U.S. Environmental Protection Agency, Research Triangle Park, North Carolina 27711

INTRODUCTION

There is an increasing body of evidence to show that a significant proportion of human disease may be caused by exposure to environmental chemicals. Among the adverse health effects produced by these chemicals, genotoxic insult is perhaps the most serious. An exposure to toxic chemicals may produce genetic alterations in somatic and/or in germ cells. Genetic damage in somatic cells may lead to cancer or myriad other diseases, but the effects are limited only to the individuals exposed to the harmful chemicals. Genomic changes in germ cells, on the other hand, may be transmitted to successive generations and become a legacy of human heritage. Cytologically observable chromosomal changes represent the most important category of genetically defined human diseases. A variety of chromosomal aberrations, such as deletions, duplications, inversions, translocations, and numerical alterations in chromosomes (aneuploidy), have been implicated in human diseases (Hook, 1983). Aneuploidy arising through nondisjunction during gametogenesis represents the single largest category of chromosomal aberrations leading to human birth defects (Hook, 1983). Although the role of aneuploidy in initiating carcinogenesis has not been established, it has been suggested that genomic alterations may predispose individuals to a greater risk of developing cancer (Arlett and Lehmann, 1978; Levan and Mitelman, 1977; Bloom, 1972; Tsutsui et al., 1983).

Aneuploidy in chemical mutagenesis has been emphasized relatively recently, and only a limited amount of work has been reported in this direction (Zimmermann et al., 1979). Most commonly used biological systems to detect chemically induced aneuploidy involve fungi (de Bertoldi et al., 1980), yeast (Parry et al., 1981), and insects (Liang et al., 1983; Zimmering, 1982). However, mitotic poisons are known to have different effects in lower and higher eukaryotes (Haber et al., 1972), and this difference has been demonstrated to be due to different target sites on the microtubule

(Heath 1975; Bond and Chandley, 1983). In mammalian cells, total chromosome count following exposure of cells to test chemicals has been used to assess abnormal chromosome segregation at mitosis (Hsu et al., 1983; Danford, 1984; Tsutsui et al., 1983). Such assays are hampered by the fact that cultured cell lines inherently contain a variable number of chromosomes, and it is difficult to assess the loss or gain of individual chromosomes.

In this report, we have described a test system in which aneuploidy is assessed by the abnormal segregation of a single human chromosome present in a human/mouse somatic cell hybrid. A human chromosome present in mouse cells can be easily distinguished by differential staining procedures and serves as a cytogenetic marker. Progeny of a monochromosomal hybrid, in general, should contain a single human chromosome. Cells containing 2 or 0 human chromosomes would result from nondisjunction. Thus, abnormal segregation resulting from the effect of chemicals can be detected from the frequency of cells containing 2 or 0 human chromosomes. In the present studies, we have used three chemicals--Colcemid, benomyl, and cyclophosphamide--previously known to induce aneuploidy for evaluation of the proposed system. This system may be extended in the future for the genotoxic evaluation of individual agents or those contained in complex environmental mixtures.

MATERIALS AND METHODS

Cell Line and Growth Conditions

The R3-5 cell line, a monochromosomal microcell hybrid of human and mouse cells, was employed to assess chemically induced aneuploidy. The R3-5 cells containing human chromosome 2 were produced by a two-step process involving gene transfer and microcell fusion. Details of the experiments on the production of R3-5 cells will be published elsewhere (Athwal et al., in press), while the properties pertinent to the present studies will be given here.

The human chromosome 2 present in R3-5 cells carries a selectable marker Ecogpt (Mulligan and Berg, 1981). Ecogpt is a bacterial gene cloned into a plasmid that codes for the enzyme xanthine guanine phosphoribosyltransferase (XGPRT), which is analogous to the mammalian enzyme hypoxanthine guanine phosphoribosyltransferase (HPRT). XGPRT catalyzes the formation of XMP from xanthine in the salvage pathway of purine biosynthesis, an activity normally not present in mammalian cells. The cells carrying Ecogpt are capable of growth in medium containing mycophenolic acid and xanthine, while the cells without Ecogpt cannot grow in this medium. Using this selection procedure (Mulligan and Berg, 1981), Ecogpt was first transferred to human cells (MGH-1) to produce a series of cell lines, each carrying Ecogpt integrated into a different chromosome. Cells with Ecogpt integrated into chromosome 2 were then used as microcell donors to transfer this chromosome to mouse cells (Athwal et al., in press).

The R3-5 cells were routinely cultured at 37°C in 10% CO_2 in Dulbeco's modified Eagle's medium supplemented with 10% fetal calf serum (DME) containing 25 µg/ml of mycophenolic acid and 100 µg/ml of xanthine (MX). Cultivation of R3-5 cells in medium containing mycophenolic acid and xanthine assures the presence of chromosome 2 in every cell at the time of treatment with chemical. Cells that have lost the human chromosome would not survive in the selective medium.

Experimental Protocol

A schematic outline of the experimental protocol is given in Figure 1. For each experiment, 0.5 x 10^6 cells were plated in T25 flasks in nonselective medium (DME) 24 h prior to the chemical treatment. Cultures were rinsed twice with PBS and fed with DME containing the test chemical. After exposure to a chemical for 24 h, cultures were rinsed twice with PBS, fed with fresh DME, allowed to grow for another 24 h, and then harvested for cytogenetic analysis. In experiments in which S9 fraction of rat liver microsomes was included for metabolic activation, cells were exposed to the test chemical for only 2 h.

Cytogenetic Analysis

Following the exposure to the test chemical, cells were arrested in mitosis with Colcemid (0.2 μg/ml) for 2 h. Mitotic cells harvested by shaking the flasks were swollen in hypotonic solution (0.075 M KCl) for 15 min and then fixed in three changes of fixative (methanol:acetic acid, 3:1). Metaphase spreads prepared by air drying were stained either by the Giemsa-11 (G-11) method (Bobrow and Gross, 1974) or with Hoechst 33258 (Hilwig and Gropp, 1972) to distinguish between human and mouse chromosomes.

For G-11 staining, 1 day old slides were soaked in water at 37°C for 30 min and then rinsed in sodium phosphate buffer (pH 11.3) for 1 min. Following the pretreatment, slides were stained in the staining solution prepared in Na_2HPO_4 (0.05 M) buffer (pH 11.3) for 3-4 min at 37°C. The staining solution was prepared fresh by mixing 240 μl of Azur B (1% in H_2O) and 200 μl of Eosin (0.25% in methanol) into 20 ml of buffer.

Alternatively, in some experiments the metaphase spreads were also stained with the fluorescent dye Hoechst 33258. The slides were pretreated with NaOH (0.07 N) dissolved in 30% ethanol for 1-2 min and rinsed three times in 70% ethanol to stop denaturation. Following the ethanol treatment, the slides were rinsed twice in PBS and stained for 15 min in Hoechst 33258 (0.05 μg/ml) dissolved in PBS. The stained slides were rinsed with distilled water and wet mounted in 0.08 M sodium phosphate (0.12 M) sodium citrate buffer (pH 4.1). The metaphase spreads were examined under a fluorescence microscope (Zeiss) equipped with an epi-illuminator to give an excitation and reflector combination of 365 nm and 420 nm, respectively. At least 200 metaphases were scored for each treatment of the chemical.

Chemicals and Concentrations

Three chemicals known to induce chromosome breaks or interfere with cell division--cyclophosphamide (Cy), Colcemid, and benomyl--were used as model agents to test the proposed system. Concentrations for each chemical (see Table 1), which were

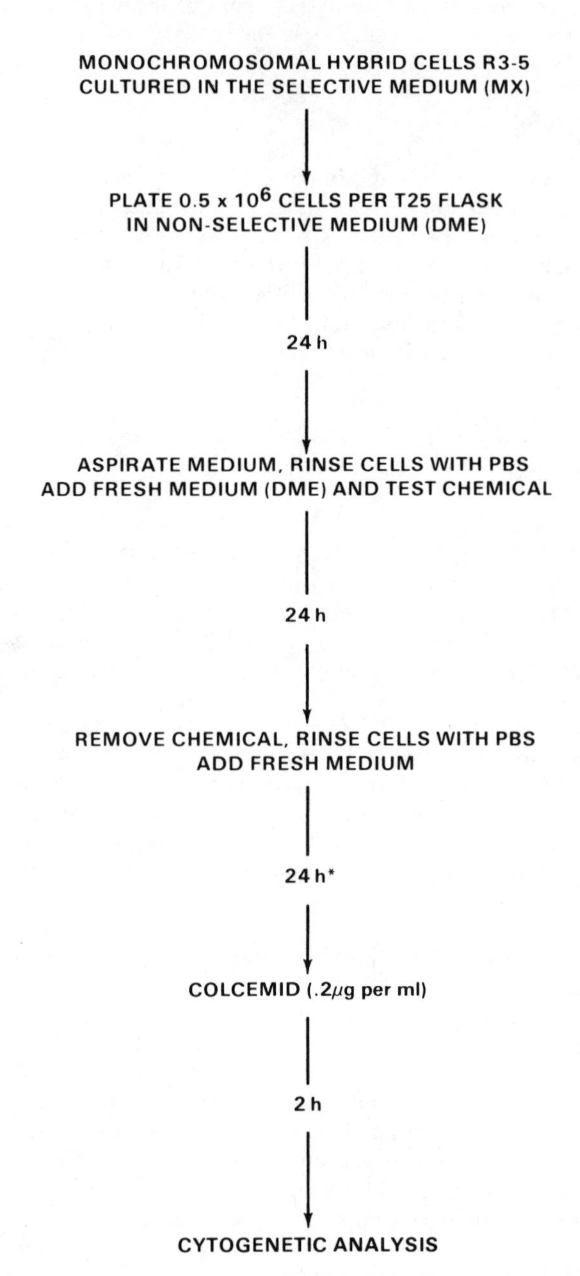

Figure 1. An outline of the experimental protocol to study chemically induced aneuploidy. *For chemicals requiring metabolic activation, exposure to the test chemical was for 2 h only.

selected on the basis of cytotoxicity, included at least one dose above and one to three doses below the 50% toxicity level. The toxicity was determined by exposing cells (10^6) seeded in T25 flasks to different concentrations of a chemical for 24 h. The inhibition of growth, determined by counting viable cells following chemical treatment for 24 h, was used as a measure of toxicity. Since the toxicity of a chemical is influenced by cell density at the time of treatment, similar experimental conditions were used for toxicity analysis as for the induction of aneuploidy.

Colcemid was purchased as a 10-µg/ml solution in Hank's BSS (Gibco) and diluted further in the medium to achieve the appropriate concentration. Cyclophosphamide (Sigma) was dissolved in water, while benomyl (du Pont de Nemours, Wilmington, DE) was dissolved in dimethylsulfoxide (DMSO).

Colcemid and benomyl do not require metabolic activation to manifest their genotoxic activity. However, both are active in the presence of S9 fraction of rat liver microsomes. Cy, on the other hand, produces genotoxic effects only when metabolically activated. Therefore, in the studies reported here, Cy was tested with and without metabolic activation using S9 fraction of rat liver microsomes (McCann and Ames, 1977). The data on Cy without metabolic activation served as a negative chemical control.

RESULTS

After chemical treatment, individual metaphases of R3-5 cells were analyzed to estimate the total number of chromosomes and to determine the number of human chromosomes present in each cell. Normally, the number of mouse chromosomes present in R3-5 cells varies from 46 to 56 (Figure 2). Following chemical treatment, R3-5 cells containing chromosome numbers within the range of the parental genome were classified as diploids, while cells with twice or more the number of R3-5 chromosomes were classified as polyploids. Only cells with a diploid number of chromosomes were scored for the segregation of human chromosomes. Human chromosomes present in mouse cells were identified by G-11 or Hoechst staining methods (Figure 2). With G-11, human chromosomes stained blue, whereas mouse chromosomes stained magenta (Figure 2). In the case of Hoechst 33258, human chromosomes present among mouse chromosomes were identified by lack of the characteristic centromeric fluorescence of the mouse chromosomes (Figure 2). The data on the induction of genomic alterations are listed in Table 1. Since the cells were analyzed for aneuploidy and polyploidy, the results for these end points are presented separately.

Aneuploidy

Normally, the progeny of a monochromosomal hybrid should contain only a single human chromosome. The cells containing 0 or 2 human chromosomes were considered to result from aberrant segregation of the human chromosome and were pooled to calculate aneuploidy (Table 1). In comparison to the control, the frequency of aneuploid cells increased in a dose-dependent fashion after treatment with Colcemid and benomyl (Table 1).

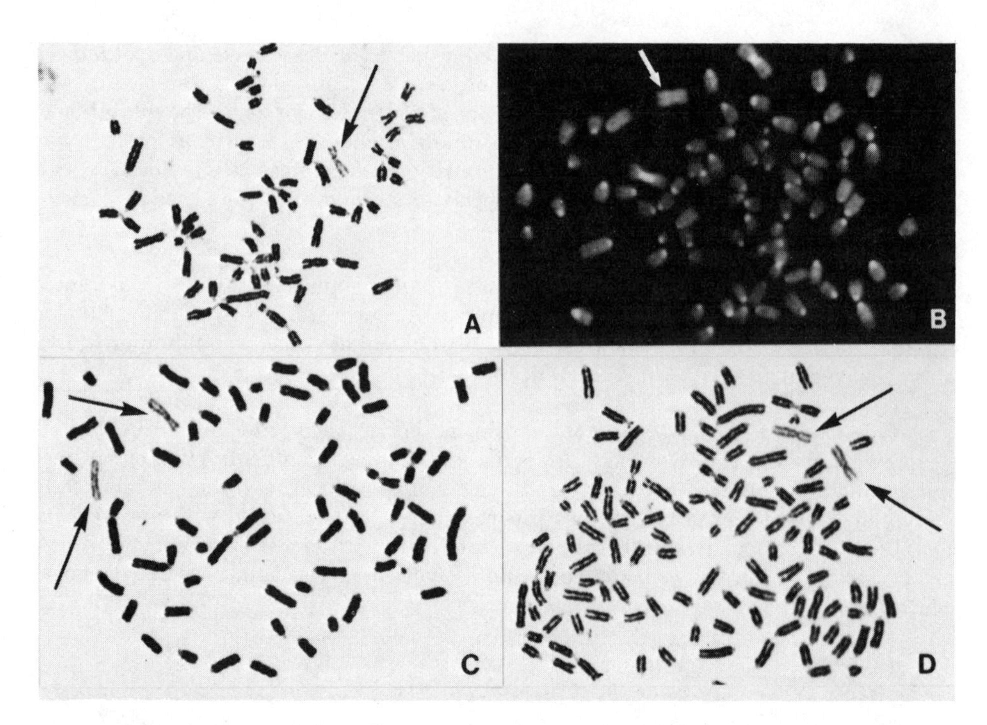

Figure 2. Metaphase spreads of monochromosomal hydrid R3-5:

A. Stained with G-11. Single human chromosome can be identified by difference in color and chromosome morphology.

B. Stained with Hoechst 33258. Human chromosome is distinguishable by the lack of intense fluorescence of centromeric heterochromatin characteristic of mouse chromosomes.

C. Metaphase of R3-5 cells containing two human chromosomes following treatment with Colcemid.

D. A polyploid cell containing two human chromosomes.

In the case of Cy, increase in the number of aneuploid cells occurred only in the presence of metabolic activation. Without metabolic activation, there was no difference between the treated and untreated cultures. Even with metabolic activation, Cy at 10 µg/ml had no effect on the frequency of aneuploid cells. The number of aneuploid cells increased from 4% in the control to 8% at 15 µg/ml of Cy and remained the same at 20 µg/ml. It was interesting to note that the increase in the number of aneuploid cells, in the case of Cy, was observed only in the class of cells that had lost the human chromosome (0-chromosome class), not in the class having an extra chromosome (2-chromosome class). On the other hand, in the cases of Colcemid and

Table 1. Numbers of Aneuploid[a] and Polyploid[a] Cells Following Treatment with Test Chemicals

Chemical and Concentration (μg/ml)	Metaphases Examined	Cells with 1, 2, and 0 Human Chromosomes			Percent Aneuploid Cells[b] (A)	Percent Polyploid Cells (P)	Ratio P/A[c]	Percent Cells with Chromosome Aberrations[d]
		1	2	0				
Cyclophosphamide (−S9)								
0	200	192	4	3	3.51	0.5	0.14	0
10	200	191	5	3	4.02	0.5	0.12	0.5
15	200	192	3	4	3.51	0.5	0.14	0
20	200	193	3	3	3.03	0.5	0.16	0.5
Cyclophosphamide (+S9)								
0	200	191	3	5	4.02	0.5	0.12	0
10	200	190	3	6	4.52	0.5	0.11	3
15	200	183	5	11	8.04	0.5	0.06	4
20	200	182	5	12	8.54	0.5	0.06	6
Colcemid (−S9)								
0	200	191	4	4	4.02	0.5	0.12	0.5
0.005	200	185	7	7	7.04	0.5	0.07	0.5
0.01	200	180	9	10	9.55	0.5	0.05	0

(continued)

Table 1. (continued)

Chemical and Concentration (μg/ml)	Metaphases Examined	Cells with 1, 2, and 0 Human Chromosomes			Percent Aneuploid Cells[b] (A)	Percent Polyploid Cells (P)	Ratio P/A[c]	Percent Cells with Chromosome Aberrations[d]
		1	2	0				
0.02	200	139	12	14	15.75	17.5	1.11	0.5
0.05	200	96	16	20	27.27	34.0	1.25	0
0.1	200	86	11	14	22.52	44.5	1.98	0.5
Benomyl (−S9)								
0	200	192	1	3	2.04	2.0	0.98	0.5
1.5	200	164	13	20	16.75	2.0	0.12	0.5
3.0	400	289	35	39	20.39	9.2	0.45	1
7.5	300	161	26	30	25.81	27.7	1.07	1
15	300	142	24	27	26.42	35.7	1.35	4

[a]In this system, aneuploidy refers to a gain or loss of the human chromosome in the R3-5 cell, while polyploidy refers to an increase in the total number of chromosomes by a factor of 2 or more.

[b]Percent aneuploid cells was calculated by adding cells with 0 and 2 human chromosomes and taking diploid and aneuploid cells (i.e., excluding polyploid cells) for the total.

[c]Ratio of polyploid/aneuploid cells was calculated from percent polyploid and percent aneuploid cells.

[d]Cells with chromosomal aberrations involving the human chromosome were counted in the scored metaphases.

benomyl, the number of cells that gained or lost the human chromosome increased simultaneously with concentration (Table 1). These results indicate that Cy may have induced aneuploidy by a mechanism different than that for Colcemid and benomyl.

Polyploidy

The data given in Table 1 show a dose-dependent increase in the frequency of polyploid cells after treatment of the hybrid cells with Colcemid and benomyl. At concentration levels up to 0.01 μg/ml of Colcemid, the number of polyploid cells did not change, while the frequency of aneuploid cells increased (Table 1). Similarly, at the lowest dose of benomyl (1.5 μg/ml), although the frequency of aneuploid cells increased from 2.04% in the control to 16.75% in the treated cells, no change in the number of polyploid cells was observed. However, at higher doses of both Colcemid and benomyl, the frequency of polyploid cells increased at a much faster rate than that of aneuploid cells. This dose-dependent difference in aneuploidy and polyploidy perhaps may be due to a relative increase in spindle disruption resulting from saturation of the target molecules at higher concentrations of Colcemid and benomyl (Table 1). The differential increase in the number of polyploid and aneuploid cells at different doses was reflected in the ratio of polyploid/aneuploid (P/A) cells. At low doses (0.01 μg/ml of Colcemid and 1.5 μg/ml of benomyl), the P/A ratio was lower than that in the control, whereas at high doses, as the number of polyploid cells increased, a corresponding shift in the P/A ratio occurred (Figure 3A and B).

Cy had no effect on the frequency of polyploid cells either with or without metabolic activation (Table 1). In comparison with the controls, no change in P/A ratio was observed for Cy without metabolic activation (Figure 3D), while a decrease in the ratio occurred when the chemical was used with metabolic activation (Figure 3C). We interpret these results to show that Cy does not interfere with cell division. Cy is a known clastogenic agent (Goetz et al., 1980), and the data given in Table 1 show that the frequency of cells with chromosome aberrations involving the human chromosome increased with dose. The increase in the frequency of cells with chromosome aberrations correlated with chromosome loss in the case of metabolically activated Cy. These results show that Cy perhaps induced aneuploidy by direct effect on the chromosome rather than by interfering with cell division.

DISCUSSION

Data presented in this paper show that the human/mouse hybrid cell line R3-5 may provide a useful tool for evaluating the potential of environmental chemicals to induce aneuploidy. The human chromosome 2 present in R3-5 cells is easily distinguishable from the mouse chromosomes, thus providing an excellent cytogenetic marker (Figure 2). Chemically induced abnormal distribution of a single chromosome to daughter cells at mitosis is used as an index for aneuploidy.

Aneuploid condition is characterized by the gain $(2n + 1)$ or loss $(2n - 1)$ of a chromosome of the complement. The gain or loss of a chromosome at mitosis usually is a complementary consequence of nondisjunction. However, loss could also occur due to the failure of a chromosome to incorporate into the daughter nucleus by other

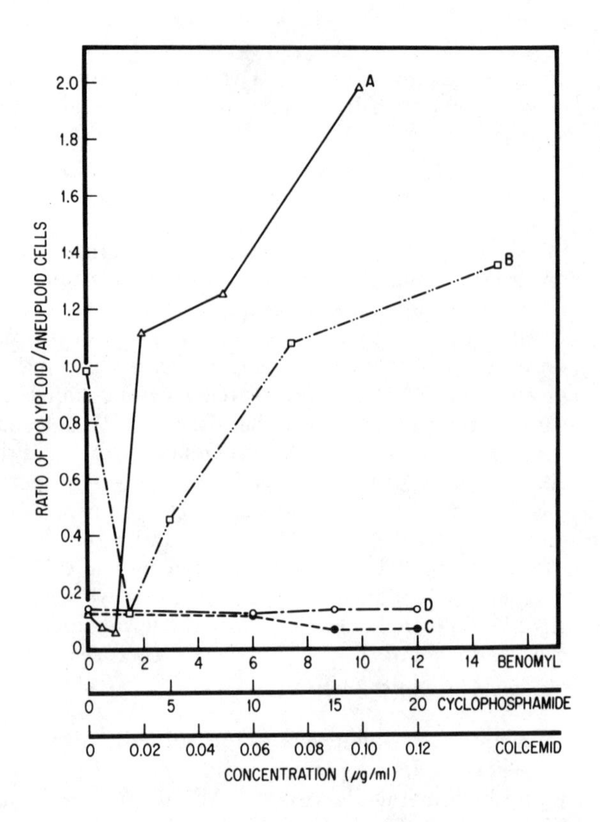

Figure 3. Ratio of polyploid/aneuploid (P/A) cells plotted against the concentration of each chemical: (A) Colcemid, (B) benomyl, (C) cyclophosphamide with metabolic activation, (D) cyclophosphamide without metabolic activation. Figure shows a dose-related increase in P/A ratio for Colcemid and benomyl, a decrease with metabolically activated cyclophosphamide, and no change with cyclophosphamide without metabolic activation.

mechanisms, such as anaphase lagging, lack of centromere separation, or chromosome breaks (Bond and Chandley, 1983; Vig, 1984). The proposed system has the capability to detect gain as well as loss of a chromosome induced by a test chemical. We would like to emphasize that this test system measures the abnormal segregation of a single chromosome rather than aneuploidy in general.

Colcemid and benomyl bind to the same target molecule (Quinlan et al., 1981), thus producing similar effects. Although the number of cells for loss and gain of the human chromosome increased simultaneously with dose of Colcemid and benomyl, the frequency of cells with chromosome loss was consistently high at all doses. These results show that more than one mechanism may account for chromosome loss, while the gain may always result from nondisjunction. In contrast, Cy induced only loss of the chromosome, which indicates a different mechanism of action for this compound. Cy is a known clastogenic agent (Goetz at al., 1980), and chromosome loss observed in this case may be due to chromosomal breaks. In fact, in our analysis, an increase in the frequency of cells with complex rearrangements involving the human chromosome corresponded with the frequency of chromosome loss with Cy (Table 1). Therefore, the aneuploid condition observed in this case is attributed to the clastogenic effect of Cy rather than interference with cell division. This notion is further supported by the fact that the frequency of polyploid cells did not increase with Cy (Table 1).

As apparent from our results, the P/A ratio remains the same in treated and control cultures when a chemical does not have any effect on cell division or chromosomes (Figure 3). An increase in the P/A ratio with the concentration would indicate interference with cell division, while a decrease indicates an effect on chromosomes rather than on cell division (Figure 3). Thus, the P/A ratio may provide an index to detect the potential of a chemical to induce aneuploidy and to determine its mode of action.

Other mammalian cell culture systems to assay for chemically induced aneuploidy include the use of continuous or early passage diploid Chinese hamster (Hsu et al., 1983; Onfelt and Klasterska, 1983; Danford, 1984; Cox, 1973) or Syrian hamster (Tsutsui et al., 1983) cells. In these systems, complete mitotic arrest (Onfelt and Klasterska, 1983), total chromosome count (Danford, 1984; Tsutsui et al., 1983), or anaphase abnormalities such as multipolar spindle and lagging chromosomes (Hsu et al., 1983) have been used as criteria to assess the ability of chemicals to interfere with cell division. Total chromosome count, in addition to being tedious and time-consuming, may not provide complete information. A cell may simultaneously lose and gain a chromosome in two different homologous pairs and still retain the parental chromosome number. In this case, although aneuploidy has occurred, the chromosome number remains unchanged.

Another system similar in principle to ours is the use of human male lymphocytes (Tenchini et al., 1983). In this system, human Y chromosome was used as a marker to assess aberrant segregation following chemical treatment. The occurrence of YY mitosis was considered to result from a nondisjunctional event. However, the frequency of YY cells was rather low. In addition, this system detected only chromosome gain and not loss.

In the hybrid cell system, easy identification of the human chromosome greatly reduces error in scoring. In addition, it eliminates the need for laborious chromosome counting or karyotypic analysis. One may contend that a human chromosome present in mouse cells may not obey the same rules at cell division as the parental mouse chromosomes, and therefore may be relatively unstable in the foreign environment. The absence of aneuploid cells in cultures treated with Cy without metabolic activation clearly shows that abnormal segregation of the human chromosome is not a random nonspecific phenomenon but a function of chemicals that interfere with cell division or cause chromosome breaks. The relative unstability of the human chromosome may, rather, be useful in making the assay more sensitive.

The future direction of this investigation is to develop a system to detect aneuploidy by selection and colony formation. Like mammalian HPRT, *E. coli* XGPRT can also transfer purine analogue 6-thioguanine (6-TG) to form the corresponding monophosphate, which is toxic to the cells. Therefore, if R3-5 cells are cultured in medium containing 6-TG, only cells that have lost the human chromosome would survive. This back selection procedure thus provides an assay system to determine the frequency of cells with chromosome loss following treatment with a test chemical.

ACKNOWLEDGMENTS

We thank Mr. M.A. Ahmad for technical assistance and Mr. S. Desai for photography. Studies included in this report were supported in part by Contract ID 5806 NAET from the U.S. Environmental Protection Agency and Grant CA 28559 from the National Cancer Institute, National Institutes of Health.

Although the research described in this article has been funded in part by the U.S. Environmental Protection Agency, it does not necessarily reflect the views of the Agency and no official endorsement should be inferred.

REFERENCES

Arlett, C.F., and A.R. Lehmann. 1978. Human disorders showing increased sensitivity to the induction of genetic damage. Ann. Rev. Genet. 12:95-115.

Athwal, R.S., M. Smarsh, B.M. Searle, and S.S. Deo. (in press). Integration of a dominant selectable marker into human chromosomes and transfer of marked chromosomes to mouse cells by microcell fusion. Mol. Som. Cell Genet.

Bloom, A.D. 1972. Induced chromosomal aberrations in man. Adv. Hum. Genet. 3:99-172.

Bobrow, M., and J. Gross. 1974. Differential staining of human and mouse chromosomes in interspecific cell hybrids. Nature 251:77-79.

Bond, D.J., and A.C. Chandley. 1983. Aneuploidy. No. 11, Oxford Monographs on Medical Genetics, Oxford University Press: Oxford, New York, Toronto.

Cox, D.M. 1973. A quantitative analysis of Colcemid induced chromosomal non-disjunction in Chinese hamster cells *in vitro*. Cytogenet. Cell Genet. 12:165-174.

Danford, N. 1984. Measurement of levels of aneuploidy in mammalian cells using a modified hypotonic treatment. Mutat. Res. 139:123-127.

de Bertoldi, M., M. Griselli, and R. Barale. 1980. Different test systems in *Aspergillus nidulans* for the evaluation of mitotic gene conversions, crossing-over and nondisjunction. Mutat. Res. 74:303-324.

Goetz, P., A.M. Malashenko, and N. Surkova. 1980. Cyclophosphamide induced chromosome aberrations in meiotic cells of male mice. Folia Biol. (Prague) 26:289-297.

Haber, J.E., J.G. Peloquin, H.O. Halvorson, and G.G. Borisy. 1972. Colcemid inhibition of cell growth and characterization of colcemid binding activity in *Saccharomyces cerevisiae*. J. Cell Biol. 55:355-367.

Health, I.B. 1975. The effect of antimicrotubule agents on the growth and ultrastructure of the fungus *Saprolegnia ferax* and their ineffectiveness in disrupting hyphal microtubules. Protoplasmia 85:147-176.

Hilwig, I., and A. Gropp. 1972. Staining of constitutive heterochromatin in mammalian chromosomes with a new fluorochrome. Exp. Cell Res. 75:122-126.

Hook, E.B. 1983. Contribution of chromosome abnormalities to human morbidity and mortality and some comments upon surveillance of chromosome mutation rate. Mutat. Res. 114:389-423.

Hsu, T.C., J.C. Liang, and L.R. Shirley. 1983. Aneuploidy induction by mitotic arrestants. Effects of diazepam on diploid Chinese hamster cells. Mutat. Res. 122:201-209.

Levan, G., and F. Mitelman. 1977. Chromosomes and the etiology of cancer. In: Chromosomes Today, Vol. 6. A. de la Chapelle and M. Sorsa, eds. Elsevier Biomedical: Amsterdam. pp. 363-371.

Liang, J.C., T.C. Hsu, and J.E. Henry. 1983. Cytogenetic assays for mitotic poisons. The grasshopper embryo system for volatile liquids. Mutat. Res. 113:467-479.

McCann, J., and B.N. Ames. 1977. The *Salmonella*/microsome mutagenicity test: predictive value for animal carcinogencity. In: Origins of Human Cancer. H.H. Hiat, J.D. Watson, and J.A. Winsten, eds. Cold Spring Harbor Laboratory: Cold Spring Harbor, NY. pp. 1431-1450.

Mulligan, R.C., and P. Berg. 1981. Selection for animal cells that express the *Escherichia coli* gene coding for xanthine-guanine phosphoribosyltransferase. Proc. Natl. Acad. Sci. U.S.A. 78:2072-2076.

Onfelt, A., and I. Klasterska. 1983. Spindle disturbances in mammalian cells. Induction of viable aneuploid/polyploid cells and multiple chromatid exchanges after treatment of V79 Chinese hamster cells with carbaryl. Modifying effects of glutathione and S9. Mutat. Res. 119:319-330.

Parry, J.M., E.M. Parry, and J.C. Barrett. 1981. Tumor promotors induce mitotic aneuploidy in yeast. Nature 294:263-265.

Quinlan, R.A., A. Roobol, C.I. Pogson, and K. Gull. 1981. A correlation between *in vivo* and *in vitro* effects of the microtubule inhibitors colchine, parbendazole and nocodazole on myxamoebae of *Physarum polycephalum*. J. Gen. Microbiol. 122:1-6.

Tenchini, M.L., A. Mottura, M. Velicogna, M. Pessina, G. Rainaldi, and L. De Carli. 1983. Double Y as an indicator in a test of mitotic nondisjunction in cultured human lymphoctyes. Mutat. Res. 121:139-146.

Tsutsui, T., H. Maizumi, J.A. McLachlan, and J.C. Berret. 1983. Aneuploidy induction and cell transformation by diethylstilbestrol: a possible chromosomal mechanism in carcinogenesis. Cancer Res. 43:3814-3821.

Vig, B.K. 1984. Sequence of centromere separation, another mechanism for the origin of non-disjunction. Hum. Genet. 66:239-243.

Zimmering, S. 1982. Induced chromosome loss following treatment of postmeiotic cells of the *Drosophila melanogaster* male with MMS and DMN and mating with repair-proficient female and the repair-deficient female mei-9 and st mus 302. Mutat. Res. 94:79-86.

Zimmermann, F.K., F.J. de Serres, and M.D. Shelby. 1979. Workshop on systems to detect induction of aneuploidy by environmental mutagens. Mutat. Res. 64:279-285.

TOXICOLOGIC RESPONSES TO A COMPLEX COAL CONVERSION BY-PRODUCT: MAMMALIAN CELL MUTAGENICITY AND DERMAL CARCINOGENICITY

Michael L. Cunningham, David A. Haugen, Frederick R. Kirchner, and Christopher A. Reilly, Jr.

Division of Biological and Medical Research, Argonne National Laboratory, Argonne, Illinois 60439

INTRODUCTION

The overall toxicity of complex mixtures cannot be predicted simply by adding the toxicities of known toxic compounds in the mixture. The toxicity of some of the components in the mixture is often unknown and synergisms and antagonisms, which cannot be predicted solely on the basis of the chemical composition, may occur. To better understand the chemical and biochemical interactions that modify the ultimate toxicity of mixtures, we are isolating and characterizing well-defined chemical classes of chemicals and determining the toxicity of these chemical class fractions with *in vivo* and *in vitro* bioassays. The need for this approach with coal conversion materials has recently been demonstrated by interactive effects observed in both *in vitro* and *in vivo* toxicity assays. For example, a mixture of polycyclic aromatic hydrocarbons from coal conversion materials inhibited the bacterial mutagencity of benzo[a]pyrene (Haugen and Peak, 1983) and potentiated the mutagenicity of 2-aminoanthracene (Pelroy and Peterson, 1979; Kawalek and Andrews, 1981) as measured in the *Salmonella*/microsome assay. In addition, tumorigenic responses to carcinogenic components of coal conversion materials were greater than those determined for the whole material (Mahlum, 1984).

Multitiered bioassay schemes have often included the Ames *Salmonella*/microsome assay (Waters et al., 1979) as the first step. This assay has been widely used to evaluate the mutagenicity and potential carcinogenicity of single compounds (McCann et al., 1975) as well as complex mixtures such as tobacco smoke condensate (Kouri et al., 1978) and coal conversion materials (Epler et al., 1978; Pelroy and Peterson, 1981; Haugen et al., 1982). The additional use of mammalian cell mutagenesis assays may enhance the value of *in vitro* cell assays as predictors of human carcinogenicity because mammalian cells have a number of components not present in bacterial cells, including a nuclear envelope, mitochondria, endoplasmic reticulum, DNA organized in chromosomes, and nucleosomes.

In the present study, we measured mutagenicity and cytotoxicity in hamster and human cells *in vitro* and tumorigenicity in mouse skin *in vivo*. The Chinese hamster ovary cell/hypoxanthine guanine phosphoribosyl transferase (CHO/HGPRT) assay has proven useful in estimating the mutagenic activity of pure compounds (O'Neill et al., 1977), but has been used only to a limited extent with complex mixtures (Hsie et al., 1978; Li et al., 1983; Wilson et al., 1984). The human teratocarcinoma cell line, designated P_3, used in these studies has recently been adapted for use in mutagenesis assays of individual compounds (Huberman et al., 1984), but has not previously been used to evaluate mutagenesis by complex mixtures. In this report, we compare the responses of the hamster and human cell lines and the mouse skin to chemical class fractions of a complex organic by-product condensate (tar) of coal gasification. The composition of this complex tar is chemically similar to that of petroleum-derived tars and products of fossil fuel combustion.

By testing basic, acidic, and neutral chemical class fractions of the complex tar, we demonstrated that the predominant genotoxic components were present in the neutral fraction, as measured both in the CHO/HGPRT and dermal carcinogenicity assays. The P_3 cells were less sensitive than the rodent cells for mutagenesis and cytotoxicity. Furthermore, fractionation and bioassay provided evidence for interactive effects that indicate the importance of combining chemical characterization and toxicologic evaluation of complex mixtures.

MATERIALS AND METHODS

Test Materials

The test material was a condensate tar formed as a by-product of coal gasification in a pilot plant (slagging, fixed-bed) operated at the University of North Dakota (Wilzbach et al., 1981). Results of chemical analyses have shown this tar to be generically similar to the complex mixtures produced during a variety of coal conversion processes. The basic, acidic, and neutral fractions were prepared by liquid partitioning, as described previously (Haugen et al., 1983). Briefly, acidic components were extracted with aqueous NaOH from a methylene chloride solution of the tar, and basic components were subsequently extracted with HCl in aqueous methanol. The use of methanol enhanced the recovery of larger aromatic bases (Boparai et al., 1983). The remaining material was designated the neutral fraction. The three fractions were qualitatively and quantitatively analyzed by capillary column gas chromatography and gas chromatography-mass spectrometry. In addition, high-performance liquid chromatography and gas chromatography-mass spectrometry were used as previously described (Haugen et al., 1982, 1983) to isolate and identify principal mutagenic components.

Cytotoxicity and Mutagenicity Assays

The CHO-K1 cell line was purchased from the American Type Culture Collection, Rockville, MD. After removing spontaneous 6-thioguanine-resistant mutants by subculturing once in medium containing hypoxanthine, aminopterin, and thymidine (Dutchland Laboratories, Denver, PA), cells were routinely passaged in growth medium consisting of antibiotic-free Hams F12 medium (Gibco Laboratories, Grand Island, NY) containing 5% fetal calf serum (Flow Laboratories, McLean, VA). The subclone P_3 was kindly supplied by Drs. C.A. Jones and E. Huberman (Argonne National Laboratory). These cells were cultured in RPMI-1640 growth medium (Gibco Laboratories) supplemented with 20% fetal calf serum, penicillin (100 units/ml), and streptomycin (100 µg/ml).

For mutagenicity and cytotoxicity assays, cells were plated at 5×10^5 cells/25-cm^2 flask. After 24 h, the growth medium was replaced with 5 ml of serum-free medium containing an NADPH-generating system consisting of glucose-6-phosphate (1.5 mg/ml) and NADP (3.0 mg/ml). Test agents were then added in 50 µl of dimethyl sulfoxide (DMSO). For some assays, 40 µl of 9,000 x g supernatant (S9) prepared from the livers of rats treated with Aroclor 1254 was also added (McCann et al., 1975). After a 1-h incubation, the medium was removed, the cells were washed once with serum-free medium, and 5 ml of growth medium was added. After a 24-h incubation, cells were removed with 0.05% trypsin-ethylenediaminetetraacetic acid. An aliquot was removed for quantification of viability (as determined by trypan blue exclusion), expressed as percentage survival relative to DMSO-treated controls. The remaining cells were assayed for mutant frequency as described by Hsie et al. (1978). Briefly, cells were subcultured three times in growth medium to allow an optimal 7- to 8-d period for expression of mutagenesis. For mutant selection, a total of 10^6 cells from each exposure group or DMSO control were plated in growth medium containing 20 µM 6-thioguanine (Sigma, St. Louis, MO) at a density of 2×10^5 cells/75 cm^2. Simultaneously, 400 cells in triplicate were plated in growth medium without 6-thioguanine for determination of cloning efficiency. Mutant frequency was expressed as the number of 6-thioguanine-resistant mutants per 10^6 surviving cells.

Treatment of Animals

Doses of the test agent (unfractionated tar or the acidic, basic, or neutral fraction) dissolved in 50 µl of acetone were applied to the backs of young adult SKH (hairless) mice three times per week on their dorsal skin. The dose of the unfractionated tar was 25 mg per application. For the neutral, acidic, and basic fractions of the tar, the amounts per application were equivalent to their fractional amounts in 25 mg tar, i.e., 9.0, 4.3, and 0.45 mg, respectively. Control mice were treated only with acetone. Each treatment group (experimental and control) contained 50 mice (25 of each sex). Animals were observed weekly during the 26-wk treatment period and for 52 wk thereafter.

RESULTS

Composition of Tar Fractions

The neutral fraction constituted 36% w/w of the tar and was largely composed of aliphatic hydrocarbons, hydroaromatic compounds, and polycyclic aromatic hydrocarbons (2-6 rings), with relatively smaller amounts of S- and N-heterocyclic aromatics and unidentified more-polar compounds. The acidic fraction (17% w/w) was composed primarily of hydroxyaromatic compounds, predominantly phenols. The basic fraction (2% w/w) was composed of 1- to 5-ring aromatic amines and azaarenes; the latter predominated. The remaining part of the tar consisted of water, insoluble material, and volatile organic chemicals. For each class of compounds present in the neutral, acidic, and basic fractions, alkyl homologs of the 1- to 3-ring members were also present in significant proportions relative to the unsubstituted compounds.

Cytotoxicity of Tar Fractions

Cells were exposed to four concentrations of the tar or its fractions (4, 20, 100, and 200 μg/ml) for 1 h; after 24 h, cell viability was measured. The toxicity of the materials was expressed as the concentration that resulted in 50% lethality, LC_{50}, as determined from concentration-toxicity curves. For the CHO cells in the absence of S9, the neutral fraction was most toxic, with an LC_{50} value about fivefold less than that for the unfractionated tar (Table 1). The toxicities of the acidic and basic fractions were similar to that for the crude tar. For the CHO cells, the presence of S9 decreased the

Table 1. Cytotoxicity of Tar and Fractions in CHO and P_3 Cells

	LC_{50} (μg/ml)[a]			
	CHO		P_3	
Fraction	−S9	+S9	−S9	+S9
Tar	60	180	240	≫400
Neutral	12	72	280	>400
Acidic	70	70	140	140
Basic	68	126	>400	>400

[a]LC_{50} is the concentration at which 50% lethality occurs after a 1-h exposure in the presence or absence of 9,000 x g supernatant (S9) from the livers of rats treated with Aroclor 1254 and an NADPH-generating system. Each value is the mean for two determinations. LC_{50} values greater than 200 μg/ml (the highest dose tested) were estimated by extrapolation of the dose-response curves.

toxicity of the crude tar as well as that of the neutral and basic fractions. However, the presence of S9 did not affect the response to the acid fraction.

The P_3 cells were less sensitive to the cytotoxic effects of the materials than the CHO cells. In the absence of S9, the LC_{50} values were twofold (acid fraction) to greater than 20-fold (neutral fraction) higher for the P_3 cell line than for the CHO cells (Table 1). In the P_3 cells, as in the CHO cells, the presence of S9 decreased the cytotoxicity of all fractions except the acidic.

Mutagenicity of Tar Fractions

None of the material was mutagenic in the absence of S9. In the presence of S9, the responses for the two lowest concentrations (4 and 20 μg/ml) tested in the CHO/HGPRT or P_3/HGPRT assays were at or near background levels (2 mutants per 10^6 surviving cells). Significant mutant frequencies (up to 15 mutants per 10^6 surviving cells) were seen for the tar and its fractions at the 100 μg/ml test concentration, but the mutant frequency was lower for the P_3 cells than for the CHO cells (Table 2). At concentrations of 200 μg/ml, the neutral and basic fractions were found to be significantly more mutagenic for the CHO cells than the crude tar or its acid fraction. In contrast to its high mutagenicity in CHO cells, the neutral fraction produced very low mutant frequencies at both the 100- and 200-μg/ml test concentrations in the P_3 cells.

Dermal Carcinogenicity

In all treatment groups except the acetone control, growths developed on the site of the exposure; these growths were judged to be tumors by visual observations. These observations were confirmed by histopathological analysis that indicated these growths were squamous cell carcinomas. The crude tar and its neutral and basic fractions induced at least one tumor in more than 90% of the animals tested. The acid fraction induced tumors in only 40% of the animals by the end of the experiment. The most active materials were distinguished by the mean time to tumor (50% incidence) values: 9 wk for the neutral fraction, 16 wk for the crude tar, and 43 wk for the basic fraction (Figure 1).

DISCUSSION

Recently, strategies have been presented to evaluate the hazards associated with exposures to complex mixtures (Neal, 1983; Murphy, 1983). Among the recommendations was the suggestion that combining analytical identification of the chemical constituents in mixtures with toxicological measurements would allow a more accurate estimate of risk of exposure to the mixtures. The present study demonstrates the usefulness of this approach, i.e., combining chemical fractionation and characterization with *in vivo* and *in vitro* bioassays. The identification and toxicologic characterization of compounds contained within complex mixtures is important for estimating potential human health hazard, but information about individual

Table 2. Mutagenicity of Tar and Fractions in CHO and P_3 Cells[a]

	Mutant Frequency[b]			
	CHO		P_3	
Fraction	100 µg/ml	200 µg/ml	100 µg/ml	200 µg/ml
Tar	8	22	22	NT[c]
Neutral	16	77	3	6
Acidic	4	10	1	7
Basic	13	62	12	24
DMSO	2	2	2	2

[a]Assays were conducted in the presence of Aroclor-induced S9 and an NADPH-generating system. Cells were exposed for 1 h at either 100 or 200 µg/ml.
[b]Mutant frequency expressed as mutants per 10^6 survivors.
[c]NT = not tested.

compounds must be supplemented with data from bioassays to detect interactions that may amplify or reduce the biological hazards.

The isolation of the neutral, acidic, and basic fractions was advantageous for both chemical and toxicologic characterization. Because the bases are only minor components of the complex tar, their detection would be difficult without fractionation. Enrichment of the amines in the basic fraction also allowed more definitive interpretation of the toxicologic data for these compounds whose metabolic activation via N-oxidation is unique among the components in the tar. Additional chromatographic fractionation, chemical analysis, and Ames *Salmonella*/microsome mutagenicity assays have shown that 3- and 4-ring primary aromatic amines are principally responsible for the relatively high mutagenicity of the basic fractions of coal conversion materials (Guerin et al., 1980; Wilson et al., 1980; Haugen et al., 1982), including the tar examined in the present study (Haugen et al., unpublished data).

The acidic fraction of complex coal conversion materials is composed primarily of hydroxyaromatic (phenolic) compounds. Phenolic compounds are well known for their cytotoxicity and their corrosive effect on tissues. However, the predominant 1- to 3-ring

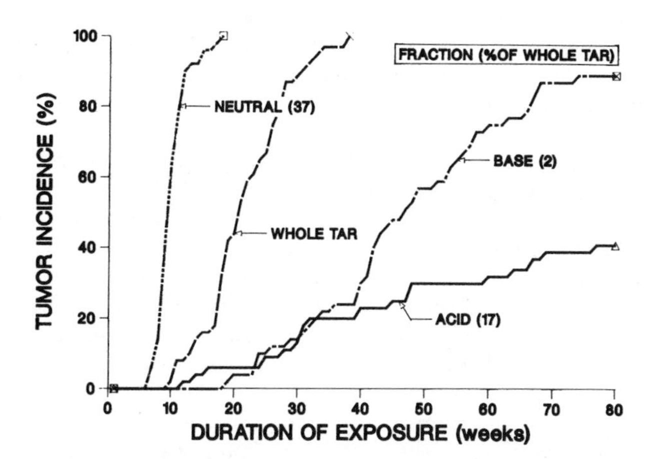

Figure 1. Tumor incidence (% mice bearing squamous cell carcinoma) in SKH hairless mice chronically exposed to fractions of a coal gasification tar. The data for each weekly point indicate the cumulative percentages of animals with at least one tumor. Values in parentheses indicate the amounts (%) of the fraction applied relative to the amount of the whole tar, reflecting their relative abundance.

phenolic compounds in the acid fraction are not individually known as mutagens. The acidic fractions have negligible mutagenic activity as measured in the *Salmonella*/microsome assay (Stamoudis et al., 1981). Isolation of these relatively polar constituents of the tar allows their chemical characterization by optimal chromatographic procedures and prevents their interference with the characterization of other components. Furthermore, separation of the acid fraction simplifies interpretation of toxicologic responses to the remaining materials both for *in vitro* and *in vivo* assays and increases the potential for detection of any less abundant genotoxic acidic components.

In contrast to the narrowly defined basic (two chemical classes) and acidic (one chemical class) fractions, the neutral fraction contains chemicals from a variety of classes. Although interpretation of toxicologic responses to the neutral fraction is facilitated by the absence of the basic and acidic components, further chemical class fractionation is necessary to identify the toxicologically-active components. The neutral fractions of the coal conversion materials we have examined account for a major proportion of the mutagenic activity of the tar as measured with bacteria (Stamoudis et al., 1981). Additional fractionation and chemical characterization have shown that the bacterial mutagenicity is due primarily to 4- to 6-ring polycyclic aromatic hydrocarbons including cyclopenta[cd]pyrene, benz[a]anthracene, benzofluoranthenes, and benzo[a]pyrene as well as alkyl and cycloalkyl homologs of these and other compounds (Haugen et al., 1983; Haugen et al., unpublished data). It is likely that these compounds are also principally responsible for the mutagenesis in mammalian cells observed in the present study.

The crude tar described here was less cytotoxic to CHO cells in the presence of rat liver S9 than were the acid, base, and neutral fractions (Table 1). This observation suggests that inhibition of toxicity occurs when the components are combined in the mixture. Exposure of CHO cells to the tar and its fractions produced higher cytotoxicity in the absence of rat liver S9 than in the presence of S9 (Table 1). This protection by S9, also observed by Li et al. (1983), may be due to the ability of the S9 system to detoxify chemicals or to physically sequester chemicals by binding them to proteins or lipid membranes. The P_3 cell line was less sensitive than the CHO cells to the cytotoxic effects of the crude tar and its fractions (Table 1). If the P_3 cell line is indeed more representative of human cells *in vivo* than are hamster cells, these results suggest that risk resulting from acute exposure to these materials may be overestimated by tests in which rodent cells are exclusively used.

Similarly, the P_3 cells exhibited lower mutant frequencies than the CHO cells after exposure to the fractions of the complex mixture (Table 2). The low response of the human cells to the neutral fraction was unexpected, because this fraction is enriched in mutagenic polycyclic aromatic hydrocarbons. Recent studies have demonstrated that benzo[a]pyrene can induce DNA repair, as measured by unscheduled DNA synthesis, in human fibroblasts but not in rat hepatocytes (Strom et al., 1981). Induced repair could account for the low mutagenicity we observed. Cells capable of repairing DNA damage produced by chemicals are less likely to undergo mutation. In addition, Strom et al. (1981) demonstrated that human cells are less able than rat hepatocytes to repair 2-acetylaminofluorene-induced DNA damage. This finding is consistent with the high mutant frequency we observed in the P_3 cells exposed to the base fraction, which is enriched with aromatic amines (Table 2). The P_3 cell line is not capable of activating promutagens (Huberman et al., 1984) and was not sensitive to the promutagens in our samples in the absence of an exogenous activation system; however, the levels of detoxification enzymes in the P_3 cell line have not been established. Thus, the relatively low sensitivity of the P_3 cells could also be due to higher levels of detoxification enzymes.

Comparison of the data in Figure 1 and Table 2 show that the relative potencies of the tar and its acidic and neutral fractions are similar for the CHO mutagenesis assay and the dermal tumorigenesis assay in that for both (a) the acidic fraction was relatively inactive and (b) the neutral fraction was more potent than the unfractionated tar, even though it was applied in an amount equivalent to its amount in the tar. Tumors not only arose sooner in mice exposed to the neutral fraction, but histopathological examination of the tumors revealed that the incidence of squamous cell carcinoma was 100% (Kirchner et al., unpublished data). The incidence of squamous cell carcinoma was only 46% in mice treated with the whole tar. For mice treated with the acidic and basic fractions, the incidences of squamous cell carcinoma were 24% and 21%, respectively. Thus, it appears that fractionation removed components from the mixture that could diminish the response to the components in the neutral fraction. The specific mutagenicity of the basic fraction was comparable to that for the predominant neutral fraction (Table 2) in CHO cells. However, the low abundance of the basic fraction in the tar suggests that its contribution to the mutagenic activity of the tar is also small. The even smaller concentrations of known dermal carcinogens in the basic fraction are consistent with the observation that its dermal carcinogenicity was much less than for the neutral fraction. These results underscore the need for testing both fractionated and unfractionated materials. The limited experience using human cells in mutation assays makes comparison of these results with those for rodent dermal tumorigenesis studies uncertain.

Complex mixtures require fractionation before reliable chemical analyses and bioassays can be performed. In our laboratory, we have combined chemical fractionation and characterization with bioassays of mutagenesis in bacterial, rodent, and human cells as well as with tests for dermal carcinogenicity in mice. This multitiered approach provides information about the identity of the principal genotoxic components and, in addition, provides the basis for a better understanding of the interactions associated with toxicologic responses to complex mixtures.

ACKNOWLEDGMENTS AND DISCLAIMER

This research was sponsored by the U.S. Department of Energy under contract number W-31-109-ENG-38. We thank Vernon Pahnke, Katherine Suhrbier, and Suzanne S. Dornfeld for expert technical assistance.

REFERENCES

Boparai, A.S., D.A. Haugen, K.M. Suhrbier, and J.F. Schneider. 1983. An improved procedure for extraction of aromatic bases from synfuel materials. In: Advanced Techniques in Synthetic Fuels Analysis. C.W. Wright, W.C. Weimer, and W.D. Felix, eds. Report Number PNL-8A-11552 (CONF-811160) (DE83015528). Technical Information Center, U.S. Department of Energy: Virginia. pp. 3-11.

Epler, J.L., J.A. Young, A.A. Hardigree, T.K. Rao, M.R. Guerin, J.B. Rubin, C.-H. Ho, and B.R. Clark. 1978. Analytical and biological analyses of test materials from the synthetic fuel technologies. I. Mutagenicity of crude oils determined by the *Salmonella typhimurium*/microsomal activation system. Mutat. Res. 57:265-276.

Guerin, M.R., C.-H. Ho, T.K. Rao, B.R. Clark, and J.L. Epler. 1980. Polycyclic aromatic primary amines as determinant chemical mutagens in petroleum substitutes. Environ. Res. 23:42-53.

Haugen, D.A., and M.J. Peak. 1983. Mixtures of polycyclic aromatic compounds inhibit mutagenesis in the Salmonella/microsome assay by inhibition of metabolic activation. Mutat. Res. 116:257-269.

Haugen, D.A., M.J. Peak, K.M. Suhrbier, and V.C. Stamoudis. 1982. Isolation of mutagenic aromatic amines from a coal conversion oil by cation exchange chromatography. Anal. Chem. 54:32-37.

Haugen, D.A., V.C. Stamoudis, M.J. Peak, and A.S. Boparai. 1983. Isolation of mutagenic polycyclic aromatic hydrocarbons from tar produced during coal gasification. In: Polynuclear Aromatic Hydrocarbons: Formation, Metabolism and Measurement. M. Cooke and A.J. Dennis, eds. Battelle Press: Columbus. pp. 607-620.

Hsie, A.W., J.P. O'Neill, J.R. San Sebastian, D.B. Couch, P. Brimer, W.N.C. Sun, J.C. Fuscoe, N.L. Forbes, R. Machanoff, J.C. Riddle, and M.H. Hsie. 1978. Quantitative mammalian cell genetic toxicology: Study of the cytotoxicity and mutagenicity of seventy individual environmental agents related to energy technologies and three subfractions of a crude synthetic oil in the CHO/HGPRT system. In: Application of Short-Term Bioassays in the Fractionation and Analysis of Complex Environmental Mixtures. M.D. Waters, S. Nesnow, J.L. Huisingh, S.S. Sandhu, and L. Claxton, eds. Plenum Press: New York. pp. 291-315.

Huberman, E., C.K. McKeown, C.A. Jones, D.R. Hoffman, and S. Murao. 1984. Induction of mutations by chemical agents at the hypoxanthine guanine phosphoribosyl transferase locus in human epithelial teratoma cells. Mutat. Res. 130:127-137.

Kawalek, J.C., and A.W. Andrews. 1981. Effect of aromatic hydrocarbons on the metabolism of 2-aminoanthracene to mutagenic products in the Ames assay. Carcinogenesis 2:1367-1369.

Kouri, R.E., K.R. Brandt, R.G. Sosnowski, L.M. Schechtman, and W.F. Benedict. 1978. In vitro activation of cigarette smoke condensate materials to their mutagenic forms. In: Application of Short-Term Bioassays in the Fractionation and Analysis of Complex Environmental Mixtures, M.D. Waters, S. Nesnow, J.L. Huisingh, S.S. Sandhu, and L. Claxton, eds. Plenum Press: New York. pp. 497-512.

Li, A.P., A.L. Brooks, C.R. Clark, R.L. Shimizu, R.L. Hanson, and J.S. Dutcher. 1983. Mutagenicity testing of complex environmental mixtures with Chinese hamster ovary cells. In: Short-Term Bioassays in the Analysis of Complex Environmental Mixtures III. M.D. Waters, S.S. Sandhu, J. Lewtas, L. Claxton, N. Chernoff, and S. Nesnow, eds. Plenum Press: New York. pp. 183-196.

Mahlum, D.D. 1984. Expression of benzo(a)pyrene and 6-aminochrysene initiating activity in complex mixtures from coal. Toxicologist 4:207.

McCann, J., E. Choi, E. Yamaski, and B.N. Ames. 1975. Detection of carcinogens as mutagens in the *Salmonella*/microsome test: Assay of 300 chemicals. Proc. Natl. Acad. Sci. U.S.A. 72:5135-5139.

Murphy, S.D. 1983. General principles in the assessment of toxicity of chemical mixtures. Environ. Health Perspect. 48:141-144.

Neal, R. 1983. Protocol for testing the toxicity of chemical mixtures. Environ. Health Perspect. 48:137-139.

O'Neill, J.P., P.A. Brimer, R. Machanoff, G.P. Hirsch, and A.W. Hsie. 1977. A quantitative assay of mutation induction at the hypoxanthine-guanine phosphoribosyl transferase locus in Chinese hamster ovary cells: Development and definition of the system. Mutat. Res. 45:91-101.

Pelroy, R.A., and M.R. Peterson. 1979. Use of Ames test in evaluation of shale oil fractions. Environ. Health Perspect. 30:191-203.

Pelroy, R.A., and M.R. Peterson. 1981. Mutagenic characterization of synthetic fuel materials by the Ames/Salmonella assay system. Mutat. Res. 90:309-320.

Stamoudis, V.C., S. Bourne, D.A. Haugen, M.J. Peak, C.A. Reilly, J.R. Stetter, and K. Wilzbach. 1981. Chemical and biological characterization of high-BTU coal gasification (The HYGAS Process). I. Chemical characterization of mutagenic fractions. In: Coal Conversion and the Environment. D.D. Mahlum, R.H. Gray, and W.D. Felix, eds. CONF801039 (DE82000105). Technical Information Center, U.S. Department of Energy: Virginia. pp. 67-95.

Strom, S., A.D. Kligerman, and G. Michalopoulos. 1981. Comparisons of the effects of chemical carcinogens in mixed cultures of rat hepatocytes and human fibroblasts. Carcinogenesis 2:709-715.

Waters, M.D., V.F. Simmon, A.D. Mitchell, T.A. Jorgenson, and R. Valencia. 1979. A phased approach to the evaluation of environmental chemicals for mutagenesis and presumptive carcinogenesis. In: In Vitro Toxicity of Environmental Agents Part B. A.R. Kolber, T.K Wong, L.D. Grant, R.S. DeWoskin, and T.J. Hughes, eds. Plenum Press: New York. pp. 417-441.

Wilson, B.W., R. Pelroy, and J.T. Cresto. 1980. Identification of primary aromatic amines in mutagenically active subfractions from coal liquefaction materials. Mutat. Res. 79:193-202.

Wilson, B.W., R.A. Pelroy, D.D. Mahlum, M.E. Frazier, D.W. Later, and C.W. Wright. 1984. Comparative chemical composition and biological activity of single- and two-stage coal liquefaction process streams. Fuel 63:46-55.

Wilzbach, K.E., J.R. Stetter, C.A. Reilly, Jr., and W.G. Willson. 1981. Environmental Research Program for Slagging Fixed-Bed Coal Gasification. Status Report ANL/SER-1 (Argonne National Laboratory, November), U.S. Department of Energy: Virginia. pp. 1-34.

INCREASED VIRUS TRANSFORMATION BY FORMALDEHYDE OF HAMSTER EMBRYO CELLS PRETREATED WITH BENZO(A)PYRENE

G.G. Hatch and T.M. Anderson

In Vitro Toxicology Division, Northrop Services, Inc.--Environmental Sciences, Research Triangle Park, North Carolina 27709

INTRODUCTION

The etiology of human cancers is most likely related to the combined effects of diverse classes and physical forms of complex mixtures of environmental carcinogens. There is increasing evidence that the development of human cancers is a multistage process (Slaga, 1978, 1980, in press; Selikoff, in press; Weinstein, in press) and is heavily linked to exposure to diverse environmental agents (Doll and Peto, 1981; International Labour Office, 1982; National Research Council, 1982; Tomatis et al., 1982). It has been suggested that a single or a simple unifactorial relationship between cancer cause and consequence may be the exception rather than the rule (Hecker et al., 1982). This idea is supported by the fact that although some 700 "solitary" carcinogens have been reported, only about 30 have been firmly linked to the etiology of human cancers (IARC, 1982b). Co-carcinogenesis theories (Sivak, 1979; Williams, in press) are useful in providing a conceptional framework for putative multifactorial and multistage etiologies of environmental cancer. Their value has been confirmed by an increasing documentation of environmental cancers by epidemiological data in human populations (Doll and Peto, 1981; Tomatis et al., 1982; Yuspa and Harris, 1982). Examples include smoking and the subsequent high incidence of lung cancer (Doll and Peto, 1981; Kuschner, in press) and the influence of dietary factors on increased cancers at selected sites (Schoental and Conners, 1981; National Research Council, 1982). Additional epidemiological evidence documents increased cancer risk by the additive and/or synergistic effects of cigarette smoking and occupational exposure to asbestos (Selikoff, in press), uranium (Archer, in press), or arsenic (Doll and Peto, 1981).

The concept of co-carcinogenesis by multiple agents was initially put forth by Rous and Kidd (1941) and Berenblum (1941). These early investigators demonstrated that subcarcinogenic doses of polycyclic hydrocarbons could produce skin tumors if followed by applications of promoters.

Studies of additive or promoting effects as general phenomena in the carcinogenic process were extended by additional observations in cultured cells. Hecker (1968), Lasne et al. (1974), and Mondal et al. (1976) demonstrated that purified phorbol esters (12-0-tetradecanoylphorbol-13-acetate (TPA)) produced increased transformation of cultured cells treated with polycyclic hydrocarbons. Similar findings were documented *in vitro* for ultraviolet or X-radiation and TPA (Mondal and Heidelberger, 1976; Kennedy et al., 1978, 1980) and for chemical carcinogens or ultraviolet radiation preceded by X-ray or methyl methanesulfonate treatment (DiPaolo and Donovan, 1976; DiPaolo et al., 1976; Doniger and DiPaolo, 1980). Weinstein (in press) and others (Slaga, 1980, in press; Hecker et al., 1982) recently reviewed current cellular and molecular mechanisms of multistage chemical carcinogenesis and co-carcinogenesis.

Large numbers of people are exposed in industrial, urban, and home environments to the ubiquitous toxicants benzo(a)pyrene (B[a]P) and formaldehyde. They represent different physical forms of carcinogens present in complex mixtures with long-term chronic exposures to diverse populations. In this study, B(a)P and formaldehyde were selected as relevant compounds to determine combined transformation effects in primary embryo cells.

Benzo(a)pyrene commonly occurs as a product of incomplete combustion and is a recognized animal (IARC, 1983) and suspect human carcinogen (U.S. Department of Health and Human Services, 1983) that induces many genotoxic effects (IARC, 1983). An estimated 1.8 million lb/yr are released from stationary sources. Human exposure can occur from the presence of B(a)P in air pollution, cigarette smoke, and food sources.

Formaldehyde is a nasal carcinogen in long-term rodent inhalation bioassays (Swenberg et al., 1980; Albert et al., 1982; IARC, 1982a; Kerns et al., 1983). Its worldwide production approaches 12 million tons. Industrial, occupational, and personal exposures to formaldehyde are of concern because of their widespread nature, relatively high exposure levels, and high chemical reactivity (National Research Council, 1981). Formaldehyde's genotoxicity, carcinogenicity, and epidemiology have been extensively reviewed (Auerbach et al., 1977; Boreiko et al., 1982; IARC, 1982a; Clary et al., 1983; Gibson, 1983; Swenberg et al., 1983).

In the environment, formaldehyde exposure occurs commonly in the form of a vapor and may be combined in a complex mixture containing other recognized or putative carcinogens and/or mutagens. Its presence in gasoline and diesel exhausts, cigarette smoke, and incinerator wastes are prime examples. The possibility of additive or synergistic genotoxic activity with other chemicals, particularly B(a)P, is a logical but largely untested possibility.

To investigate the combined effects of B(a)P and formaldehyde, a short-term *in vitro* assay for mutagens and carcinogens (Casto 1973, 1981a; Casto et al., 1973, 1974; Hatch et al., 1982a, c, d, 1983a, b; Hatch and Anderson, in press) that measures the ability of chemicals of many diverse classes to enhance the adenovirus (SA7) transformation of Syrian hamster embryo (SHE) cells was selected. Effects of complex mixtures (Casto et al., 1981) and multiple agents (Hatch and Anderson, 1984, Hatch et al., unpublished results) have also been detected in this bioassay. Methodology most suitable for evaluation of the vapor phase of formaldehyde has previously been developed in this culture system (Hatch et al., 1982b, 1983a) and utilized to evaluate other volatile liquids and gases, including halogenated chemicals (Hatch et al., 1982b, 1983b) and ethylene oxide (Hatch et al., 1982c).

This virus enhancement assay reflects the capability of a chemical to damage cell DNA by either direct or indirect means (Casto, 1981a) and provides a rapid mammalian cell bioassay system for identifying suspect genotoxic environmental chemicals (Casto, 1981a; Hatch et al., 1983a, b; Hatch and Anderson, in press). Positive test results in this system are generally concordant with those obtained in bacterial and mammalian cell mutagenesis assays (Casto 1981b; Hatch et al., 1982c, 1983a, b) and *in vivo* carcinogenesis assays (Casto et al., 1978; Hatch et al., 1982c, 1983a, b; Heidelberger et al., 1983).

In this study, cells were treated with non-toxic and toxic concentrations of B(a)P and then subsequently exposed to selected concentrations of formaldehyde vapors. Increased SA7 virus transformation frequencies (TF) were produced by sequential B(a)P and formaldehyde treatment.

MATERIALS AND METHODS

Cell Cultures and Virus Stocks

Primary SHE cells were prepared by trypsinization of eviscerated and decapitated 13- to 14-d hamster embryos (Charles River Laboratories, Wilmington, MA), as described previously (Hatch et al., 1983b). After 3 d, the number of cells was approximately 5×10^6 per plate.

Vero (ATCC:CCL-81) cell cultures were inoculated with Simian adenovirus SA7 for preparation of virus stocks, as previously described (Hatch et al., 1983b).

Chemical Treatment

A fresh stock solution of B(a)P (Aldrich Gold Label; 99+% pure, Aldrich Chemical Co., Milwaukee, WI) in acetone was further diluted in acetone and added to medium to give the desired final concentrations. Final acetone concentration was 0.5%.

Formaldehyde solution (37%) was obtained from Fisher Scientific (Fairlawn, NJ), and dilutions were made in sterile deionized distilled water.

Cells were cultured for 72 h, refed approximately 8 h prior to treatment, and then treated with medium containing appropriate concentrations of B(a)P. After 20 h, cells were rinsed with complete medium immediately prior to formaldehyde treatment.

A complete description of the methodology utilized for treatment with volatile liquids and gases including formaldehyde has been reported (Hatch et al., 1982b, 1983a, b). In brief, cells in minimal amounts of culture medium were exposed to formaldehyde vapors in closed treatment chambers on rocker platforms. After 2 h, the vapor was exhausted and the cells assayed for viability and enhancement of viral transformation.

Transformation and Survival Assays

Detailed procedures have been described previously (Hatch et al., 1983b). SA7 virus was added to chemically-treated SHE cells at approximately 4×10^7 plaque forming units per dish and was adsorbed for 3 h. Cells were replated into 60-mm dishes at 2×10^5 cells per dish for the transformation assay and 700 cells per dish for the survival assay. Transformation assay plates were maintained in low (0.1 mM) $CaCl_2$ medium (Freeman et al., 1967) and Bactoagar (BBL Microbiology Systems, Cockeysville, MD) (0.3 g/100 ml). Survival and transformation assay plates were fixed and stained at approximately 9 and 28 d, respectively, from the beginning of the experiment.

Determination of Enhancement

Detailed methods for determining surviving fractions, TF, and enhancement ratios have been reported (Hatch et al., 1983b). The fraction of cells surviving chemical treatment was determined from five assay plates seeded with approximately 700 cells per plate. The TF, i.e., the number of SA7 foci per 10^6 surviving cells, was calculated by dividing the number of SA7 foci from 10 transformation plates by two and by the surviving fraction of treated cells. Enhancement was expressed as the ratio between the TF of treated cells and the TF of control cells. Statistical significance was determined using a table of ratios (Casto et al., 1973) derived from the Lorenz (1962) table that is based on the Poisson distribution. The increased TF was considered statistically significant at the 5 or 1% confidence level if the enhancement ratio exceeded the appropriate value obtained from the table of critical ratios.

RESULTS

The effects of B(a)P on formaldehyde were measured by cytotoxicity and enhancement of SA7 virus transformation in hamster embryo cells (Table 1). A 20-h exposure to 0.03, 0.06, and 0.12 µg/ml of B(a)P resulted in concentration-related cytotoxicity. Significant increases (approximately threefold as indicated by the enhancement ratios) in TF were produced at 0.06 and 0.12 µg/ml of B(a)P. A 2-h treatment with formaldehyde alone produced significant cytotoxicity (62% survival) at the highest concentration of 6.0 ppm; lower concentrations were non-toxic. Significant increases in TF (approximately sixfold and twofold) were produced by 6.0 and 3.0 ppm, respectively. These concentrations produced a twofold to threefold increase in the absolute number of virus foci observed, demonstrating that the increased TF produced by formaldehyde was not dependent on cytotoxicity for significance. Both B(a)P and formaldehyde produced concentration-related enhancement of virus transformation at two concentrations when cells were treated with either chemical alone.

Figure 1 shows the survival and transformation produced by combined B(a)P and formaldehyde treatment. SHE cells were pretreated for 20 h with 0.03 or 0.06 µg/ml of B(a)P. Replicate dishes from each B(a)P treatment group were exposed for 2 h to five concentrations (0.4 to 6.0 ppm) of formaldehyde vapor. The cytotoxicity observed with combined treatment at both B(a)P concentrations seemed to be primarily related to the B(a)P-induced cytotoxicity.

Table 1. Enhancement of SA7 Transformation by Benzo(a)pyrene or Formaldehyde

Chemical	Treatment Concentration[a]	% Survival[b]	SA7 Foci[c]	Trans-formation Fre-quency[d]	Enhance-ment Ratio[e]
Benzo(a)pyrene	0.12[f] μg/ml	11	12	53	3.4**
	0.06	34	37	54	3.5**
	0.03	76	30	20	1.3
	0	100	31	15	1.0
Formaldehyde	6.0 ppm	62	108	86	5.6**
	3.0	102	66	32	2.1**
	1.5[g]	107	41	19	1.2
	0	100	31	15	1.0

[a]Mass cultures of hamster embryo cells were treated for 20 h with benzo(a)pyrene dilutions in culture medium or for 2 h with formaldehyde vapors in individual treatment chambers. Cells were rinsed and SA7 virus was adsorbed for 3 h. Cells were then transferred for survival (700 cells per dish) and transformation (200,000 cells per dish) assays.

[b]Determined from plates receiving 700 cells. The number of colonies from virus- and chemically-treated cells was divided by the number of colonies from virus-inoculated control cells to give the surviving fraction (survival values are shown as a percent). Plating efficiency of control was 15%.

[c]Number of foci from 2 x 10^6 plated cells.

[d]Transformation frequency per 10^6 surviving cells was determined by dividing SA7 foci by two and by the surviving fraction at each dose.

[e]Enhancement ratio was determined by dividing the transformation frequency of treated cells by that of control cells. Statistical significance was detemined using a table of ratios (Casto et al., 1973) derived from the Lorenz (1962) table that is based on the Poisson distribution. The increased transformation frequency was considered statistically significant if the enhancement ratio exceeded the appropriate value obtained from the table of ratios. A statistically significant ($p \leq .01$) enhancement ratio is indicated by **.

[f]Higher doses were completely toxic.

[g]Lower doses were non-toxic.

A concentration-related increase of TF was observed with formaldehyde in both B(a)P pretreated and non-pretreated cells (Figure 1). Those increases in TF that were statistically significant are identified with asterisks in Figure 1. Transformation frequencies significantly higher than the non-treated control were produced by

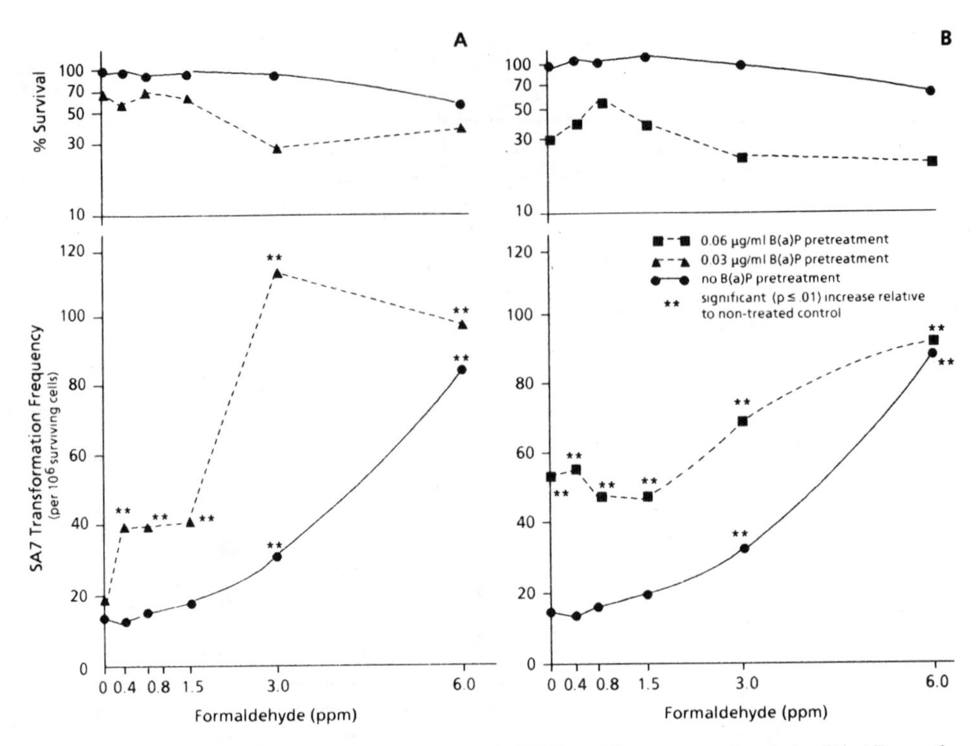

Figure 1. SA7 virus transformation of SHE cells treated with B(a)P and formaldehyde. Panel A: 0.03 µg/ml B(a)P. Panel B: 0.06 µg/ml B(a)P.

formaldehyde at five concentrations (0.4, 0.8, 1.5, 3.0, and 6.0 ppm) in cells pretreated with 0.03 or 0.06 µg/ml B(a)P. Increased TF in 0.03 µg/ml B(a)P pretreated cells were produced in the absence of overt cytotoxicity, demonstrating that the increased TF were not dependent on cell killing for significance.

The highest concentration of formaldehyde or B(a)P alone that did not produce a significant effect on TF was 1.5 ppm and 0.03 µg/ml, respectively (Table 1). However, the combined treatment of SHE cells with formaldehyde (0.4, 0.8, or 1.5 ppm) and B(a)P (0.03 µg/ml) produced TF that were significantly higher than those observed with either chemical alone.

Figure 1 suggests that the effects on transformation of sequential B(a)P and formaldehyde treatment were approximately additive. The TF of cells treated with both 0.03 µg/ml B(a)P and 0.4, 0.8, 1.5, and 6.0 ppm formaldehyde (41, 40, 42, and 99) were approximately the sum of the TF of cells treated with formaldehyde (14, 16, 19, and 86) and B(a)P (20) alone (Figure 1A). However, pretreatment with 0.03 µg/ml of B(a)P and 3.0 ppm formaldehyde may have produced a synergistic effect on enhancement of SA7 transformation. The TF of cells treated with 0.03 µg/ml B(a)P and 3.0 ppm formaldehyde (115) was greater than the sum of the TF of cells treated with each chemical alone (20 and 32, respectively) (Figure 1A). Pretreatment with 0.06 µg/ml B(a)P and formaldehyde produced effects on TF that were less than additive (Figure 1B).

DISCUSSION

The enhancement of virus transformation by physical or chemical agents has been successfully employed in mammalian cells to study the interactions between virus and either single chemical carcinogens or complex mixtures. Experimental data suggest that discrete changes in cell DNA resulting from chemical treatment can be detected by observing a quantitative increase in virus transformation (Casto et al., 1979). Increases in virus transformation have been demonstrated in hamster, rat, and human cells treated with a variety of carcinogenic agents (Casto, 1981a; Hatch et al., 1983a, b; Hatch and Anderson, 1984, in press). The virus enhancement assay reflects the capacity of a chemical to damage cell DNA by either direct or indirect means. It has been shown that carcinogen treatment results in an increased incorporation of virus DNA into cellular DNA (Casto et al., 1979). In addition, a quantitative relationship between B(a)P-DNA adduct formation and biological effect as measured by enhanced SA7 virus transformation frequency has been demonstrated (Theall et al., 1982). Data from the bioassay of approximately 130 test chemicals from 29 different chemical classes using the enhancement of virus transformation assay have been reported (Casto, 1981a; Hatch et al., 1982a, c, d, 1983a, b).

Relevant, quantitative cell culture transformation systems can provide an increased understanding of the nature of the neoplastic process and the relative effects of the combined action of two or more carcinogens. In this report, we demonstrate increased sensitivity and enhanced transformation of cultured hamster cells by combined treatment with two chemical agents. The sequential chemical treatment protocol utilized in these experiments (i.e., exposure to formaldehyde of cells pretreated with B(a)P) was designed to minimize cytotoxicity, demonstrate quantitative transformation data, and allow for assessment of the relative effect of each compound.

Formaldehyde treatment produced significant transformation at an eightfold lower concentration (0.8 ppm) in cells pretreated with a non-toxic and nontransforming concentration of B(a)P (0.03 µg/ml). The TF of B(a)P-pretreated cells increased with increasing doses of formaldehyde. The data demonstrating increased transformation by both agents is most consistent with additive rather than synergistic effects.

Additional studies have documented that reversal of the treatment order, i.e., formaldehyde pretreatment followed by B(a)P exposure, also produces increased virus transformation (unpublished data). Increased transformation by both chemicals is, therefore, independent of the sequential treatment order. However, increasing the time interval between treatment with the first and second chemical decreases transformation produced by combined treatment (data not shown).

It is significant that increased transformation by combined chemical treatment can be demonstrated independently of cytotoxicity. The use of vapor phase formaldehyde treatment allows utilization of lower treatment concentrations that permit increased detection of genotoxic effects (Hatch et al., 1983a) and is consistent with previously reported results for chlorinated hydrocarbons (Hatch et al., 1983b). The increased efficiency of vapor phase treatment may be related to the extent of solvation in the vapor state and effects on cell permeability.

The increased transformation produced in primary hamster embryo cells by formaldehyde in combination with low concentrations of B(a)P is consistent with other published reports documenting formaldehyde's ability to transform continuous cell lines of mouse (Ragan and Boreiko, 1981; Brusick, 1983) and hamster (Ashby and Lefevre, 1983) origin and to produce mutation or DNA damage in a variety of cultured mammalian cells, including human cells (Goldmacher and Thilly, 1983; Grafstrom et al., 1983).

Swenberg et al. (1983) have speculated that formaldehyde could accelerate carcinogenesis *in vivo* in the nasal passage. Weak promoting activity has been reported in mouse embryo cell cultures initiated with N-methyl-N'-nitro-N-nitrosoguanidine (Frazelle et al., 1983).

This laboratory has previously reported increased adenovirus transformation of hamster embryo cells treated with formaldehyde (Hatch et al., 1983a). We have also reported enhanced transformation by treatment with B(a)P singly and in combination with diverse agents in complex mixtures from sources including coke ovens, roofing tars, diesel and gasoline emissions, and cigarette smoke condensate (Casto et al., 1981). The activity produced by some of these complex mixtures was in excess of that predicted by their B(a)P content, suggesting additive and/or synergistic effects.

Previous collaborative studies (Theall et al., 1982) have documented a linear relationship between B(a)P-DNA binding (micromoles B(a)P per mole of DNA) and B(a)P concentration (0.1-1.0 μg B(a)P/ml) in primary SHE cell cultures. However, there was no increase in virus TF until a binding level of 8-10 μmol B(a)P/mol DNA was reached. After this "threshold" level was achieved, there was a linear increase in TF as a function of the extent of B(a)P-DNA binding.

The mechanisms by which B(a)P and other carcinogens enhance virus-induced cell transformation are not completely known, but this reported "threshold" suggests a critical level of DNA damage before chemical enhancement of virus transformation is demonstrated. If formaldehyde treatment is given while the effects of DNA damage from B(a)P are still present, then cumulative genotoxic effects might reasonably be expected. Previous studies with this bioassay system demonstrated a correlation between enhancement of virus transformation and persistence of DNA damage by both alkylating agents and polycyclic hydrocarbons, including B(a)P (Casto et al., 1976). For B(a)P, enhanced virus transformation persisted 72-96 h following chemical treatment.

Similarly, the length of the treatment time with formaldehyde and the time interval before subsequent bioassay are critical for optimal activity. Previous studies indicated that a 2-h treatment time is optimal. Longer exposure periods putatively allow for repair of damage after the initial insult, and increased transformation effects decrease significantly by 20 h (Hatch et al., 1983a).

In summary, increased transformation effects by sequential treatment with B(a)P and formaldehyde have been documented in a well-validated mammalian embryo cell system. The enhanced transformation produced by formaldehyde in SHE cells pretreated with B(a)P may provide a relevant model for determining the additive and/or synergistic effects of multiple genotoxic agents.

ACKNOWLEDGMENTS

We would like to thank Drs. Stephen Nesnow, Raymond Tennant, and Donald Gardner for helpful comments and suggestions in the preparation of this manuscript. The excellent technical assistance of Kathy Warn and Gail Wyatt is appreciated. This research was supported by the National Institute of Environmental Health Sciences under contract NO1-ES-15796. Portions of this work were presented at the 1984 annual meeting of the Environmental Mutagen Society.

REFERENCES

Albert, R.E., A.R. Sellakumar, S. Laskin, M. Kuschner, N. Nelson, and C.A. Snyder. 1982. Gaseous formaldehyde and hydrogen chloride induction of nasal cancer in the rat. J. Natl. Cancer Inst. 68:597-603.

Archer, V.E. (in press). Enhancement of lung cancer by cigarette smoking in uranium and other miners. In: Cancer of the Respiratory Tract: Predisposing Factors. M. Mass, D. Kaufman, J. Siegfried, V. Steele, and S. Nesnow, eds. Raven Press: New York.

Ashby, J., and P. Lefevre. 1983. Genetic toxicology studies with formaldehyde and closely related chemicals including hexamethylphosphoramide (HMPA). In: Formaldehyde Toxicity. J.E. Gibson, ed. Hemisphere Publishing Corp.: New York. pp. 85-97.

Auerbach, C., M. Moutschen-Dahman, and J. Moutschen. 1977. Genetic and cytogenetical effects of formaldehyde and related compounds. Mutat. Res. 39:317-362.

Berenblum, I. 1941. The cocarcinogenic action of croton resin. Cancer Res. 1:44-48.

Boreiko, C.J., D.B. Couch, and J.A. Swenberg. 1982. Mutagenic and carcinogenic effects of formaldehyde. Environ. Sci. Res. 25:353-367.

Brusick, D. 1983. Genetic and transforming activity of formaldehyde. In: Formaldehyde Toxicity. J.E. Gibson, ed. Hemisphere Publishing Corp.: New York. pp. 72-84.

Casto, B.C. 1973. Enhancement of adenovirus transformation by treatment of hamster cells with ultraviolet irradiation, DNA base analogs, and dibenz(a,h)anthracene. Cancer Res. 33:402-407.

Casto, B.C. 1981a. Detection of chemical carcinogens and mutagens in hamster cells by enhancement of adenovirus transformation. In: Advances in Modern Environmental Toxicology, Vol. 1. N. Mishra, V. Dunkel, and M. Mehlman, eds. Senate Press: Princeton. pp. 241-271.

Casto, B.C. 1981b. Mutagenesis and carcinogenesis testing *in vitro*. Biotechnol. Bioeng, Vol. XXIII. pp. 2659-2671.

Casto, B.C., W.J. Pieczynski, and J.A. DiPaolo. 1973. Enhancement of adenovirus transformation by pretreatment of hamster cells with carcinogenic polycyclic hydrocarbons. Cancer Res. 33:819-824.

Casto, B.C., W.J. Pieczynski, and J.A. DiPaolo. 1974. Enhancement of adenovirus transformation by treatment of hamster embryo cells with diverse chemical carcinogens. Cancer Res. 34:72-78.

Casto, B.C., W.J. Pieczynski, N. Janosko, and J.A. DiPaolo. 1976. Significance of treatment interval and DNA repair in the enhancement of viral transformation by chemical carcinogens and mutagens. Chem.-Biol. Interact. 13:105-125.

Casto, B.C., N. Janosko, J. Meyers, and J.A. DiPaolo. 1978. Comparison of *in vitro* tests in Syrian hamster cells for detection of carcinogens. Proc. Am. Assoc. Cancer Res. 19:83.

Casto, B.C., M. Miyagi, J. Meyers, and J.A. DiPaolo. 1979. Increased integration of viral genome following chemical and viral treatment of hamster embryo cells. Chem.-Biol. Interact. 25:255-269.

Casto, B.C., G.G. Hatch, S.L. Huang, J.L. Huisingh, S. Nesnow, and M.D. Waters. 1981. Mutagenic and carcinogenic potency of extracts of diesel and related environmental emissions: *In vitro* mutagenesis and oncogenic transformation. Environ. Int. 5:403-410.

Clary, J.J., J.E. Gibson, and A.S. Waritz, eds. 1983. Formaldehyde: Toxicology, Epidemiology and Mechanisms. Marcel Dekker: New York. 280 pp.

DiPaolo, J.A., and P.J. Donovan. 1976. *In vitro* morphologic transformation of Syrian hamster cells by UV-irradiation is enhanced by X-irradiation and unaffected by chemical carcinogens. Int. J. Radiat. Biol. 30:41-53.

DiPaolo, J.A., P.J. Donavan, and N.C. Popescu. 1976. Kinetics of Syrian hamster cells during X-irradiation enhancement of transformation *in vitro* by chemical carcinogen. Radiat. Res. 66:310-325.

Doll, R., and R. Peto. 1981. The causes of cancer: quantitative estimates of avoidable risks of cancer in the United States today. J. Natl. Cancer Inst. 66:1191-1308.

Doniger, J., and J.A. DiPaolo. 1980. Excision and postreplication DNA repair capacities, enhanced transformation, and survival of Syrian hamster embryo cells irradiated by ultraviolet light. Cancer Res. 40:582-587.

Freeman, A.E., P.H. Black, R. Wolford, and R.J. Huebner. 1967. Adenovirus type 12-rat embryo transformation system. J. Virol. 1:362-367.

Frazelle, J.H., D.J. Abernethy, and C.J. Boreiko. 1983. Weak promotion of C3H/10T1/2 cell transformation by repeated treatments with formaldehyde. Cancer Res. 43:3236-3239.

Gibson, J.E., ed. 1983. Formaldehyde Toxicity. Hemisphere Publishing Corp.: New York. 312 pp.

Goldmacher, V.S., and W.G. Thilly. 1983. Formaldehyde is mutagenic for cultured human cells. Mutat. Res. 116:417-422.

Grafstrom, R.C., A.J. Fornace, Jr., H. Autrup, J.F. Lechner, and C.C. Harris. 1983. Formaldehyde damage to DNA and inhibition of DNA repair in human bronchial cells. Science 220:216-218.

Hatch, G.G., and T.M. Anderson. 1984. Increased sensitivity and enhanced transformation by formaldehyde of hamster embryo cells pretreated with benzo(a)pyrene. Environ. Mutagen. 6:474.

Hatch, G.G., and T.M. Anderson. (in press). Assays for enhanced DNA viral transformation of primary Syrian hamster embryo (SHE) cells. In: Evaluation of Short Term Tests for Carcinogens: Report of the International Programme on Chemical Safety Collaborative Study on *In Vitro* Assays. J. Ashby, F. de Serres, M. Draper, M. Ishidate, B.E. Margolin, B. Matter, and M. Shelby, eds. Elsevier/North Holland: Amsterdam.

Hatch, G.G., P. Conklin, T. Anderson, M. Waters, and S. Nesnow. 1982a. Enhanced viral transformation produced by treatment of hamster embryo cells with pesticides. J. Cell Biol. 95(2):452a.

Hatch, G.G., P.D. Mamay, M.L. Ayer, B.C. Casto, and S. Nesnow. 1982b. Methods for detecting gaseous and volatile carcinogens using cell transformation bioassays. In: Genotoxic Effects of Airborne Agents. R.R. Tice, D.L. Costa, and K.M. Schaich, eds. Plenum Press: New York. pp. 75-90.

Hatch, G.G., P.D. Mamay, C.C. Christensen, R. Langenbach, C.R. Goodhart, and S. Nesnow. 1982c. Enhanced viral transformation of primary Syrian hamster embryo cells and mutagenesis of Chinese hamster lung cells (V79) exposed to ethylene oxide in sealed treatment chambers. Proc. Am. Assoc. Cancer Res. 23:74.

Hatch, G.G., P.D. Mamay, and S. Nesnow. 1982d. Enhancement of viral transformation of hamster embryo cells by pretreatment with 4-chloromethylbiphenyl. Mutat. Res. 100:229-233.

Hatch, G.G., P.M. Conklin, C.C. Christensen, B.C. Casto, and S. Nesnow. 1983a. Synergism in the transformation of hamster embryo cells treated with formaldehyde and adenovirus. Environ. Mutagen. 5:49-57.

Hatch, G.G., P.D. Mamay, M.L. Ayer, B.C. Casto, and S. Nesnow. 1983b. Chemical enhancement of viral transformation in Syrian hamster embryo cells by gaseous and volatile chlorinated methanes and ethanes. Cancer Res. 43:1945-1950.

Hecker, E. 1968. Cocarcinogenic principles from the seed oil of croton tigluim and from other euphorbraceae. Cancer Res. 28:2332-2349.

Hecker, E., N.E. Fusenig, W. Kunz, F. Marks, and H.W. Thielmann, eds. 1982. Carcinogenesis. A Comprehensive Survey, Volume 7: Cocarcinogenesis and Biological Effects of Tumor Promoters. Raven Press: New York. 664 pp.

Heidelberger, C., A.E. Freeman, R.J. Pienta, A. Sivak, J.S. Bertram, B.C. Casto, V.C. Dunkel, M.W. Francis, T. Kakunaga, J.B. Little, and L.M. Schechtman. 1983. Cell transformation by chemical agents--a review and analysis of the literature. Mutat. Res. 114:283-385.

International Agency for Research on Cancer. 1982a. IARC Monographs on the Evaluation of the Carcinogenic Risk of Chemicals to Humans. Vol. 29. Some Industrial Chemicals and Dyestuffs. International Agency for Research on Cancer: Lyon, France. pp. 345-389.

International Agency for Research on Cancer. 1982b. IARC Monographs on the Evaluation of the Carcinogenic Risk of Chemicals to Humans, Supplement 4. Chemicals, Industrial Processes and Industries Associated with Cancer in Humans. Volumes 1 to 29. International Agency for Research on Cancer: Lyon, France. 292 pp.

International Agency for Research on Cancer. 1983. IARC Monographs on the Evaluation of the Carcinogenic Risk of Chemicals to Humans. Vol. 32. Polynuclear Aromatic Compounds, Part 1, Chemical, Environmental and Experimental Data. International Agency for Research on Cancer: Lyon, France. pp. 69-90.

International Labour Office. 1982. Occupational Safety and Health Series. Vol. 46. Prevention of Occupational Cancer--International Symposium. International Labour Office: Geneva, Switzerland. 658 pp.

Kennedy, A.R., S. Mondal, C. Heidelberger, and J.B. Little. 1978. Enhancement of X-ray transformation by 12-O-tetradecanoylphorbol-13-acetate in a cloned line of C3H mouse embryo cells. Cancer Res. 38:439-443.

Kennedy, A.R., G. Murphy, and J.B. Little. 1980. The effect of time and duration of exposure to 12-O-tetradecanoylphorbol-13-acetate (TPA) on x-ray transformation of C3H 10T1/2 cells. Cancer Res. 40:1915-1920.

Kerns, W.D., K.L. Pavkov, D.J. Donofrio, E.J. Gralla, and J.A. Swenberg. 1983. Carcinogenicity of formaldehyde in rats and mice after long-term inhalation exposure. Cancer Res. 43:4382-4392.

Kuschner, M. (in press). The relationship of underlying pathologic disease states of infectious and noninfectious etiologies to predisposition to the development of lung cancer. In: Cancer of the Respiratory Tract: Predisposing Factors. M. Mass, D. Kaufman, J. Siegfried, V. Steele, and S. Nesnow, eds. Raven Press: New York.

Lasne, C., A. Gentil, and I. Chouroulinkov. 1974. Two-stage malignant transformation of rat fibroblasts in tissue culture. Nature 274:490-491.

Lorenz, R.J. 1962. Zur statistik des plaque-testes. Arch. Gesamte Virus forsch. 12:108-137.

Mondal, S., and C. Heidelberger. 1976. Transformation of C3H/10T1/2 Cl 8 mouse embryo fibroblasts by ultraviolet irradiation and a phorbol ester. Nature 260:710-711.

Mondal, S., D.W. Brankow, and C. Heidelberger. 1976. Two-stage chemical oncogenesis in cultures of C3H/10T1/2 cells. Cancer Res. 36:2254-2260.

National Research Council. 1981. Formaldehyde and Other Aldehydes. National Academy Press: Washington, DC. 340 pp.

National Research Council. 1982. Diet, Nutrition, and Cancer. National Academy Press: Washington, DC.

Ragan, D.L., and C.J. Boreiko. 1981. Initiation of C3H/10T1/2 cell transformation by formaldehyde. Cancer Lett. 13:325-331.

Rous, P., and J.G. Kidd. 1941. Conditional neoplasms and subthreshold neoplastic states. A study of the tar tumors of rabbits. J. Exp. Med. 73:365-390.

Schoental, R., and T.A. Conners, eds. 1981. Dietary Influences on Cancer: Traditional and Modern. CRC Press: Boca Raton. 260 pp.

Selikoff, I. (in press). Asbestos workers and the synergistic effects of cigarette smoking. In: Cancer of the Respiratory Tract: Predisposing Factors. M. Mass, D. Kaufman, J. Siegfried, V. Steele, and S. Nesnow, eds. Raven Press: New York.

Sivak, A. 1979. Cocarcinogenesis. Biochem. Biophys. Acta 560:67-89.

Slaga, T.J., ed. 1978. Carcinogenesis. A Comprehensive Survey, Vol. 2: Mechanisms of Tumor Promotion and Cocarcinogenesis. Raven Press: New York. 605 pp.

Slaga, T.J., ed. 1980. Carcinogenesis. A Comprehensive Survey, Vol. 5: Modifiers of Chemical Carcinogenesis. Raven Press: New York. 285 pp.

Slaga, T.J. (in press). Cellular mechanisms for tumor promotion and enhancement. In: Cancer of the Respiratory Tract: Predisposing Factors. M. Mass,

D. Kaufman, J. Siegfried, V. Steele, and S. Nesnow, eds. Raven Press: New York.

Swenberg, J.A., W.D. Kerns, R.E. Mitchell, E.J. Gralla, and K.L. Pavkov. 1980. Induction of squamous cell carcinomas of the rat nasal cavity by inhalation exposure to formaldehyde vapor. Cancer Res. 40:3398-3402.

Swenberg, J.A., E.A. Gross, H.W. Randall, and C.S. Barrow. 1983. The effect of formaldehyde exposure on cytotoxicity and cell proliferation. In: Formaldehyde: Toxicology, Epidemiology and Mechanisms. J.J. Clary, J.E. Gibson, and R.S. Wartz, eds. Marcel Dekker: New York. pp. 225-236.

Theall, G., I.B. Weinstein, D. Grunberger, S. Nesnow, and G. Hatch. 1982. Quantitative relationships between DNA adduct formation and biological effects. In: Banbury Report 13: Indicators of Genotoxic Exposure. B.A. Bridges, B.E. Butterworth, and I.B. Weinstein, eds. Cold Spring Harbor Laboratory: New York. pp. 231-243.

Tomatis, L., N.E. Breslow, and H. Bartsch. 1982. Experimental studies in the assessment of human risk. In: Cancer Epidemiology and Prevention. D. Schottenfeld and J.F. Fraumeni, eds. Saunders: Philadelphia. pp. 44-73.

U.S. Department of Health and Human Services. 1983. Third Annual Report on Carcinogens, Summary. U.S. Department of Health and Human Services: Washington, DC. 229 pp.

Weinstein, I.B. (in press). Molecular mechanisms of multistage chemical carcinogenesis. In: Cancer of the Respiratory Tract: Predisposing Factors. M. Mass, D. Kaufman, J. Siegfried, V. Steele, and S. Nesnow, eds. Raven Press: New York.

Williams, G.M. (in press). Types of enhancement of carcinogenesis and influences on human cancer. In: Cancer of the Respiratory Tract: Predisposing Factors. M. Mass, D. Kaufman, J. Siegfried, V. Steele, and S. Nesnow, eds. Raven Press: New York.

Yuspa, S.H., and C.C. Harris. 1982. Molecular and cellular basis of chemical carcinogenesis. In: Cancer Epidemiology and Prevention. D. Schottenfeld and J.F. Fraumeni, eds. Saunders: Philadelphia. pp. 23-43.

ENHANCEMENT OF ANCHORAGE-INDEPENDENT SURVIVAL OF RETROVIRUS-INFECTED FISCHER RAT EMBRYO CELLS FOR DETERMINING THE CARCINOGENIC POTENTIAL OF CHEMICAL AND PHYSICAL AGENTS

William A. Suk, Joan E. Humphreys, E. Perry Hays, and John D. Arnold

Laboratory of Cellular and Molecular Oncology, *In Vitro* Toxicology Program, Northrop Services, Inc.--Environmental Sciences, Research Triangle Park, North Carolina 27709

INTRODUCTION

Proliferation of cells in an anchorage-independent state in semisolid medium (agar, methyl cellulose) has been considered a confirmation of neoplastic transformation (Barrett et al., 1979; Freedman and Shin, 1974; Kakunaga, 1978; Shin et al., 1975). The ability of transformed cells to aggregate and/or survive in liquid media above a base layer of solid agar is an extension of their anchorage independence and has been used to confirm transformation of a variety of cells exposed to either viruses or chemicals (Cho et al., 1976; Steuer and Ting, 1976; Steuer et al., 1977). Attachment-independent survival has recently been used in an assay for initiators and promoters of carcinogenesis (Eker and Sanner, 1983). Anchorage-independent survival as an indicator of neoplastic transformation has been measured and applied to Rauscher leukemia virus-infected Fischer rat embryo (RIFRE) cells (Traul et al., 1979, 1981b). Recently, a number of modifications have been made in the assay that have increased its sensitivity and reproducibility (Suk et al., 1983) and have enabled the RIFRE cells to respond to carcinogen treatment dose dependently. Moreover, recent data show that cells exhibiting carcinogen-mediated enhancement of anchorage-independent survival subsequently progressed to a transformed phenotype (unpublished data). The modified system has been proven to be useful in the evaluation of the transforming potential of a number of agents selected by the World Health Organization's International Program for Chemical Safety (Suk and Humphreys, in press) and of ultraviolet (UV) irradiation (Suk and Arnold, 1984).

MATERIALS AND METHODS

Cells

The RIFRE cell cultures ($2FR_450$) used in this assay have been described previously (Traul et al., 1981a). Frozen ampules of the $2FR_450$ cells were received from Dr. K. Traul via the American Type Culture Collection. The cells at passage levels 7 and 15 were propagated and prepared for cryopreservation by suspending 3×10^6 cells/ml in Eagle's minimum essential medium (EMEM) with 10% fetal bovine serum (FBS) and 7.5 - 10% dimethyl sulfoxide (DMSO). Cultures described in this work were used between passage 17 and 35. The cultures were maintained by passaging at 7-10 d intervals, seeding at 2.67×10^4 cells/cm^2, and refeeding once between passages. The cells had a population doubling time of approximately 42 h, reaching confluency within 72 h. Cells were grown in 37°C incubators with a water-saturated atmosphere of 5% CO_2 in air. The cells were cultivated in EMEM with 10% FBS. No antibiotics were used. Trypsin (Enzar-T) (Reheis, Tarrytown, NY) diluted 1:40 was used routinely for dissociating the cells during serial subpassaging.

Retrovirus infection of the cell was confirmed by immunofluorescence using antiviral serum (Tennant et al., 1973), reverse transcriptase assay (Grandgenett et al., 1972), and XC plaque-induction (Rowe et al., 1970). The cells were found to be free of mycoplasma contamination in tests performed at Biotech Research Laboratories, Inc. (Rockville, MD).

Cytotoxicity Test

A cytotoxicity test for determination of test doses was performed prior to the anchorage-independent survival assay. RIFRE cells were seeded into 24-well multiwell plates at 5.2×10^4 cells per well. After incubating for 24 h, the cells were treated with a range of chemical doses; each dose was administered in triplicate. Media and solvent controls were also included in the test. After treatment, the cells were incubated for an additional 72 h and then washed twice with Earle's balanced salt solution (EBSS) and refed with EMEM. After incubating 72 h, the cells were fixed and stained with methylene blue. Microscopic and macroscopic examination of the stained cultures, comparing test chemical doses to the media and solvent controls, led to dosage selection for the short-term bioassay.

Anchorage-Independent Survival Assay

Anchorage-independent survival of the RIFRE cells was determined using a modification of a previously described procedure (Traul et al., 1981b). Figure 1 schematically represents the modified assay procedure. The RIFRE cells were plated at 6.66×10^3 cells/cm^2, 500,000 cells/75-cm^2 flask (1-3 flasks per condition), and incubated for 24 h. Plating medium was then aspirated and replaced with medium containing the preselected test chemical doses. Control cells received EMEM with or without solvent.

One to three flasks were then trypsinized and counted using trypan blue exclusion to obtain the relative plating efficiency at the time of chemical treatment. Following a

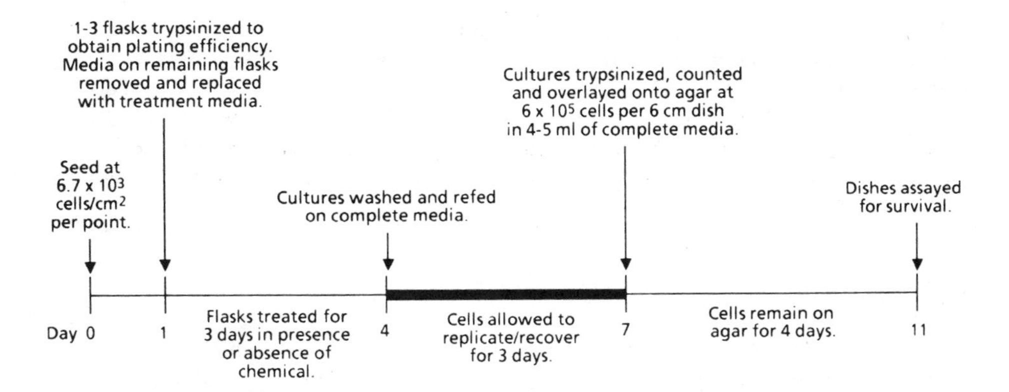

Figure 1. Schematic representation of the RIFRE cell anchorage-independent survival assay. The details of the procedure are described in Materials and Methods.

treatment incubation period of 72 h, the experimental treatment media were removed, the cells were washed twice with EBSS, and the cells were then refed with EMEM without chemical. The cultures were then allowed to incubate and replicate (recover) an additional 72 h. After dissociating with trypsin, cells treated with the same chemical dose were pooled and resuspended in EMEM. Viable cell counts were made for each test dose, including medium and solvent controls. Six hundred thousand viable cells were seeded onto a 60-mm dish (3 dishes/test dose) in 4 ml of EMEM over a solid agar base: 1% agar Noble medium was made by combining 2X concentrations of EMEM with 2% agar at a 1:1 ratio. (The agar dishes were prepared 3 d before cells were added.) The dishes were incubated for 4 d. Cells from one to three dishes for each set of controls and treatment points were harvested by decanting the suspended cells into a 15-ml conical centrifuge tube, washing the dishes twice with EBSS, and decanting the wash into the tube. The tubes were centrifuged at 600-1200 x g at 4°C for 10 min, the supernatant was removed, and the cell pellet was resuspended in 3 ml of EBSS. Cells for cell counting were taken directly from these tubes. Cells were counted using trypan blue exclusion to determine total viable cell count; three hemacytometer readings were performed for each treatment point and the number was averaged, thereby reducing the amount of variability inherent in the counting procedure.

An enhanced response was expressed as the number of anchorage-independent surviving viable cells induced by carcinogen treatment relative to the solvent or media control cells. This was computed as

$$\%\ \text{Enhanced Survival} = \frac{\text{Treated VCN} - \text{Control VCN}}{\text{Control VCN}} \times 100$$

where VCN equals viable cell number (Traul et al., 1981b).

Treatment with ultraviolet irradiation. To test the effects of UV irradiation on the RIFRE cells, the assay procedure was modified. The cells were seeded at

6.66×10^3 cells/cm^2 in 100-mm petri dishes. Cultures were incubated for 24 h, at which time one to three plates were trypsinized to obtain a relative plating efficiency. The medium was removed from the experimental dishes and the cultures were exposed to UV light (254 nm) at an intensity of 100 µJ/cm^2/s. Cultures were immediately refed with EMEM and incubated for 72 h. Unexposed control cultures were also refed with medium. From this point, the procedure followed that of the anchorage-independent survival assay.

Determination of the Transformed Phenotype

In some experimental conditions, after the viable cells from each set of control and treatment points from the survival assay were counted, they were added to 25-cm^2 flasks to propagate and establish new cell lines. These cell lines upon reaching confluence were carried vertically/horizontally in mass culture in a modification of previously described procedures (Freeman et al., 1973; Traul et al., 1981a). The passage number of those cells replated following anchorage-independent survival was arbitrarily set at passage (P) + 1. The cultures were incubated an additional 7 d after reaching confluency, at which time they were subcultured into three 25-cm^2 flasks. Fourteen days later, one flask from each point was washed with EBSS and fixed and stained with methylene blue, the second set of flasks was refed with complete media for an additional two weeks and subsequently stained, and the third set of flasks was subcultured to provide three new flasks per point. Although cell counts were made at each passage level, cultures were routinely subcultured 1:3. All cultures were refed with EMEM two times per week. This cycle was repeated every 2 wk. Cultures that were not subdivided were designated as the horizontal series; subdivided cultures were called the vertical series. Each stained flask was examined for the presence of darkly stained foci lacking polar orientation and contact inhibition.

Cell cultures derived from anchorage-independent survival cells were seeded into semisolid medium at 5×10^5 cells per dish according to the procedure of MacPherson (1969) with these modifications: the addition of 1% each of sodium pyruvate, nonessential amino acids, essential amino acids, L-glutamine, and 0.12% bactopeptone to the base layer and top layer of the agar medium. Colonies were counted 28 d later.

RESULTS

Anchorage-Independent Survival Studies

To evaluate this short-term bioassay for its ability to detect chemical and physical carcinogenic agents, we began our testing process with known carcinogens, including polycyclic hydrocarbons, nitrosamines, and heterocyclic compounds as well as known noncarcinogens. As the research progressed and the results became more consistent, we incorporated coded compounds into the testing system. A cytotoxicity test, as described in Materials and Methods, was performed on each agent before it was tested in the anchorage-independent assay. Due to the preselected test doses, problems associated with extreme toxicity were rarely encountered.

In general, anchorage-dependent media control and solvent control cells showed a rapid decline in cell survival; however, cells that had been treated with carcinogen did not undergo the destructive process that took place in control cells, indicating specificity. When RIFRE cells were treated with increased concentrations of a carcinogen, the number of viable anchorage-independent surviving cells was significantly higher than solvent-treated control cells.

Dose-response relationships were achieved with five "model" carcinogens using our modified assay (Figure 2 a-e). The polycyclic hydrocarbons 7,12-dimethylbenz(a)anthracene (DMBA) and benzo(a)pyrene (B[a]P) were responsive in their ability to induce enhanced anchorage-independent survival as a function of increasing concentration. The nitrosamines N-nitrosodimethylamine (DEN) and N-methyl-N'-nitro-N-nitrosoguanidine (MNNG) and the heterocyclic compound 4-nitroquinoline-N-oxide (4NQO) also produced concentration-dependent responses. Although at first DEN does not appear to meet the criteria of a positive enhanced survival response, this agent is indeed positive when represented as mean number of cells plus or minus their standard deviation. Of the carcinogens tested, DMBA, B(a)P, and DEN are known to require metabolic activation and MNNG and 4NQO are direct acting. Therefore, anchorage-independent survival when used as a marker for *in vitro* transformation can assess known carcinogens in a dose-dependent manner.

Several known noncarcinogens were also tested for enhanced anchorage independence in the assay. As indicated in Table 1, anthracene, pyrene, benzo(e)pyrene (B[e]P), sodium azide (NaN_3), caprolactam, and benzoin showed little to no enhancement compared to the untreated or solvent controls and were considered to be negative in the assay. The results affirm the capacity of RIFRE cells to differentiate between chemicals with similar structures when determining the carcinogenicity of an agent.

A physical agent, UV irradiation, was also tested in the anchorage-independent survival assay. Figure 3 represents a dose-response relationship achieved after exposing RIFRE cells to increasing doses of UV irradiation. As the results indicate, the anchorage-independent survival assay can be used to test physical agents as well as chemical agents.

Studies to Determine the Transformed Phenotype

To determine if those cells that exhibited enhanced cell survival in the assay subsequently expressed morphologically transformed and neoplastic phenotypes, additional experiments were performed. Cells taken directly from the assay after the 4-d agar phase were plated into 25-cm^2 flasks to establish cell cultures. These cultures were passaged biweekly and examined macroscopically and microscopically for changes in their cellular morphology. Figure 4 compares a flask of control RIFRE cells with a flask of B(a)P-treated cells. A higher magnification of each monolayer is also shown. When stained, the control cells appear to be homogenous; the treated cells form darkly stained foci. The higher magnifications show the sharp contrast between the monolayer of the control culture and the foci of the treated culture. The control cells are normal in their appearance, but the B(a)P-transformed cells show loss of contact inhibition and polar orientation.

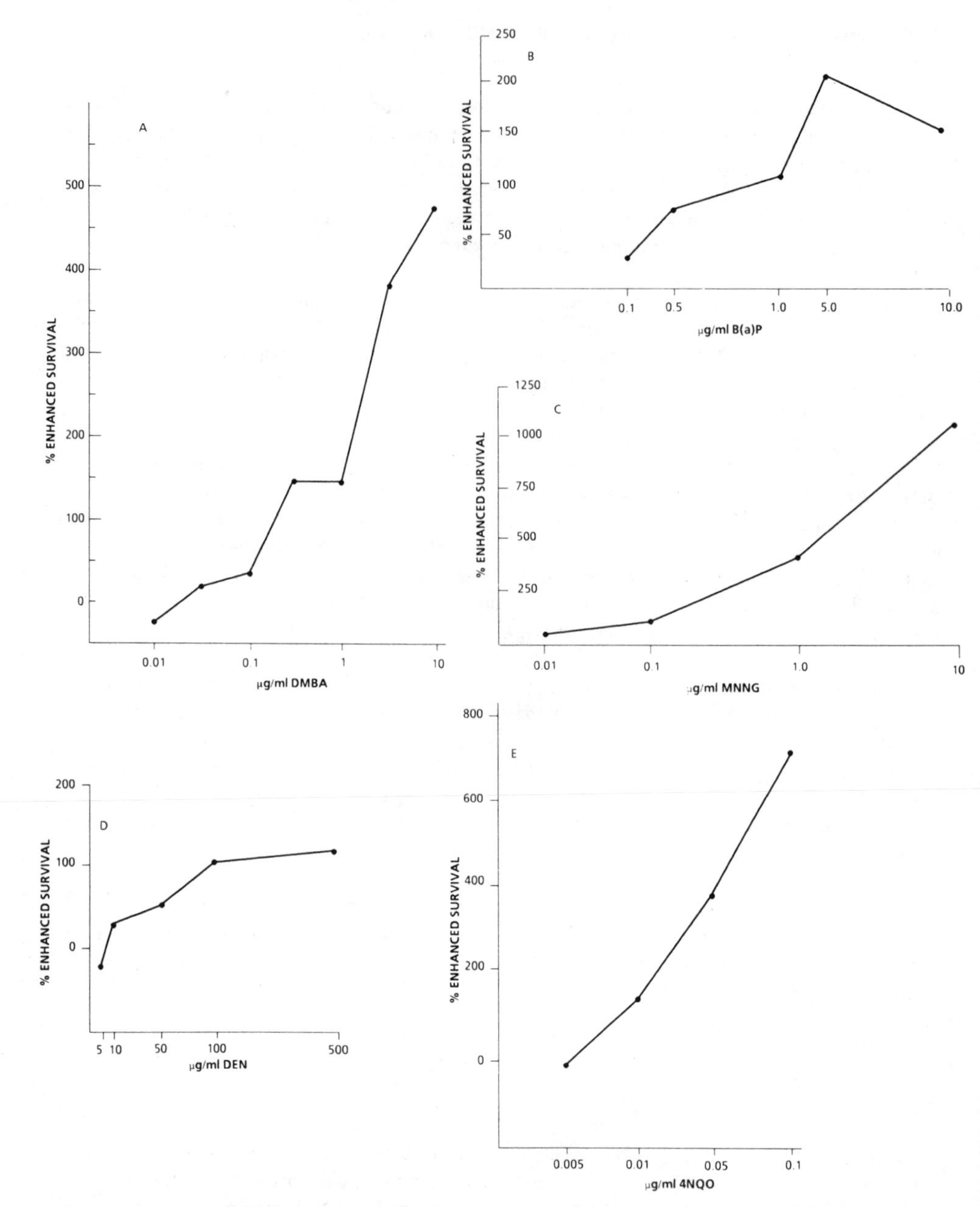

Figure 2. Enhancement of anchorage-independent survival as a function of concentration. Monolayer cultures were exposed to increasing concentrations of: A, DMBA; B, B(a)P; C, MNNG; D, DEN; and E, 4NQO. The cultures were subsequently assayed as described in Materials and Methods. The data are expressed as percent survivability of the number of viable cells with respect to the DMSO control (except DEN, which is expressed in terms of the medium control) obtained with each chemical following 4 d in medium above an agar base. All values were determined in triplicate for two or three dishes per treatment point in a single experiment.

Table 1. Effect of Noncarcinogen Treatment on the Anchorage-Independent Survival of RIFRE Cells

Treatment (µg/ml)[a]	Experiment 1 Mean No. Cells ± S.D. (x 10^3)[b]	Experiment 1 % Enhanced Survival[c]	Treatment (µg/ml)	Experiment 2 Mean No. Cells ± S.D. (x 10^3)	Experiment 2 % Enhanced Survival
Medium	60.0 ± 1.6	-29	Medium	41.8 ± 2.8	27
DMSO (0.1%)	85.3 ± 1.5	0	DMSO (0.2%)	32.8 ± 2.5	9
DMBA (0.5)[d]	213.3 ± 4.9	150	B(a)P (10)[d]	69.9 ± 8.2	113
Anthracene (10)	101.3 ± 2.6	19	Pyrene (20)	42.2 ± 3.9	29
Anthracene (5)	69.0 ± 1.0	-19	Pyrene (10)	33.0 ± 1.5	0
Anthracene (1)	61.3 ± 1.5	-28	Pyrene (5)	33.4 ± 5.4	2
Anthracene (0.5)	73.3 ± 1.0	-14			
Anthracene (0.1)	62.7 ± 0.7	-26			

Treatment (µg/ml)	Experiment 3 Mean No. Cells ± S.D. (x 10^3)	Experiment 3 % Enhanced Survival	Treatment (µg/ml)	Experiment 4 Mean No. Cells ± S.D. (x 10^3)	Experiment 4 % Enhanced Survival
Medium	37.5 ± 3.7	6	Medium	27.0 ± 3.4	-25
DMSO (0.1%)	35.3 ± 2.9	0	DMSO (0.1%)	36.0 ± 3.7	0
DMBA (0.1)	105.0 ± 7.2	197	DMBA (0.1)[d]	91.0 ± 8.4	153
B(e)P (10)	26.3 ± 3.1	-25	NaN_3 (50)	30.0 ± 3.2	-17
B(e)P (1)	37.0 ± 4.4	5	NaN_3 (25)	34.3 ± 7.8	-5

Treatment (µg/ml)	Experiment 5 Mean No. Cells ± S.D. (x 10^3)	Experiment 5 % Enhanced Survival	Treatment (µg/ml)	Experiment 6 Mean No. Cells ± S.D. (x 10^3)	Experiment 6 % Enhanced Survival
Medium	25.5 ± 1.1	-26	Medium	14.0 ± 7.0	-52
DMSO (0.1%)	34.0 ± 9.0	0	DMSO (0.1%)	29.2 ± 1.4	0
DMBA (1)	179.2 ± 3.1	427	DMBA (1)	149.2 ± 4.1	411
Caprolactam (50)[e]	20.8 ± 7.0	-39	Benzoin (50)[e]	20.0 ± 1.1	-31
Caprolactam (5)	20.8 ± 5.0	-39	Benzoin (5)	26.0 ± 1.2	-11
Caprolactam (0.5)	33.2 ± 1.8	-2	Benzoin (0.5)	19.2 ± 1.6	-34

[a] RIFRE cells were treated with varying concentrations of known agent for three days, washed and refed for an additional three days. They were subsequently trypsinized, adjusted for cytotoxicity, and overlaid at 6 x 10^5 cells per 6 cm dishes that contained 1% agar Noble medium. Viable cell counts were made following four days in suspension above agar.

[b] The numbers represent the mean and standard deviation of the viable number of cells counted based on 1-3 dishes.

[c] The percent (%) survival for each treatment point is calculated based on the solvent control, 0.1 - 0.2% DMSO.

[d] DMBA or B(a)P were used as positive controls in most experiments.

[e] Caprolactam and benzoin were received from the World Health Organization and are part of a collaborative study (Suk and Humphreys in press).

In correlative studies, these same cells were plated into semisolid agar. The anchorage-independent phenotype was selected because of the relationship between malignant potential and the ability of cells to grow in semisolid agar (Shin et al., 1975). The results of these correlative studies are shown in Table 2. The frequency of expression of growth in semisolid agar correlated with the appearance of morphological transformation and with enhanced cell survival that had been shown in the assay performed some weeks before. Those cultures that expressed an altered morphology and an enhanced cell survival also grew in semisolid agar, although the frequencies were lower. (Altered morphology, as seen at higher magnifications, describes cells that

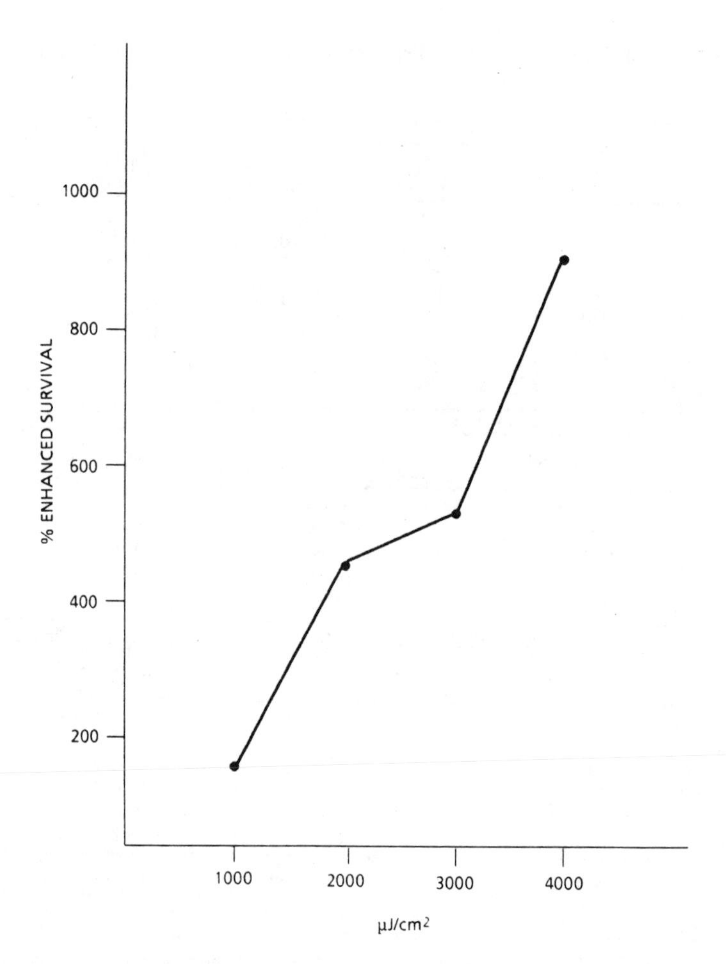

Figure 3. Enhancement of anchorage-independent survival as a function of increased UV irradiation exposure. Monolayer cultures were exposed to increasing doses of UV irradiation (254 nm). The cultures were subsequently assayed as described. The data are expressed as in Figure 2.

are clearly not homogenous like control cells and do not appear as darkly stained foci as in the transformed cultures.) Control and noncarcinogen (pyrene, anthracene) cultures exhibited no enhanced survival in the assay, no morphologically-altered cells or transformed cells, and no growth in semisolid agar.

Results similar to those shown in Table 2 were achieved when the RIFRE cells were exposed to UV irradiation (unpublished data). Ultraviolet-exposed cells showing enhancement of anchorage-independent survival were plated into 25-cm^2 flasks, were established in culture, and showed subsequent changes in cellular morphology as well as growth in semisolid agar.

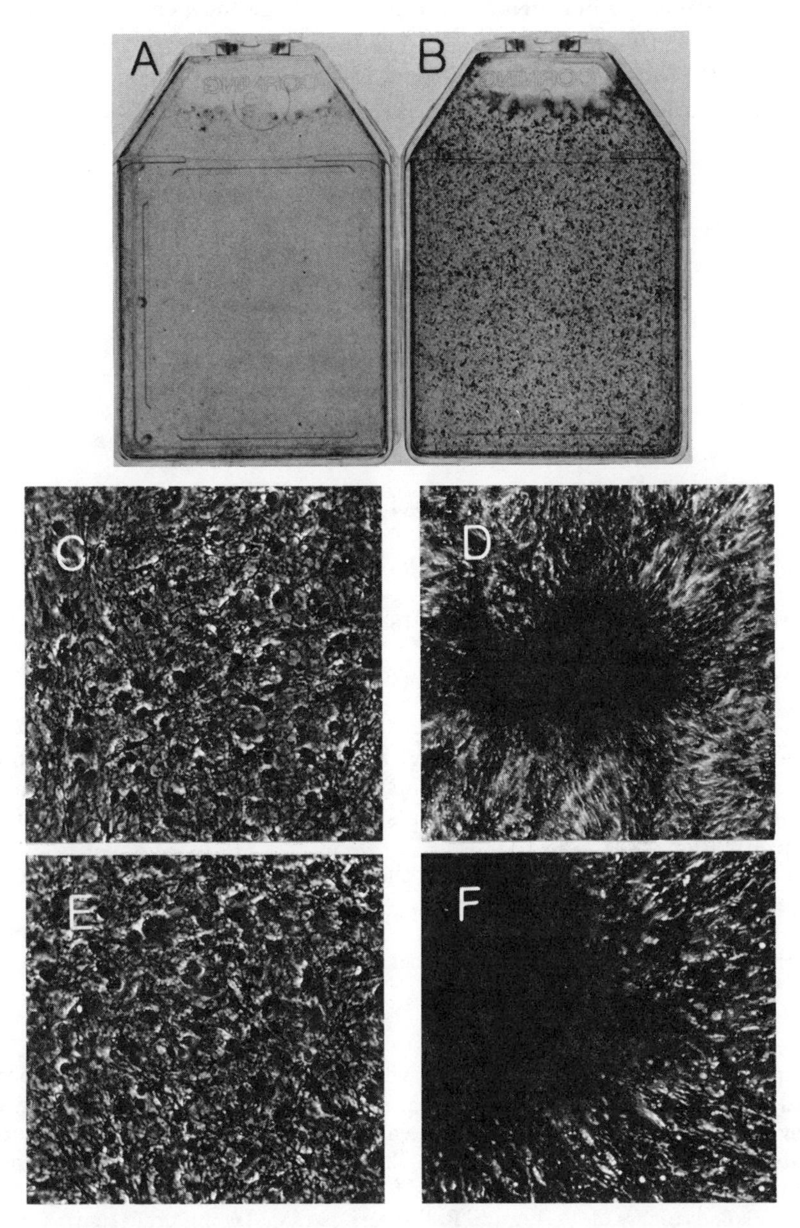

Figure 4. Morphologic transformation of RIFRE cells that exhibit enhanced anchorage-independent survival. Stained flasks of: A, control and B, B(a)P-treated cells (1 μg/ml) carried in culture (1X), P15 + 13 + 7; C, the appearance of the monolayer of the control cells at 40X magnification; D, same for treated cells; E, 100X magnification of the control cells; F, same for treated cells.

Table 2. Correlation Between Enhanced Anchorage-Independent Survival of Carcinogen-Treated RIFRE Cells and Their Cellular Morphology and Growth in Semisolid Agar

Treatment (μg/ml)	Mean Number Cells ± SD (x10^3)	Enhanced Survival (%)	Cellular Morphology (P+)[a,b]	Growth in Semisolid Agar[c] (%)
None	42.4 ± 2.8	6	Normal (+8)	0
DMSO (0.1%)	40.0 ± 3.2	0	Normal (+8)	0
DMBA (10)	425.6 ± 9.9	964	Transformed (+7)	3.8
DMBA (1)	282.4 ± 6.7	606	Altered (+6)	0.8
B(a)P (10)	70.4 ± 4.3	76	Altered (+7)	0.2
B(a)P (1)	97.6 ± 4.4	144	Transformed (+6)	2.2

[a] Cells were counted for enhanced survival and subsequently added to 25-cm^2 flasks to propagate and establish new cell lines. These cell lines were then carried in culture as described in the Materials and Methods section and screened for transformation.

[b] The number in parentheses indicates the passage level at which the described cellular morphology was first observed.

[c] 5 x 10^5 cells/6-cm dish were put into semisolid agar at P + 6; the numbers represent the percent of cells forming foci based on the mean of two dishes.

Experiments were performed to test whether the cells expressing enhanced cell survival were transplantable in syngeneic rats. Preliminary results indicate that carcinogen-mediated enhanced surviving cells subsequently carried in culture and inoculated subcutaneously into newborn Fischer rats induced the development of malignant lymphomas of thymic origin in five of seven animals; the control cultures failed to develop tumors (J. Poiley and R. Raineri, personal communication). Because of the unusual pathology of the developed neoplasia, further experiments are in progress. In a previous experiment in which carcinogen-mediated enhanced cell survival was correlated with growth in semisolid agar, cells inoculated into newborn rats caused nodule formation; no nodules were formed from control cultures. However, the nodules regressed within 3 wk of appearance, possibly due to host immunological response to murine leukemia virus antigens (Holden and Herberman, 1981; Sobis et al., 1980) and to suppression of growth by a bystander cell population (Nagai et al., 1983).

DISCUSSION

The ability to detect hazardous chemical and physical agents is rapidly becoming more important with increased production of new agents. The ability to test these two types of agents together is also important for detecting their combined or possibly synergistic effects. We feel that the anchorage-independent survival assay may be employed to test a combination of agents. The present study supports the hypothesis that enhanced anchorage-independent survival is a preneoplastic event associated with progression to neoplastic transformation and is relevant to the carcinogenic process.

The RIFRE cell focal assay (for review see Price and Mishra, 1980) has been used to determine the *in vitro* transformation potential of DEN (Freeman et al., 1970); 3-methylcholanthrene (Price et al., 1971); extract of city smog (Freeman et al., 1971); DMBA (Rhim et al., 1971); (-)-trans-Δ^9-tetrahydrocannabinol (Price et al., 1972); and over 30 polycyclic hydrocarbons, azo dyes, aromatic amines, and miscellaneous chemicals (Freeman et al., 1973). A variety of other agents were also tested (Auletta and Suk, 1977; Hetrick and Kos, 1973; Kouri et al., 1975; Price et al., 1975; Price et al., 1978a, b; Suk and Price, 1982). When the supply of these cells was ultimately exhausted, Traul et al. (1981a) derived a new RIFRE cell line, $2FR_450$, for use in assessing chemical carcinogenicity. The evaluation of this system was then extended to more than 75 compounds covering many classes or types of chemical carcinogens (Dunkel et al., 1981; Traul et al., 1981a).

Traul et al. (1979, 1981b) applied measurement of anchorage-independent survival as an indicator of neoplastic transformation to RIFRE cells. There are a number of studies that have evaluated malignant transformation *in vitro* and have correlated it with anchorage-independent survival (Eker and Sanner, 1983; Steuer and Ting, 1976; Suk et al., 1983; Traul et al., 1981b). In this way, cellular transformation is measured by survival in an aggregate/anchorage-independent form when the cells are suspended in liquid medium above an agar base. A number of modifications to the assay system have improved the usefulness of the RIFRE cells as target cells and have led to the detection of carcinogens in a dose-dependent manner (Suk et al., 1983). Although there is still some variation in the enhanced survival response of the cells to the chemicals, the modifications have increased the sensitivity and reproducibility of the system. In addition, all the chemicals so far have been tested without the addition of an exogenous metabolic activation system; the positive response to known carcinogens that are metabolized to reactive intermediates suggests that the RIFRE cells have the capacity to perform these same functions.

Anchorage independence of retrovirus-infected rat embryo cells has been shown to be an event preceding morphological transformation (Mishra and Ryan, 1973). This phenotypic marker correlates very well with neoplastic potential and is the most reliable *in vitro* index of tumorigenicity (Shin et al., 1975). However, a certain skepticism prevails with regard to anchorage-independent survival as a conclusive marker for neoplastic transformation. Therefore, a number of experiments have been designed to answer the query as to whether carcinogen-induced anchorage-independent survival precedes the expression of neoplastic transformation. Thus far, the results indicate that RIFRE cells exhibiting enhanced cell survival plated back onto a solid substrate and carried in culture express transformation-related changes in their cellular morphology and produce macroscopic foci when suspended in semisolid agar.

Preliminary tumorigenicity studies indicate that these cells have the ability to induce tumors in syngeneic and immunologically-suppressed animals; more definitive tumorigenicity studies are presently in progress. These observations support our hypothesis that this assay system measures the effects of agents indicative of their capacity to induce neoplasia.

ACKNOWLEDGMENTS

The authors thank Dr. Raymond W. Tennant for his continued interest and support; Drs. Judith Poiley and Ronald Raineri for sharing preliminary tumorigenicity data; and Drs. Judson Spalding and Anton Steuer for their helpful discussions. This research was sponsored by NIEHS/NTP Contract NO1-ES-15798.

REFERENCES

Auletta, A.E., and W.A. Suk. 1977. Transformation of the Fischer rat embryo cell system by two carcinogenic mutagens. Proc. Am. Assoc. Cancer Res. 18:547.

Barrett, J.C., B.D. Crawford, L.O. Mixter, L.M. Schechtman, P.O.P. Ts'o, and R. Pollock. 1979. Correlation of *in vitro* growth properties and tumorigenicity of Syrian hamster cells lines. Cancer Res. 39:1504-1510.

Cho, H., E. Cutchins, J. Rhim, and R. Huebner. 1976. Revertants of human cells transformed by murine sarcoma virus. Science 194:951-953.

Dunkel, V.C., R.J. Pienta, A. Sivak, and K.A. Traul. 1981. Comparative neoplastic transformation responses of Balb/3T3 cells, Syrian hamster embryo cells, and Rauscher murine leukemia virus-infected Fischer 344 rat embryo cells to chemical carcinogens. J. Natl. Cancer Inst. 67:1303-1315.

Eker, P., and T. Sanner. 1983. Assay for initiators and promoters of carcinogenesis based on attachment-independent survival of cells in aggregates. Cancer Res. 43:320-323.

Freedman, V.H., and S. Shin. 1974. Cellular tumorigenicity in nude mice: Correlation with cell growth in semi-solid medium. Cell 3:355-359.

Freeman, A.E., P.J. Price, H.J. Igel, J.C. Young, J.M. Maryak, and R.J. Huebner. 1970. Morphological transformation of rat embryo cells induced by diethylnitrosamine and murine leukemia viruses. J. Natl. Cancer Inst. 44:65-78.

Freeman, A.E., P.J. Price, R.J. Bryan, R.J. Gordon, R.V. Gilden, G.J. Kelloff, and R.J. Huebner. 1971. Transformation of rat and hamster embryo cells by extracts of city smog. Proc. Natl. Acad. Sci. U.S.A. 68:445-449.

Freeman, A.E., E.K. Weisburger, J.H. Weisburger, R.G. Wolford, J.M. Maryak, and R.J. Heubner. 1973. Transformation of cell cultures as an indication of the carcinogenic potential of chemicals. J. Natl. Cancer Inst. 51:799-808.

Grandgenett, D.P., G.F. Gerard, and M. Green. 1972. Ribonuclease H: A ubiquitous activity in virions of ribonucleic acid tumor viruses. J. Virol. 10:1136-1142.

Hetrick, F.M., and W.L. Kos. 1973. Transformation of Rauscher virus-infected cell cultures after treatment with Hycanthone and Lucanthone. Pharmacol. Exp. Therapeutics 186: 425-429.

Holden, H.T., and R.B. Herberman. 1981. Immune mechanisms involved in the antitumor response to murine sarcoma virus-induced tumors. In: Mechanisms of Immunity to Virus-Induced Tumor. J.W. Blasecki, ed. Marcel Dekker, Inc.: New York. pp. 1-67.

Kakunaga, T. 1978. Neoplastic transformation of human diploid fibroblasts by chemical carcinogens. Proc. Natl. Acad. Sci. U.S.A. 75:1334-1338.

Kouri, R.E., S.A. Kurtz, P.J. Price, and W.F. Benedict. 1975. 1-β-D-arabinofuranosylcytosine-induced malignant transformation of hamster and rat cells in culture. Cancer Res. 35:2413-2419.

MacPherson, I. 1969. Agar suspension culture for quantitation of transformed cells. In: Fundamental Techniques in Virology. K. Abel and N.P. Salzman, eds. Academic Press: New York. p. 274.

Mishra, N.K., and W.L. Ryan. 1973. Effect of 3-methylcholanthrene and dimethylnitrosamine on anchorage dependence of rat fibroblasts chronically infected with Rauscher leukemia virus. Int. J. Cancer 11:123-130.

Nagai, A., B. Zbar, N. Terata, and J. Hovis. 1983. Rejection of retrovirus-infected tumor cells in Guinea pigs: Effect on bystander tumor cells. Cancer Res. 43:5783-5788.

Price, P.J., and N.K. Mishra. 1980. The use of Fischer rat embryo cells as a screen for chemical carcinogens and the role of the nontransforming type "C" RNA tumor viruses in the assay. In: Advances in Modern Environmental Toxicology. Vol. 1, Mammalian Cell Transformation by Chemical Carcinogens. N. Mishra, V. Dunkel, and M. Mehlman, eds. Senate Press, Inc.: New Jersey. pp. 213-239.

Price, P.J., A.J. Freeman, W.T. Lane, and R.J. Huebner. 1971. Morphological transformation of rat embryo cells by the combined action of 3-methylcholanthrene and Rauscher leukemia virus. Nature (London) New Biol. 230:144-146.

Price, P.J., W.A. Suk, G.J. Spahn, and A.E. Freeman. 1972. Transformation of Fischer rat embryo cells by the combined action of murine leukemia virus and (-)-trans-Δ^9-tetrahydrocannabinol. Proc. Soc. Exp. Biol. Med. 140:454-456.

Price, P.J., W.A. Suk, P.C. Skeen, L.A. Chirigos, and R.J. Huebner. 1975. Transforming potential of the anti-cancer drug adriamycin. Science 187:1200-1201.

Price, P.J., C.M. Hassett, and J.I. Mansfield. 1978a. Transforming potential of trichloroethylene and proposed industrial alternatives. In Vitro 14:290-293.

Price, P.J., W.A. Suk, A.E. Freeman, W.T. Lane, R.L. Peters, M.L. Vernon, and R.J. Huebner. 1978b. In vitro and in vivo indications of the carcinogenicity and toxicity of food dyes. Int. J. Cancer 21:361-367.

Rhim, J.S., W. Vass, H.Y. Cho, and R.J. Huebner. 1971. Malignant transformation induced by 7,12-dimethylbenz(a)anthracene in rat embryo cells infected with Rauscher leukemia virus. Int. J. Cancer 7:65-74.

Rowe, W.P., W.E. Pugh, and J.W. Hartley. 1970. Plaque assay technique for murine leukemia viruses. Virology 42:1136-1139.

Shin, S., V.N. Freedman, R. Risser, and R. Pollack. 1975. Tumorigenicity of virus-transformed cells in nude mice is correlated specifically with anchorage-independent growth *in vitro*. Proc. Natl. Acad. Sci. U.S.A. 11:4435-4439.

Sobis, H., L. Van Hovem, H. Heremaus, M. Deley, A. Billian, and M. Vandeputte. 1980. Induction of immune reaction against rat embryonal carcinoma by activation of viral genome. Int. J. Cancer 26:93-99.

Steuer, A.F., and R.C. Ting. 1976. Formation of larger cell aggregates by transformed cells: An *in vitro* index of cell transformation. J. Natl. Cancer Inst. 56:1279-1280.

Steuer, A.F., P. Hentosh, L. Diamond, and R.C. Ting. 1977. Survival differences exhibited by normal and transformed rat liver epithelial cell lines in the aggregate form. Cancer Res. 37:1864-1867.

Suk, W.A., and J.D. Arnold. 1984. Ultraviolet irradiation induces enhanced anchorage-independent survival in retrovirus-infected Fischer rat embryo cells. Proc. Am. Assoc. Cancer Res. 25:528.

Suk, W.A., and J.E. Humphreys. (in press). Transforming potential of selected WHO/IPCS chemicals as determined by their enhancement of anchorage-independent survival of retrovirus-infected Fischer rat embryo cells. In: Evaluation of Short Term Tests for Carcinogens: Report of the International Programme on Chemical Safety Collaborative Study on *In Vitro* Assays. J. Ashby, F. deSerres, M. Draper, M. Ishdate, Jr., B. Margolin, B. Matter, and M.S. Shelby, eds. Elsevier/North Holland: Amsterdam.

Suk, W.A., and P.J. Price. 1982. Transformation of rat embryo cells by the antituberculosis drug, para-aminosalicylic acid. In Vitro 18:112.

Suk, W.A., J.E. Humphreys, J.D. Arnold, and E.P. Hays. 1983. Increased expression of anchorage independence as an indicator of neoplastic transformation. Proc. Am. Assoc. Cancer Res. 24:401.

Tennant, R.W., J.G. Farrelly, J.M. Ihle, B.C. Kenney, and A. Brown. 1973. Effects of polyadenylic on murine RNA tumor virus functions. J. Virol. 12:1216-1225.

Traul, K.A., V. Kachevsky, and J.S. Wolff. 1979. A rapid *in vitro* assay for carcinogenicity of chemical substances in mammalian cells utilizing an attachment-independence endpoint. Int. J. Cancer 23:193-196.

Traul, K.A., R.J. Hind, J.S. Wolff, and W. Korol. 1981a. Chemical carcinogenesis *in vitro*. An improved method for chemical transformation in Rauscher leukemia virus-infected rat embryo cells. J. Appl. Toxicol. 1:32-37.

Traul, K.A., K. Takayama, V. Kachevsky, R.J. Hink, and J.S. Wolff. 1981b. A rapid *in vitro* assay for carcinogenicity of chemical substances in mammalian cells utilizing an attachment-independent endpoint. J. Appl. Toxicol. 1:190-195.

CORRELATION BETWEEN *IN VIVO* TUMORIGENESIS AND *IN VITRO* CYTOTOXICITY IN CHO AND V79 CELLS AFTER EXPOSURE TO MINERAL FIBERS

L.D. Palekar,[1] B.G. Brown,[1] and D.L. Coffin[2]

[1]Northrop Services, Inc.--Environmental Sciences, Research Triangle Park, North Carolina 27709, and [2]Health Effects Research Laboratory, U.S. Environmental Protection Agency, Research Triangle Park, North Carolina 27711

INTRODUCTION

It is well known that occupational exposure to asbestos constitutes a major public health hazard. It is also recognized that asbestos already in the environment such as in sprayed accoustical ceilings in schools (Nicholson et al., 1979), emissions from building demolition, and reintrainment of dust from asbestos waste dumps presents a possible threat to the general population (Nicholson and Pundsack, 1973). Furthermore, chronic pleural fibrosis and even tumors of the lungs and pleura are known to be associated with asbestos and asbestos-like minerals in the natural environment (Kiviluoto, 1965; Baris et al., 1978). Substitutes for asbestos are currently being sought, but unfortunately the very properties (durable thin fibers) that provide insulation from high temperature are also likely to enhance the substitute's tumorigenesis.

There is a need to predict the potential health hazards after exposure to dust in the environment and to asbestos substitutes. Some available animal models such as inhalation exposure, intrapleural inoculation, and intratracheal instillation ascertain the tumorigenic potential of the mineral fibers. The time and cost required to conduct these studies, however, make the large-scale application of these models impractical. It is, therefore, of utmost importance that short-term bioassays, which may predict potential health hazards to humans, be developed.

Several *in vitro* systems are being exploited in the search for a short-term predictive model. Some of the methods often used are sheep erythrocyte hemolysis, rat and rabbit alveolar macrophage cytotoxicity, Chinese hamster ovary (CHO) cell cytotoxicity, and Chinese hamster lung cell (V79) cytotoxicity. While these systems reflect the cellular interaction with mineral fibers, none of them have been sufficiently studied to judge their utility as predictors of tumorigenesis.

This paper presents data on the evaluation of the cytotoxicity of several selected mineral fibers to CHO and V79 cells. The minerals were selected according to the degree of mesothelioma induction after intrapleural inoculations in rats. In these studies, the degree of cytotoxicity in both cell lines is compared to the degree of tumorigenesis.

In addition, to ascertain whether the cytotoxic reactions of the mixtures are additive, antagonistic, or synergistic, four mixtures of known asbestos minerals were also tested using the V79 cell cytotoxicity test.

MATERIALS AND METHODS

Cell Cultures

V79 cells. The cells were maintained in William's Medium (GIBCO, Grand Island, NY) supplemented with 5% v/v fetal bovine serum (GIBCO), 300 μg/ml L-glutamine, 2.5 μg/ml fungizone, and 30 μg/ml gentamicin.

CHO cells. The cells were received from the American Type Culture Collection. They were maintained in Nutrient Mixture F-12, HAM (GIBCO) supplemented with 10% fetal bovine serum, 2.5 μg/ml fungizone, and 30 μg/ml gentamicin.

Exposure to Test Samples

Most mineral samples are composed of particles of various sizes, causing variability in their rate of sedimentation in the liquid nutrient medium. It is, therefore, difficult to ensure uniform contact between cells and particles of all sizes when the cells are already attached to the bottom of the culture vessel. Two methods of exposure of mineral fibers to cells have been employed to ascertain whether a method of exposure can alter the degree of cytotoxicity.

Method I. Approximately 200 cells were plated in 60-mm diameter petri dishes. The cells were allowed to attach for 24 h, after which the test material was added. The concentrations of test material used were 10, 20, 40, 60, 80, and 100 μg/ml. Each concentration including the control was done in five replicates. Each experiment was repeated three times. The cells were allowed to grow and form colonies for 6 d. After the 6-d exposure period, the attached colonies were fixed with a 10% formol-saline solution and then stained with 0.04% crystal violet. Stained colonies were counted and compared to controls. Colony survival was expressed as a percent of control.

Method II. Approximately 200 cells were plated in 60-mm petri dishes. The test material was added immediately, with no attachment period. The concentrations of test material used were the same as in method I. The cells were allowed to grow and were processed in the same manner as in method I.

Mineral Samples

The test asbestos samples used were standard UICC samples of crocidolite, chrysotile, and amosite. Other samples were erionite (a fibrous zeolite from Oregon), ferroactinolite (a fibrous sample from a rock near a mine in Minnesota), nonfibrous grunerite (a nonfibrous sample from a rock in the same area), min-u-sil, fiber glass, and quartz.

RESULTS

For a comparison between method I and method II, four mineral samples, UICC amosite, UICC crocidolite, UICC chrysotile, and erionite, were selected. For each cell line, mineral, and concentration, a two sample t-test was performed to compare exposure methods I and II. The mean response versus dose using both methods in the CHO cell cytotoxicity assay is illustrated in Figure 1. Excluding chrysotile and the highest concentration of amosite, the data indicate method I yielded higher survival ($p < 0.05$). The exceptions are not surprising because at the points where the data were not significant both chrysotile and amosite were extremely cytotoxic.

A comparison between cytotoxicity of the same mineral samples using methods I and II in the V79 cell cytotoxicity assay is shown in Figure 2. The results were similar to those observed with CHO cells, except that the cytotoxicity due to chrysotile was comparable to that observed for the other three samples. In all cases, method I yielded higher survival ($p < 0.05$) than method II, except at very low or very high concentrations. Since method II rendered higher reactivity than method I for all minerals, method II was selected for further experiments.

A concentration-response relationship using the CHO cell cytotoxicity assay on a variety of fibers is depicted in Figure 3. A comparison was made between cytotoxicity due to tumorigenic mineral fibers (amosite, crocidolite, chrysotile, erionite, and ferroactinolite) and nontumorigenic samples (quartz, fiber glass, nonfibrous grunerite, and min-u-sil). The data indicate that with the exception of ferroactinolite, four tumorigenic samples were more cytotoxic than nontumorigenic mineral samples. A concentration at which there were 50% or more surviving colonies was considered noncytotoxic; the concentration at which there were less than 50% surviving colonies was considered cytotoxic (Chamberlain et al., 1982).

Based on these criteria, erionite, amosite, and chrysotile were cytotoxic in the CHO system at a low concentration of 20 μg/ml and crocidolite was cytotoxic at 40 μg/ml. Quartz, fiber glass, and min-u-sil were considered to be cytotoxic at 60, 80, and 100 μg/ml, respectively. Ferroactinolite and nonfibrous grunerite were found to be noncytotoxic.

To validate these methods as a possible model for screening tumorigenic minerals, sensitivity and specificity values were calculated as recommended by Cooper et al. (1979) and MacMahon and Pugh (1970). The sensitivity is the measurement of the proportion of tumorigenic minerals that are also cytotoxic; the specificity is the measurement of the nontumorigenic minerals that are also noncytotoxic. The accuracy of the assay is defined as the proportion of correct predictions in all samples tested.

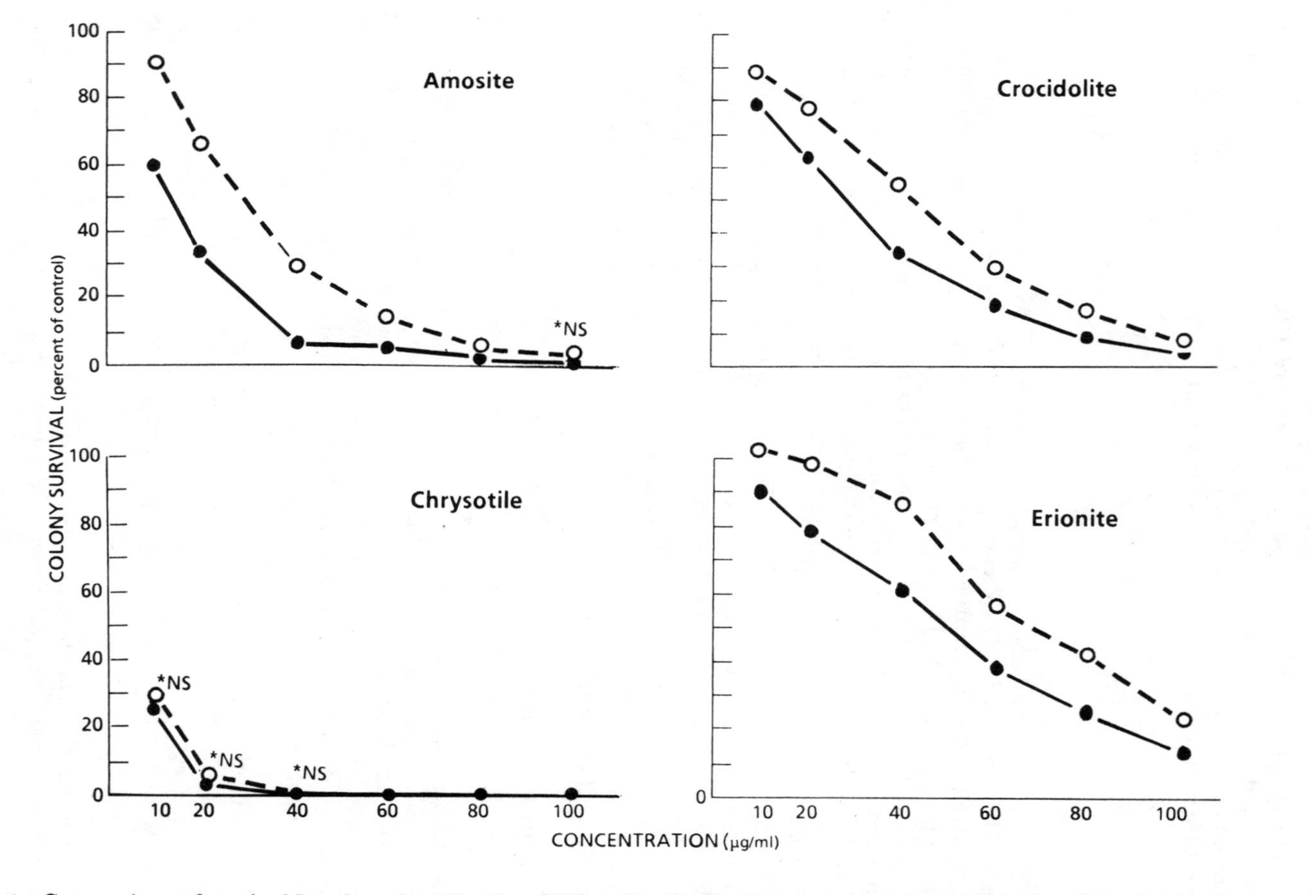

Figure 1. Comparison of method I and method II using CHO cells. Method I (o---o), samples added after 24 h of cell attachment; method II (●—●), samples added simultaneously with the cells. Nonsignificant values (NS).

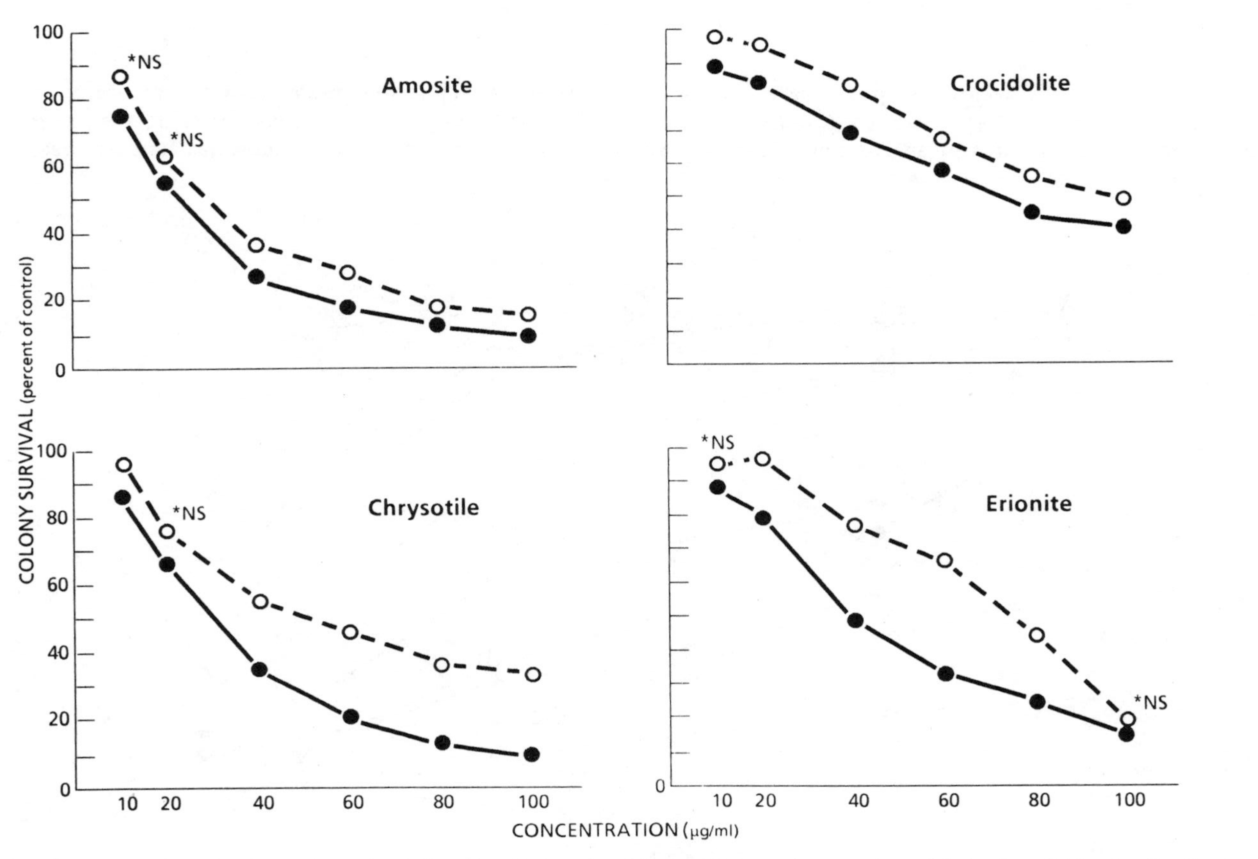

Figure 2. Comparison of method I and method II using V79 cells. Method I (o---o), samples added after 24 h of cell attachment; method II (●—●), samples added simultaneously with the cells. Nonsignificant values (NS).

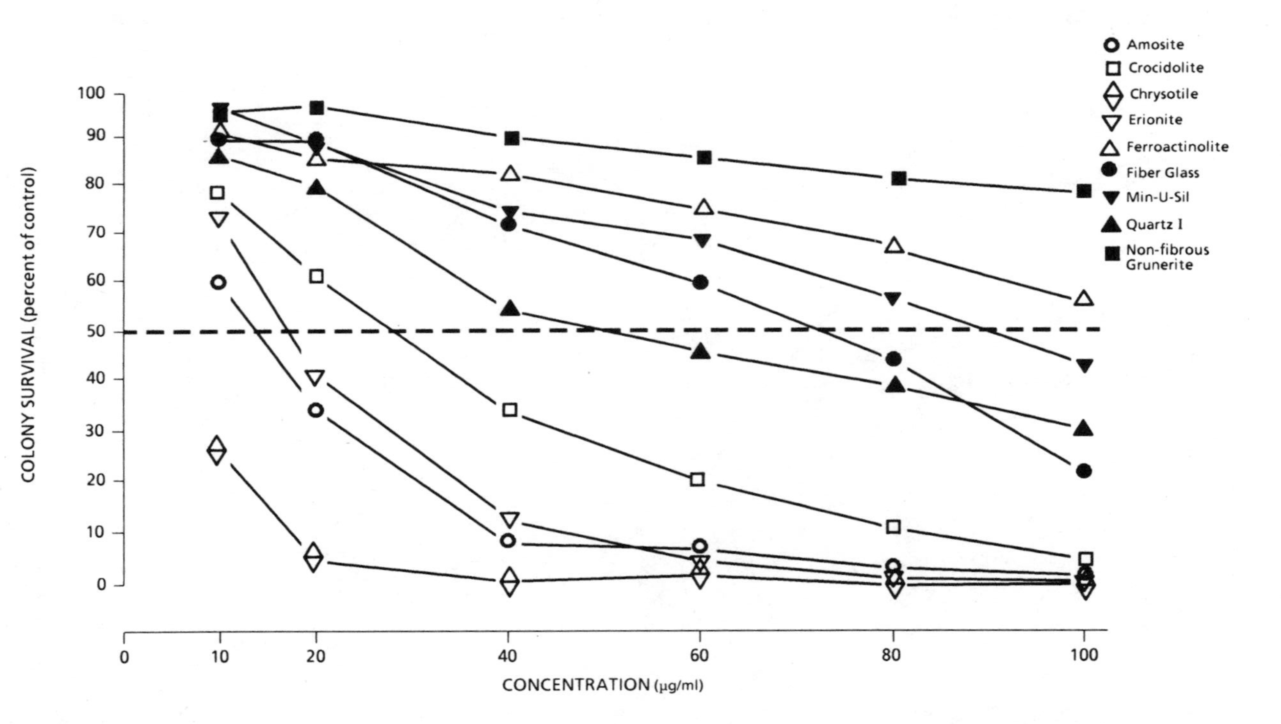

Figure 3. Dose-response relationship in CHO cell cytotoxicity assay. Open symbols represent tumorigenic mineral samples. Closed symbols represent nontumorigenic mineral samples. The dotted line represents 50% survival. Values below the dotted line are considered toxic and those above are nontoxic. For the significance of the values see Table 1.

Table 1 shows the validation measures for the CHO cell cytotoxicity bioassay. While sensitivity was only 20% at the lowest concentration, it increased to 80% at all concentrations above 40 μg/ml. The specificity, on the other hand, was 100% at lower concentrations but decreased at 60, 80, and 100 μg/ml to 75, 50, and 25%, respectively. The highest accuracy, 89%, was achieved at 40 μg/ml, but it was much less at other lower and higher concentrations.

A dose-response relationship using the V79 cytotoxicity assay is shown in Figure 4, and the validation measures are given in Table 2. In this system, cytotoxicity of erionite, chrysotile, and amosite was apparent at 40 μg/ml. Crocidolite cytotoxicity was not shown until a concentration of 100 μg/ml. The other mineral samples, fibrous grunerite, fiber glass, quartz, min-u-sil, and ferroactinolite, were noncytotoxic. The sensitivity ranged from 60 to 80% at higher concentrations; the specificity was 100% at all concentrations. Accuracy increased with increasing concentrations, reaching 89% at the highest concentration, 100 μg/ml.

Another comparison of tumorigenicity and cytotoxicity for each fiber is shown in Table 3. The concentration selected for *in vitro* cytotoxicity ranged from 10-100 μg/ml. The selected tumorigenic samples, chrysotile, crocidolite, amosite, and erionite, were cytotoxic in both the CHO and V79 cell systems. Other nontumorigenic samples, nonfibrous grunerite, quartz, and fiber glass, were noncytotoxic in the V79 cell system, but showed cytoxicity in the CHO cell system. Min-u-sil was noncytotoxic in both the CHO and V79 cell systems. With the exception of ferroactinolite in the V79 cell system, all tumorigenic samples were cytotoxic and all nontumorigenic samples were

Table 1. Validation of the CHO Cell Cytotoxicity Bioassay's Ability to Predict Tumorigenicity[a]

	Dose (μg/ml)					
	10	20	40	60	80	100
Sensitivity[b]	0.2	0.6	0.8	0.8	0.8	0.8
Specificity[c]	1.0	1.0	1.0	0.75	0.5	0.25
Accuracy[d]	0.56	0.78	0.89	0.78	0.67	0.56

[a]Multiply tabulated values by 100 to express as percentages.
[b]Proportion of tumorigenic samples that were also cytotoxic.
[c]Proportion of nontumorigenic samples that were also noncytotoxic.
[d]Proportion of correct predictions in all samples.

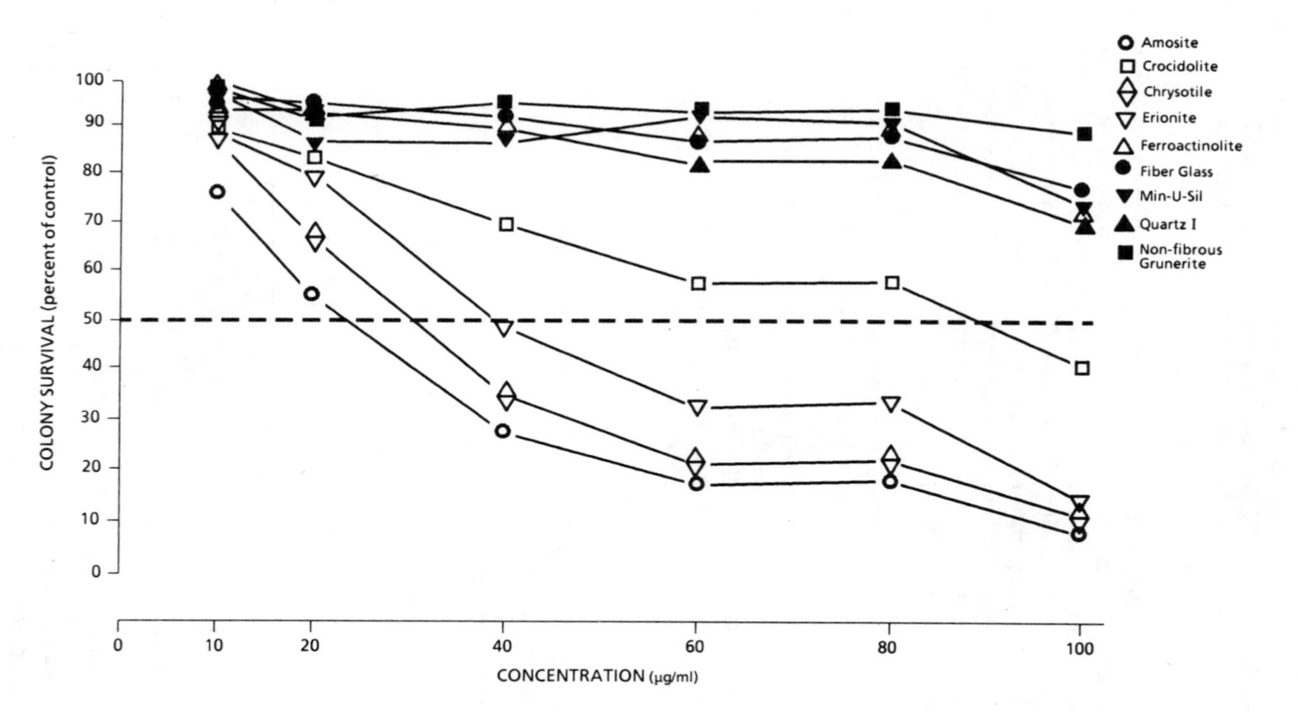

Figure 4. Dose-response relationship in V79 cell cytotoxicity assay. Open symbols represent tumorigenic mineral samples. Closed symbols represent nontumorigenic mineral samples. The dotted line represents 50% survival. Values below the line are considered toxic and those above are nontoxic. For the significance of these values see Table 2.

Table 2. Validation of the V79 Cell Cytotoxicity Bioassay's Ability to Predict Tumorigenicity[a]

	Dose (μg/ml)					
	10	20	40	60	80	100
Sensitivity[b]	0	0	0.6	0.6	0.6	0.8
Specificity[c]	1.0	1.0	1.0	1.0	1.0	1.0
Accuracy[d]	0.45	0.45	0.78	0.78	0.78	0.89

[a]Multiply tabulated values by 100 to express as percentages.
[b]Proportion of tumorigenic samples that were also cytotoxic.
[c]Proportion of nontumorigenic samples that were also noncytotoxic.
[d]Proportion of correct predictions in all samples.

noncytotoxic. In the CHO cell system, the correlation between the tumorigenic potential and cytotoxicity of the mineral fibers was not as strong; some nontumorigenic samples, quartz, fiber glass, and nonfibrous grunerite, were found to be cytotoxic.

Since the results from V79 cell cytotoxicity assays were more comparable to the tumorigenic potential of mineral samples, only V79 cells were used to determine the effects of mixtures of cytotoxicity. Four mixtures of two samples each were selected: mixture I, amosite plus crocidolite; mixture II, erionite plus crocidolite; mixture III, erionite plus amosite; and mixture IV, chrysotile plus crocidolite. The concentrations were chosen by estimating the amount required to yield 10, 25, and 40% cytotoxicity by each mineral.

The expected cytotoxicity of the mixtures was computed assuming the independent action of the two minerals on a probability basis. The following formula was used: the probability that the mixture of component A and component B is cytotoxic is equal to

$$P(A) + P(B) - P(A)\,P(B)$$

where $P(A)$ = probability that component A acting alone is cytotoxic and $P(B)$ = probability that component B acting alone is cytotoxic. Multiplication by 100 yields percent toxicity (Parzen, 1964).

Table 3. Comparison of *In Vivo* and *In Vitro* Results

Sample	Tumorigenesis[a]	CHO Cell Cytotoxicity[b]	V79 Cell Cytotoxicity
Chrysotile	+	+	+
Crocidolite	+	+	+
Amosite	+	+	+
Erionite	+	+	+
Ferroactinolite	+	–	–
Nonfibrous Grunerite	–	–	–
Quartz	–	–	–
Fiber glass	–	+	–
Min-u-sil	–	+	–

[a] "+" indicates that mesotheliomas were produced in animals after intrapleural inoculations.
[b] "+" for CHO and V79 indicates that cytotoxicity was observed for at least one of the concentrations used that ranged from 10-100 µg/ml.

For example, if the probability of cytotoxicity for compound A acting alone is 10% and that of compound B acting alone is 10%, the probability of the cytotoxicity of the mixture is not 20% but 19%. Similarly, the expected mixture cytotoxicity for 25% and 25% will be 44% and that for 40% and 40% will be 64%.

The actual cytotoxicity was compared to the calculated expected cytotoxicity. The statistical significance of the difference between the two values was determined by a t-test at $p \leq 0.05$ (Table 4). The data indicate that at all three concentrations amosite and crocidolite act independently, i.e., no significant difference was observed between the expected and the actual values. The mixture of erionite and crocidolite, however, showed significantly higher cytotoxicity than expected at all three concentrations, thus exhibiting a synergistic reaction. A mixture of erionite and amosite also showed significantly higher cytotoxicity, although only at the expected values of 44 and 64%. No significant difference was observed at the expected value of 19%. When chrysotile and crocidolite were combined, there was also a significant difference between the expected and actual values. It was, however, noted that at lower concentrations the

Table 4. Comparison of Expected and Observed Toxicity of Mixtures of Minerals in the V79 Cell Cytotoxicity Assay

Mixtures	Initial Calculated Cytotoxicity (%)	Expected Cytotoxicity of Mixture (%)[a]	Observed Mean Cytotoxicity (% ± SE)	Difference	Statistical Significance of Difference
I. Amosite + Crocidolite	20	19	20.3 ± 2.3	+1.3	N.S.
	50	44	41.8 ± 3.1	−2.2	N.S.
	80	64	63.3 ± 3.5	−0.7	N.S.
II. Erionite + Crocidolite	20	19	28.8 ± 2.7	+9.8	<0.05
	50	44	51.6 ± 2.5	+7.6	<0.05
	80	64	69.8 ± 1.9	+5.8	<0.05
III. Erionite + Amosite	20	19	23.2 ± 2.6	+4.2	N.S.
	50	44	51.6 ± 2.5	+7.6	<0.05
	80	64	70.8 ± 3.1	+6.8	<0.05
IV. Crocidolite + Chrysotile	20	19	12.3 ± 1.7	−6.7	<0.05
	50	44	54.9 ± 1.6	+10.9	<0.05
	80	64	71.8 ± 2.6	+7.8	<0.05

[a]The expected cytotoxicity was based on the independent action of the two minerals on the probability basis by using the following formula: $P(A) + P(B) - P(A)\,P(B)$ where $P(A)$ = probability that component A acting alone is cytotoxic and $P(B)$ = probability that component B acting alone is cytotoxic (multiplication by 100 yields percent toxicity).

actual cytotoxicity was lower than the expected value, indicating an antagonism between the minerals. At higher concentrations, the minerals seemed to be synergistic.

DISCUSSION

The results indicate that in most cases in which the cells and mineral fibers were added simultaneously the cytotoxicity was significantly ($p < 0.05$) higher than when the cells were plated prior to the addition of mineral fibers. This suggests that the initial contact of cells with mineral fibers before the attachment increases the cytotoxicity. This may be due either to the fact that small fibers, which would ordinarily stay in suspension and not come in contact with the attached cells, cause the additional cytotoxicity or to the fact that unattached cells are more vulnerable to mineral fibers in suspension. It is, however, quite certain that long and narrow fibers, with low sedimentation rate, are biologically active (Stanton and Wrench, 1972). It is more likely that the difference in cytotoxicity observed here is due to the proper exposure of mineral fibers to the cells. Moreover, the concentration-response relationship observed with the two methods also shows a parallel response. This seems to be an indication of a uniform reaction that is perhaps proportional to the number of small fibers at each concentration.

The data presented here indicate that in the V79 cell cytotoxicity assay method II was more sensitive for amosite, crocidolite, erionite, and chrysotile. These data are not quite in agreement with those observed by Brown and Chamberlain (1980), who reported that the cytotoxicity was higher for amphiboles but lower for chrysotile when cells and mineral samples were added together. In another study, Chamberlain et al. (1982) reported that erionite was not cytotoxic in the V79 assay. Data presented here indicate that cytotoxicity of erionite was higher than crocidolite but lower than amosite. There is no explanation for this discrepancy. It must be realized, however, that the samples used in the two laboratories were from different lots and that this may constitute a source of variability.

Although the standard UICC asbestos and erionite samples were found to be very cytotoxic in the CHO cell cytotoxicity bioassay, the nontumorigenic samples, min-u-sil, quartz, and fiber glass, were also cytotoxic at higher concentrations. In the V79 cell cytotoxicity assay, a better correlation between cytotoxicity and tumorigenesis was achieved, with the exception of ferroactinolite. Amosite, chrysotile, crocidolite, and erionite were cytotoxic; min-u-sil, fiber glass, nonfibrous grunerite, and quartz were noncytotoxic.

In order to compare the two assays, sensitivity, specificity, and accuracy were estimated at six concentrations ranging from 10-100 μg/ml by the method recommended by Cooper et al. (1979). It is evident from the data presented here that the concentration is very important when several bioassays are evaluated as potential screening tests. The data presented in Tables 1 and 2 show that the CHO cell cytotoxicity assay showed increased accuracy between tumorigenesis and cytotoxicity, indicating better correlation at a low concentration. An accuracy of the same value, however, was evident in V79 cell cytotoxicity assays only at a high concentration. On the other hand, in the CHO cell cytotoxicity bioassay, the specificity value was 100% at low concentrations, which indicates that nontumorigenic samples are also noncytotoxic. However, at high concentration, quartz, min-u-sil, and fiber glass were cytotoxic, thus lowering the specificity of the test. In comparison, the V79 cell cytotoxicity bioassay showed a 100% specificity at all concentrations. It is, however, important to note that the highest values ascertained in this investigation, 80% sensitivity, 100% specificity, and 89% accuracy, were achieved at 40 μg/ml in the CHO cell cytotoxicity assay; the same values were noted at 100 μg/ml in the V79 cell cytotoxicity bioassays.

The highest value of sensitivity of both assays was 80% because the tumor-producing ferroactinolite was noncytotoxic in both systems. The tumorigenicity of each mineral was determined by its capability to produce pleural tumors in F344 animals by intratracheal and intrapleural inoculations (Coffin et al., 1982). Although the fiber size and fiber number analysis of the ferroactinolite indicated that there were fewer fibers than in UICC amosite at the time of inoculation, the number of fibers recovered after one year of residence in the lung of animals was ten times higher than in the animals treated with amosite for the same period (Cook et al., 1982). This increase in number was due to the longitudinal splitting of the ferroactinolite fibers during their residence in the lung. The authors suggest that this "*in vivo*" splitting was responsible for a greater degree of tumorigenesis in the animals. It is likely that this sample would be nontumorigenic if a normal clearance with no longitudinal splitting had occurred. The increase in number of fibers by longitudinal splitting is a function of time. Since *in vitro* bioassays are short term, there is insufficient time to experience this phenomenon and, therefore, the *in vitro* exposure was not quite comparable to the *in vivo* exposure for this particular mineral sample.

The response to the mixture of mineral fibers in the V79 cell cytotoxicity bioassay was not the same for all mixtures. The data indicate that amosite and crocidolite act independently in the systems; when crocidolite or amosite are mixed with erionite, they act synergistically. Although erionite has been recognized as a potent tumorigenic mineral since the reports of endemic mesotheliomas in the population of Karain district in Turkey (Artvenli and Baris, 1979; Kiviluoto, 1965), its mineralogy is not properly understood. More information is needed in order to properly explain the interactions of erionite with amosite or crocidolite. A possible explanation for the interaction of chrysotile and crocidolite, antagonistic at lower concentrations and synergistic at higher concentrations, may be that chrysotile is positively charged whereas amosite and crocidolite are negatively charged (Light and Wei, 1977).

The search for short-term bioassays that may aid in the evaluation of minerals and particles with respect to their potential hazard to human health is important because long-term assays are often impractical. We are in agreement with Bridges (1976), who proposed a multitier test system for such evaluations in which the *in vitro* bioassays would be used for screening and selection of mineral samples for further evaluation by other *in vivo* and epidemiological studies.

ACNOWLEDGMENTS AND DISCLAIMER

The authors would like to thank Dr. Bernard Most for his valuable suggestions of the statistical methods for the evaluation of the data.

This paper has been reviewed by the U.S. Environmental Protection Agency and approved for publication. Mention of trade names or commercial products does not constitute endorsement or recommendation for use.

REFERENCES

Artvenli, M., and Y.I. Baris. 1979. Malignant mesotheliomas in a small village in the Antolian region of Turkey: An epidemiological study. J. Natl. Cancer Inst. 63:17-22.

Baris, Y.I., A.A. Sahin, M. Ozesmi, I. Kerse, E. Ozen, B. Kolacan, M. Altinörs, and A. Göktepeli. 1978. An outbreak of pleural mesothelioma and chronic fibrosing pleurisy in the village of Karain/Urgüp in Anatolia. Thorax 33:181-192.

Bridges, B. 1976. Use of a three-tier protocol for evaluation of long-term toxic hazards particularly mutagenicity and carcinogenicity. In: Screening Tests in Chemical Carcinogenesis. IARC Scientific Publication No. 12. International Agency for Research on Cancer: Lyon, France. pp. 549-559.

Brown, R.C., and M. Chamberlain. 1980. The activity of two types of asbestos in tissue culture. In: Biological Effects of Mineral Fibers, Volume I. J.C. Wagner, ed. Inserm Symposia Series Volume 92. International Agency for Research on Cancer: Lyon, France. pp. 40-407.

Chamberlain, M., R. Davies, R.G. Brown, and D.M. Griffiths. 1982. In vitro tests for the pathogenicity of mineral dusts. In: Inhaled Particles. V. W.H. Walton, ed. Pergamon Press: Oxford. pp. 583-592.

Coffin, D.L., L.D. Palekar, and P.M. Cook. 1982. Tumorigenesis by ferroactinolite minerals. Toxicol. Lett. 13:143-150.

Cook, P.M., L.D. Palekar, and D.L. Coffin. 1982. Interpretation of the carcinogenicity of amosite asbestos and ferroactinolite on the basis of retained fiber dose and characteristics. Toxicol. Lett. 13:151-158.

Cooper, J.A., II, R. Saracci, and P. Cole. 1979. Describing the validity of carcinogen screening tests. Br. J. Cancer 39:87-89.

Kiviluoto, R. 1965. Pleural plaque and asbestos: Further observation on endemic and other non-occupational asbestos. Ann. N.Y. Acad. Sci. 132:235-239.

Light, W.G., and E.T. Wei. 1977. Surface charge and hemolytic activity of asbestos. Environ. Res. 13:135-145.

MacMahon, B., and T.F. Pugh. 1970. Epidemiology Principles and Methods. Little, Brown and Co.: Boston. p. 261.

Nicholson, W.J., and F.L. Pundsack. 1973. Asbestos in the environment. In: Biological Effects of Asbestos. IARC Scientific Publication No. 8. International Agency for Research on Cancer: Lyon, France. pp. 126-130.

Nicholson, W.J., E.J. Sevoslowski, Jr., A.N. Porohl, J.D. Todaro, and A. Adams. 1979. Asbestos contamination in the U.S. state schools from use of asbestos surfacing material. Ann. N.Y. Acad. Sci. 330:587-596.

Parzen, E. 1964. Probability and theory on the study of mathematical models of random phenomena. In: Modern Probability Theory and Its Application. John Wiley and Sons: New York. pp. 1-31.

Stanton, M.F., and C. Wrench. 1972. Mechanisms of mesothelioma induction with asbestos and fibrous glass. J. Natl. Cancer Inst. 48:797-816.

QUALITY ASSURANCE AND GOOD LABORATORY PRACTICE REGULATIONS FOR SHORT-TERM BIOASSAYS OF COMPLEX MIXTURES

Robert S. DeWoskin

Quality Assurance Unit, Chemistry and Life Sciences, Research Triangle Institute, Research Triangle Park, North Carolina 27709

INTRODUCTION

Quality assurance monitoring of the data generated in health effects testing has increased as a result of recent government regulatory agency rulings and increased competition in the private sector. Quality is evaluated by comparing selected attributes of a study to a set of reference standards (ASQC, 1973; Juran and Gryna, 1980). Reference standards in highly technical or rapidly growing research areas may or may not be published and are frequently revised subject to a changing consensus. The use of different standards or different study attributes to evaluate the quality of a study often results in inconsonant conclusions. To formalize the evaluation for the quality of health effects studies, the U.S. Food and Drug Administration in 1978 (U.S. FDA, 1978) and more recently the Environmental Protection Agency (U.S. EPA, 1983a, 1983b) have published reference standards known as the Good Laboratory Practice standards (GLPs). The GLPs from both agencies were prompted by the discovery that poor quality and in some cases fraudulent data (Crawford, 1979; Richards, 1983) were influencing the outcome of safety assessments.

An important understanding in implementing a quality assurance GLP compliance program is the difference between an evaluation for quality based upon the GLP standards and an evaluation for quality by peer review. In the former, the main criteria are data quality and integrity; in the latter, the main criterion is scientific merit--in other words, how well the study resolves the scientific problem under investigation. Data integrity is possible in a study that is lacking in scientific merit, but scientific merit is unlikely for a study where the data are of poor quality.

In research or methods development studies, standardized practices are often impractical so that data quality and integrity have, for the most part, remained the responsibility of the principal investigator. The GLP standards do not remove this responsibility from the principal investigator (Gardner, 1979; Hile, 1979). Instead, they provide a frame of reference that auditors not directly involved in the conduct of the study can use to evaluate the quality of the data. In the U.S., compliance with the GLP standards has become a prerequisite for acceptance of any study, including short-term bioassays, submitted to regulatory agencies in support of a research permit, a market permit, or an exposure standard. Government research agencies participating in the National Toxicology Program can also contractually require that extramural projects comply with the GLP standards (Grieshaber and Whitmire, 1984). Compliance with the GLPs for short-term bioassays of complex mixtures, especially in research and development studies, presents a unique challenge both to the traditional evaluation of quality based upon scientific merit and to the organizational and documentation skills of the principal investigator. This paper briefly discusses the scope of the GLPs, the provisions applicable to short-term bioassays of complex mixtures, and the inherent features of the GLPs that allow their practical application to short-term studies of a nonroutine or research orientation.

GLP COMPLIANCE REQUIREMENTS

The three GLP standards (two from EPA and one from FDA) are very similar in text and emphasis but have different applications. Table 1 briefly lists and describes each set. The FDA GLPs, the first to be published in final form (i.e., after two years of debate), apply to nonclinical laboratory studies on FDA-regulated products that support or are intended to support applications for research or marketing permits. EPA has published two sets of GLPs: one to help implement the Toxic Substances Control Act (TSCA), and one to help implement the Federal Insecticide, Fungicide, and Rodenticide Act (FIFRA). The TSCA GLPs apply to all studies relating to health effects, environmental effects, and chemical fate testing pursuant to TSCA Section 4 (Testing of Chemical Substances and Mixtures) and all data developed under TSCA Section 5 (Manufacturing and Processing Notices). The FIFRA GLPs are required for all studies that support or are intended to support applications for research or marketing permits for pesticide products regulated by EPA.

The GLP standards apply specifically to the generation of health and safety data used in regulatory decision making. The requirements are based upon the assumption that the methods of the study are routine at least to the extent that critical procedures can be targeted for quality control. Nonroutine studies, such as research and methods development studies and range-finding studies, have generally been considered exempt from the compliance requirement. Nonroutine study results, however, are sometimes included in safety assessments in the absence of more routine bioassay data. Also, it is to be expected that the data developed under government programs not listed above might be included in applications reviewed by TSCA, FIFRA, or FDA panels and could be challenged on the basis of noncompliance with the GLP standards. Therefore, to determine whether a particular study need comply with the GLPs, the principal investigator should not only consider the intended use of the results but also all the probable future uses.

The next section discusses the GLP approach to ensuring quality.

Table 1. Good Laboratory Practice Standards

Government Agency	Title	Reference	Scope
FDA	Good Laboratory Practice for Nonclinical Laboratory Studies	U.S. FDA, 1978	Applies to all nonclinical laboratory studies that support or are intended to support applications for research or marketing permits for FDA-regulated products pursuant to the Federal Food, Drug, and Cosmetic Act
EPA	Toxic Substances Control; Good Laboratory Practice Standards	U.S. EPA, 1983a	Applies to all studies relating to health effects, environmental effects, and chemical effects pursuant to Sections 4 and 5 of the Toxic Substances Control Act
EPA	Pesticide Programs; Good Laboratory Practice Standards	U.S. EPA, 1983b	Applies to studies that support or are intended to support applications for research or marketing for pesticides pursuant to the Federal Insecticide, Fungicide, and Rodenticide Act

GLP APPROACH TO ENSURING QUALITY

To understand the GLP requirements it is helpful to first understand how the GLPs attempt to ensure quality. The GLP regulations are process-oriented; that is, they attempt to ensure the quality of the data by introducing controls on study processes that are generally thought to affect the overall quality of health effects study results. Quality, however, is commonly defined in terms of the end product or service--not in terms of the process. Quality is defined as "fitness for use" (Juran and Gryna, 1980) or as "the totality of features and characteristics of a product or service that bear on its ability to satisfy a given need" (ASQC, 1973). Ensuring the quality of data by enforcing process controls will only be successful if the processes identified have a significant influence upon the fitness of the end results for their intended use. Because the GLPs are soundly based, the processes targeted for control often do directly affect the quality of the final results from health effects research.

GLP PROVISIONS APPLICABLE TO SHORT-TERM BIOASSAYS OF COMPLEX MIXTURES

Many GLP provisions are written in general terms and contain phrases like "adequate to maintain" or "as needed" or "sufficient to ensure" or "periodically." This phrasing effectively defers the responsibility for detailing the laboratory procedures to the principal investigator. The originators of the GLPs purposely included the above phrases so that the standards could be widely applied. The preamble to the GLP text, the official commentary, and regulatory agency policy statements clearly state that the "spirit" or intent of the GLPs is to ensure data quality and integrity. When the study records comply with the requirements to document methods and GLP deviations, then the regulatory agency auditor will have the information needed to evaluate the validity of a researcher's interpretation of the general provision.

Table 2 lists ten GLP subheadings that describe different aspects of a health effects study along with a few examples of specific requirements. In the FDA GLPs, there are a total of 141 operational provisions (Taylor, 1984). Many of the GLP requirements address laboratory procedures characteristic of *in vivo* animal testing because the GLPs evolved primarily from experience with chronic toxicity bioassays. Obviously, the provisions specifically addressing whole-animal care would not apply to a short-term *in vitro* bioassay. But, because of the general language used, most of the provisions would apply, and the specific procedures would depend upon the interpretation.

The initially proposed versions of the GLPs of both EPA and FDA included references to widely accepted standards like the National Institutes of Health Guidelines for Care and Use of Laboratory Animals to assist researchers in developing specific laboratory procedures. Table 3 lists some of the standards that agency auditors often use for guidance in interpreting the more ambiguous GLP provisions. The final versions of the GLPs deleted references to the standards listed in Table 3 because of the logistical and legal problems of having standards nested within standards. This in effect gave more responsibility to the research laboratory.

Referring again to Table 2, the GLP requirements that pose the greatest challenge to a compliance program for short-term bioassays of complex mixtures will be those in subheadings 5 (Operations), 6 (Protocol), and 7 (Test and Control Substances). Some of the requirements in these areas assume an *a priori* knowledge about the test organism, test substance, or probable outcome of the study. Subheading 5, for example, includes a provision requiring signed and approved Standard Operating Procedures (SOPs) for all routine phases of the study. Changes in procedures are to be documented in the form of an SOP amendment. This requirement can result in a burdensome amount of paperwork for bioassays that, although past the developmental stage, are still subject to frequent revision. Bioassay or analytical studies in the developmental stage have an even greater difficulty complying with the SOP requirements and the requirements in Subheading 6 to develop a detailed protocol that is signed and approved prior to the

Table 2. Examples of Specific FDA GLP Requirements

GLP Subheading	Examples of Specific Requirements
1. Personnel	Personnel shall be qualified. Sufficient number present. *Curriculum vitae* available on file.
2. General Facilities	Facility is of suitable size and construction. Separate areas adequate to prevent contamination.
3. Specialized Facilities	Laboratory space for receipt, housing, and treatment of animals. Separate rooms for biohazardous compounds. Facilities for disposal of waste.
4. Equipment	SOPs available for maintenance, calibration, and repair. Records kept on all equipment used to generate data.
5. Operations	Management commitment to quality assurance. SOPs. A Quality Assurance Unit monitors the study for GLP compliance.
6. Protocol	Protocol describing study is approved, signed, and dated prior to start of laboratory phase. Changes in procedures that affect quality are documented as amendments.
7. Test and Control Substances	Substances are characterized for identity, purity, strength, and stability. Storage and handling prevent degradation or contamination.
8. Records	All records are in ink, signed, and dated. All changes are signed, dated, and explained without obscuring the original entry. Integrity of the data is protected (i.e., limited access).
9. Reports	Reports contain the minimum information listed within this section. Report contains a Quality Assurance Unit statement.
10. Archives	Archive minimizes deterioration of samples and data. Limited access and easy retrieval of information.

Table 3. Safety and Specific Test Standards (Examples)

Source	Date Published	Title	Reference
EPA	1979	Standards for Development of Test Data--Chronic Health Effects	44 CFR 27334
	1979	Standards for Development of Test Data--Acute and Subchronic, Mutagenic Teratogenic/Reproductive, Metabolism Effects	44 CFR 44054
Interagency Regulatory Liaison Group	1980	Recommended Guidelines for Teratogenicity Studies in the Rat, Mouse, Hamster or Rabbit	EPA: Washington, DC
National Research Council	1981	Prudent Practices for Handling Chemicals in Laboratories	QD51.N32
Health and Human Services	1981	Occupational Health Guidelines for Chemical Hazards	DHHS (NIOSH) 81-123
	1976	Guidelines for Carcinogen Bioassay in Small Rodents	NIH 76-801
	1974	Guide for the Care and Use of Laboratory Animals	NIH 80-23
	1981	NIH Guide for the Laboratory Use of Chemical Carcinogens	NIH 81-2385
	1975	NCI Safety Standards for Research Involving Chemical Carcinogens	NIH 77-900

(continued)

Table 3. Continued

Source	Date Published	Title	Reference
Committee for the Institute of Laboratory Animal Resources, Assembly of Life Sciences, National Research Council	1976	Long Term Holding of Laboratory Rodents	ILAR News 19:L1
International Agency for Research on Cancer	1980	Long Term and Short Term Screening Assays for Carcinogens	IARC Monograph Supplement No. 2
Nuclear Regulatory Commission	1983	Handling of Radioactive Materials	Title 10 CFR

start of the laboratory phase. But undoubtedly the most difficult GLP requirements for the bioassay testing of complex mixtures are those of Subheading 7, the characterization of the test compound. Test compounds are to be characterized for identity, strength, purity, and composition prior to initiation of the study. Also, the stability of the test compound alone and in the presence of carriers should be known prior to the start of the laboratory phase. A label on the test compound container must contain an expiration date, generally the earliest date that any of the components will degrade. Of course, this information may be very difficult to obtain for complex mixtures.

QUALITY ASSURANCE PROGRAM FOR THE TESTING OF COMPLEX MIXTURES

The GLPs were designed to ensure the quality and integrity of health effects data. Deviations from specific GLP requirements are an available option in a quality assurance program as long as the deviations are well documented and do not compromise the quality of the study.

The GLP provisions on protocols and SOPs are primarily designed to prevent bias and fraud. In the absence of *a priori* knowledge, researchers developing assays can comply with the intent of the GLPs by maintaining adequate notebooks to assist an outside auditor in the reconstruction of the study and the evaluation of the integrity of the data. The GLP requirements for chemical characterization are intended to prevent test compound substitution, alteration, or degradation. A resarcher using complex mixtures should answer the following questions: (1) Has all the available information on the composition of the mixture been compiled? (2) Is there enough sample to perform additional stability analyses? (3) Do raw data records clearly describe any storage, handling, or preparation procedures that have been designed to maintain the mixture? In the testing of complex mixtures, the integrity of the test compound from collection to post-assay analysis commands particular attention. Quality controls to ensure the integrity of the test compound often include chain of custody documentation and detailed environmental controls for storage, handling, and testing.

The quality assurance program therefore properly addresses the GLP intent of ensuring data quality and integrity by encouraging compliance with the provisions of the GLPs whenever they are applicable to short-term bioassay testing of complex mixtures and, when they are not applicable, by supporting the practice of documenting alternative procedures and the reasons that the alternatives further the quality of the study.

ACKNOWLEDGMENTS

I would like to thank Dr. Bill Durham of EPA (Research Triangle Park, NC) for his insights on the relationship between peer review and GLP compliance and to acknowledge the research scientists at Research Triangle Institute for their valuable suggestions on implementing GLPs for nonroutine or research studies.

REFERENCES

ASQC. 1973. Glossary and Tables for Statistical Quality Control. American Society for Quality Control: Milwaukee, WI.

Crawford, L.M. 1979. The final order on Good Laboratory Practices. Presented at the FDA Briefing Session on Good Laboratory Practices Regulations, Washington, DC.

Gardner, S. 1979. Maintaining the creative balance. In: Quality Control in Toxicology. G.E. Paget, ed. University Park Press: Baltimore, MD.

Grieshaber, C.K., and C.E. Whitmire. 1984. Effects of Good Laboratory Practices on chemistry requirements for toxicity testing. In: Chemistry for Toxicity Testing. C.W. Jameson and D.B. Walters, eds. Butterworth Publishers: Boston.

Hile, J.P. 1979. The current regulatory status in the USA: Regulation begins. In: Good Laboratory Practice. G.E. Paget, ed. University Park Press: Baltimore, MD.

Juran, J.M., and F.M. Gryna. 1980. Quality Planning and Analysis. McGraw-Hill: New York.

Richards, B. 1983. Papers from trial of former IBT officers raise many questions on product safety. Wall Street Journal, May 13, p. 31.

Taylor, D.W. 1984. Inspection of computer-supported toxicological data submitted to the FDA. Drug Inf. J. 18:189-194.

U.S. EPA. 1983a. Toxic substances control; Good Laboratory Practice standards. Fed. Regist. 48:53922-53944.

U.S. EPA. 1983b. Pesticide programs; Good Laboratory Practice standards. Fed. Regist. 48:53946-53969.

U.S. FDA. 1978. Nonclinical laboratory studies; Good Laboratory Practice regulations. Fed. Regist. 43:59986-60020.

GENERATION AND CHARACTERIZATION OF COMPLEX GAS AND PARTICLE MIXTURES FOR INHALATION TOXICOLOGIC STUDIES

Michael T. Kleinman, Robert F. Phalen, and T. Timothy Crocker

Department of Community and Environmental Medicine, University of California, Irvine, Irvine, California 92717

INTRODUCTION

Laboratory toxicology studies of the relative potencies of complex, polluted urban atmospheres have been surprisingly slow in confirming results observed by epidemiologists, that is, that air pollution can seriously affect human health. The responses of human volunteers and laboratory animal models after exposure to single compounds and simple mixtures of compounds observed in plumes and ambient air are not sufficiently intense to explain the respiratory morbidity and mortality associated with serious air pollution exposure excursions. Yet there is little doubt that during severe episodes such as those that occurred in London in 1952 or in Donora, PA in 1948 some components present in the air were responsible for the observed respiratory-related illnesses and deaths.

Epidemiological studies have identified several compounds that are most likely associated with the observed air pollution health effects, and some of these materials might be the actual etiological agents. Significant decrements in human lung function have been attributed to environmental exposures to the combined influences of temperature, sulfur dioxide (SO_2), nitrogen dioxide (NO_2), ozone (O_3), and aerosols (Goldsmith and Friberg, 1977; Kagawa et al., 1980). Many of these compounds alone and in mixtures, with the notable exception of O_3, have exhibited minimal effects on respiratory function and symptoms in laboratory tests, even when the concentrations used were near "worst case" ambient conditions (Hackney et al., 1980; Hackney and Linn, 1981).

The lack of correspondence between effects in the laboratory and in the environment suggests that the aforementioned compounds are not themselves causal, but that they are in some way related to the actual etiologic agents (Mazumdar et al., 1982). These etiologic agents are as yet unidentified, in part because they have not been adequately monitored in air samples, or because they are so labile they may exist only as transient species. They might be rapidly converted to some end product (of lesser toxicity) either during the course of sampling or during later laboratory analysis. For example, it is well documented that SO_2 can be taken up by wet and dry aerosols and can be oxidized to acid sulfates (Dlugi et al., 1981). The presence of soot (Benner et al., 1982), fly ash (Liebsch and de Pena, 1982), marine aerosols (Hitchcock et al., 1980), or catalytic metal ions such as iron or manganese (Kaplan et al., 1981; Harrison et al., 1982) in the particulate phase all accelerate conversion from sulfur (IV) to sulfur (VI) species. During the course of the absorption, dissolution, dissociation (complexation), and oxidation processes, a number of relatively unstable intermediate reaction products may be present in or on the surface of the aerosol. These might include sulfites and bisulfites, dissolved SO_2, formaldehyde-bisulfite adducts (Dasgupta, 1981), Fe(III)-sulfite complexes (Eatough et al., 1978), and acid sulfates and their ammonia neutralization products. Compounds such as dithionic, sulfonic, hydosulfonic, and thiosulfuric acids and their respective ammonia neutralization products may also be formed in urban aerosols (Charlson et al., 1978). If nitrogen oxides (NO_x) are present, these may dissolve in aerosols to yield nitrites that can subsequently be oxidized to nitric acid (Martin et al., 1981) and that may also interact with SO_2 to form more complex species such as hydroxylamine disulfonate (Oblath et al., 1981). From the standpoint of toxicology, sulfur(IV) compounds that are chemically reactive should be more potent than sulfur(VI) species, and Last et al. (1980) have demonstrated lesions in the lungs of rodents exposed to sodium sulfite aerosols. The association of SO_2 and salt-containing liquid aerosol droplets has also been reported to enhance the irritancy of the gas. This has been demonstrated in both animals and humans (McJilton et al., 1973, 1976; Amdur and Underhill, 1968; Koenig et al., 1980, 1981). Several hypotheses, including formation of acid or the transport of adsorbed or dissolved SO_2 to more distal lung regions, have been offered to explain this enhancement, but none has been conclusively demonstrated to be correct (or most important).

Inhalation toxicology studies with atmospheres that closely simulate the complexity of those in the real world are critically important if reasonable evaluations are to be made regarding the potential risks associated with the presence of possible reaction intermediates and unstable sulfur species in community air. This article will discuss some of the techniques developed to simulate, in the laboratory, conditions in an aging, reacting plume and to characterize the resulting atmospheric components. It is hoped that such information will help others to design cost-effective, productive inhalation toxicology studies.

AIR PURIFICATION

Inhalation toxicology studies are often conducted in locations in which ambient pollutant levels are sufficiently high to mask or otherwise alter the effects of the atmosphere under study. Important environmental contaminants that are often present in sufficient concentrations to influence the inhalation studies include viable and nonviable particles, O_3, SO_2, and NO_x (i.e., NO, NO_2). Gaseous hydrocarbons are typically present at levels below those thought to significantly interfere with the inhalation studies.

In order to prevent such interference and to enhance the reliability and reproducibility of studies, chamber atmospheres should be generated using purified air as a starting point. In most applications, cost considerations prohibit the use of tanked synthetic air to supply dynamic flowing laboratory inhalation chamber systems. It is, therefore, necessary to design or purchase an appropriate air purification system. As a guideline for inhalation studies with mixtures of ambient pollutants at ambient levels or above, the following maximum levels of contaminants would provide air of sufficient purity: CO, 1.0 ppm; NO, NO_2, O_3, and SO_2, 0.01 ppm; hydrocarbons, 5.0 ppm; and condensation nuclei, 100 cm^{-3}. For maximum flexibility, it is desirable to have simulation capability for the full range of ambient humidities; however, for many purposes a range of 40 to 90% relative humidity is appropriate. The volume capacity of the air purifier will depend on the configuration of the specific exposure system and the desired number of air exchanges. For example, to service two 1-m^3 inhalation chambers, an airflow rate of 1200 slpm would be sufficient to meet purified air requirements of compressed air-driven aerosol generators and the inhalation chambers with capacity to spare. This particular system is currently in use at the Air Pollution Health Effects Laboratory, University of California, Irvine (Walters and Phalen, 1982). Ambient air is first filtered and compressd to 100 psig (67.9×10^6 dyn/cm^2) using a liquid ring unit (Nash Model CD663C). The 100-psig saturated air leaving the receiver is filtered to remove liquid water and entrained particles and passed through a 31-cm diameter x 80-cm deep bed of permanganate-impregnated alumina pellets (Purafil, Inc., Chamblee, GA). The moisture in the inlet gas stream is sorbed by the alumina, dissolving the permanganate that can then oxidize sorbed contaminants. The unit efficiently removes SO_2, NO_x, O_3, and some hydrocarbons from the pressurized air stream. Further purification and drying of the air stream is then accomplished in activated alumina-filled heatless dryer columns operated in a pressure-swing desorption regenerating mode. In this operation, high pressure air is dried by passage through one drying column; a portion of this dried air is simultaneously passed through a second idling column at atmospheric pressure to desorb moisture and contaminants, thus enabling the idling column to act as the drying column in the next cycle. Drying of the pressurized air stream to less than 100-ppm (by weight) water content promotes the adsorption performance of the downstream activated carbon bed and is essential to prevent water vapor "poisoning" of the next unit, a fixed Hopcalite bed operating at ambient temperature. The heatless dryer and Hopcalite units are part of a system supplied by Del-Tech Engineering Co. (Model No. 5N12) that is used for the purification of compressed breathing air. After passage through coconut-base activated carbon to remove heavier hydrocarbons, the air stream is filtered to remove Hopcalite and charcoal dust and brought to indoor laboratory temperature using an air/air heat exchanger. Delivery to inhalation chambers immediately follows air pressure reduction, humidification, and final high efficiency (HEPA) filtration.

A higher throughput system (Bell et al., 1980) has been in use for serveral years at Rancho Los Amigos Hospital in Downey, California. This system operates along similar principles but uses oxidative catalysts at elevated temperatures rather than at higher pressures to facilitate air cleaning.

ATMOSPHERE "AGING"

Many of the atmospheric transformations and chemical reactions (described in the Introduction) that may be of toxicologic significance can be simulated under controlled laboratory conditions. Under some circumstances, adequate simulation can be achieved using a static system such as a continuously stirred tank reactor. Often, however, inhalation toxicology studies require a dynamic system in which reactants are continuously input, products are formed, and animals are exposed to a flow of well-mixed pollutant atmosphere. By varying the throughput rate of the system, animals can be exposed to atmospheres of different average ages (Walters et al., 1982).

The Walters plug-flow aging line and exposure system has been used to simulate several important phenomena that occur in plumes as they mix and interact with compounds in the ambient atmosphere in order to study atmospheric transformations of SO_2 (Kleinman et al., 1984b) and to conduct sophisticated toxicology studies (Mannix et al., 1982). The atmospheres studied contained O_3, SO_2, and droplet aerosols, either sulfuric acid or ammonium sulfate. The droplets have contained metal ions (Fe^{+3}, Mn^{+2}) known to catalyze the conversion of SO_2 gas to particulate species.

Results of initial toxicologic studies with unaged atmospheres containing O_3 and sulfate particles suggested that, if anything, effects of the mixture were less decremental than if the same amount of O_3 had been administered alone (Phalen et al., 1980). More recently, however, studies with aged, multicomponent atmospheres have suggested an enhancement of O_3-induced lung parenchymal focal lesions by the mixture (Kleinman et al., 1984a).

Some features of the aging line system are shown in Figure 1. The figure also shows the experimental setup for the aforementioned study of SO_2 uptake by aerosols. Note that the sampling of gas-particle mixtures is far more involved than sampling for simple, one-component atmospheres. The importance of sampling is often underestimated in toxicologic reasearch. Reactive components in the gas or vapor phase can often react with sampling media or with particles collected on filters during the sampling process. Figure 1 illustrates one method available for avoiding this type of artifact: use of a "denuder." The denuder shown is an efficient diffusion collector with walls that are covered with a suitable absorbing material. For example, a basic coating such as lead oxide, sodium carbonate, or triethanolamine can be used to remove more than 99% of acid gases such as NO_2 or SO_2 from an air stream, while allowing better than 99% of aerosols to pass through. Conversely, an acidic coating like oxalic acid can be used to remove a basic gas, for instance ammonia, from the sampled air.

The aging line has been used to generate atmospheres that are more environmentally relevant than might be otherwise obtained. As an example of the usefulness of the aging line, we examined the influence of atmospheric composition and aging time of the conversion of SO_2 to particulate species such as sulfite and sulfate.

Radiolabeled $^{35}SO_2$ was used as a tracer, and the uptake of SO_2 by ammonium sulfate droplet aerosols with and without catalytic metal ions (iron and manganese sulfate) was studied. Relative humidity was maintained at 85%. Air samples were collected through SO_2 denuders onto formaldehyde-impregnated filters. Samples were collected at different locations along the aging line to represent atmospheres 10-, 20-,

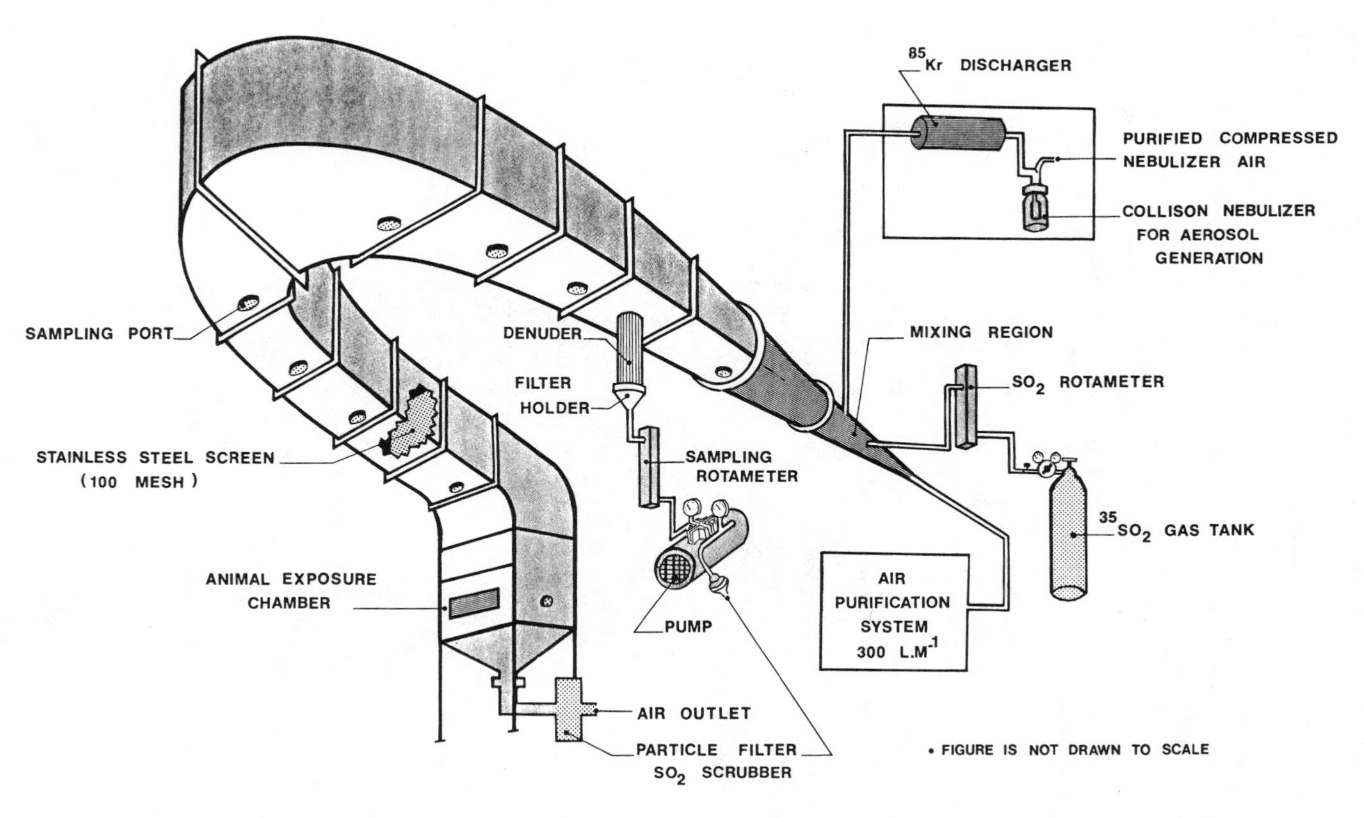

Figure 1. A schematic representation of a continuous flow aging line for inhalation toxicology studies.

and 30-min downstream of the entrance plume. The denuder terminated reactions during sample collection by stripping unreacted SO_2 from gas and aerosol phases, and the formaldehyde inhibited further oxidation of incorporated sulfur (IV) to sulfur (VI) species. Bisulfite (or sulfite), sulfate, and thiosulfate concentrations in filter extracts were determined by ion chromatography. Levels of sulfur (IV) species in collected samples were below the limit of detection of the method, 2 $\mu g/m^3$. Incorporation of SO_2 by the particulate phase was measured by liquid scintillation counting of ^{35}S in the filter sample extracts.

No SO_2 uptake was observed in ammonium sulfate aerosol without catalysts. The addition of catalytic metal ions resulted in small but measurable incorporation of SO_2 into aerosols; the rate of incorporation was equivalent to 0.02%/h.

Additional studies to determine the influence of oxidant gases (NO_2 and O_3) on $^{35}SO_2$ incorporation were also conducted. The atmospheric concentrations and experimental conditions are summarized in Table 1.

As shown in Figure 2, the presence of oxidant gases can greatly increase the rate of uptake of $^{35}SO_2$ by ammonium sulfate aerosols that contain catalytic metals (AMS*). The addition of NO_2 resulted in a rate of uptake that was about 23 times greater than that for SO_2 + AMS*. Ozone was even more effective: about 43 times more potent than SO_2 + AMS* alone. The conversion rates, expressed as percent SO_2 converted per hour, are summarized in Table 2. The conversion rates in the presence of oxidant gases approach those reported in ambient power plant plumes.

Table 1. Summary of Atmospheric Concentrations and Experimental Conditions

Relative Humidity	85%
Aging System Flow	300 L/min
Aerosol Mass Concentration	1 mg/m^3
Aerosol Size	0.5 μm MMD
Gas Concentrations	
SO_2	5 ppm
NO_2	5 ppm
O_3	0.6 ppm

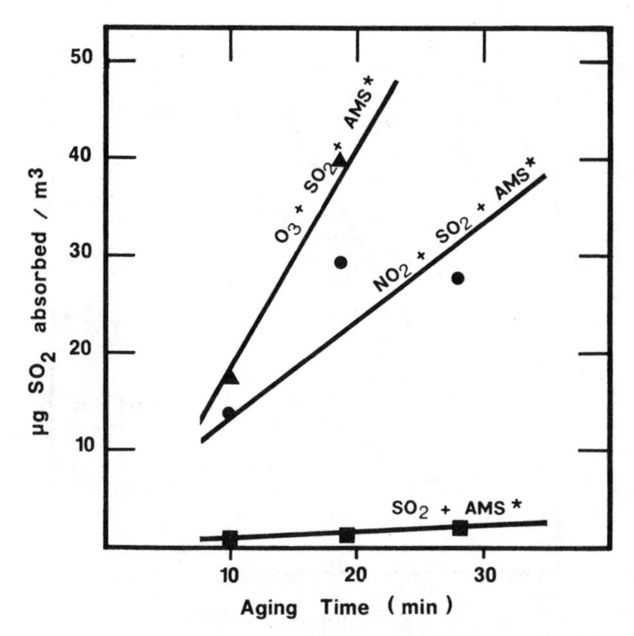

Figure 2. The influence of oxidant gases on the rate of incorporation of gaseous SO_2 into ammonium sulfate aerosols that contain catalytic iron and manganese ions (AMS*).

Table 2. Effect of Oxidant Gases on the Rate of $^{35}SO_2$ Conversion in Sulfate Aerosols

Atmosphere	Conversion Rate (%/h)
SO_2 + AMS*	0.02
NO_2 + SO_2 + AMS*	0.46
O_3 + SO_2 + AMS*	0.86

We also investigated the importance of the catalytic ions in influencing SO_2 conversion rates in the presence of an oxidant gas, O_3. As shown in Figure 3, iron and

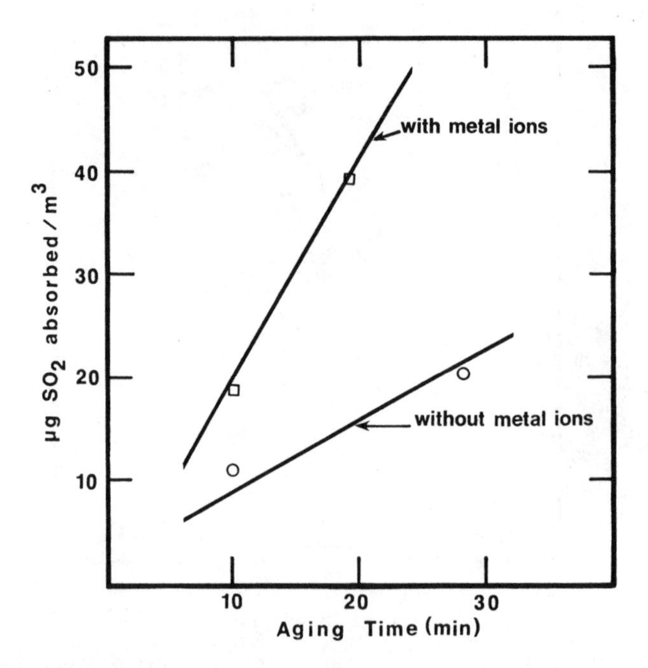

Figure 3. The influence of catalytic metal ions (iron and manganese) on the rate of SO_2 incorporation into ammonium sulfate aerosols.

manganese ions in the aerosol phase markedly increase the rate of uptake of SO_2 by ammonium sulfate aerosol in an O_3-containing atmosphere.

CONCLUSIONS

We have summarized some of the key considerations that should be accounted for in the generation and characterization of complex gas and particle mixtures used in inhalation toxicology studies. The risk of deriving erroneous toxicological inferences from inhalation studies of atmospheres that do not adequately reflect the complexity of the real world are real and pertinent. Investigators should use the available technology to increase the level of sophistication of inhalation toxicology studies so that we can obtain a better understanding of the nature of environmental hazards.

ACKNOWLEDGMENTS

We would like to thank R. Walters, R. Mannix, M. Azizian, B. Nordenstam, and T. Nguyen for their technical assistance. These studies are parts of programs supported by Southern California Edison, Contract C1022901; Electric Power Research Institute, Contract RP 1962-1; the U.S. Environmental Protection Agency, Grant R808267-01-0; and the California Air Resources Board (AO-128-32).

REFERENCES

Amdur, M.A., and D. Underhill. 1968. The effects of various aerosols on the responses of guinea pigs to sulfur dioxide. Arch. Environ. Health 16:460-468.

Bell, K.A., E.L. Avol, R.M. Bailey, M.T. Kleinman, D.A. Landis, and S.L. Heisler. 1980. Design, operation and dynamics of aerosol exposure facilities for human subjects. In: Generation of Aerosols and Facilities for Exposure Experiments. Klaus Willeke, ed. Ann Arbor Science Press: Ann Arbor, MI. pp. 475-491.

Benner, W.H., R. Brodzinski, and T. Novakov. 1982. Oxidation of SO_2 in droplets which contain soot particles. Atmos. Environ. 16:1333-1339.

Charlson, R.J., D.S. Covert, T.V. Larson, and A.P. Waggoner. 1978. Chemical properties of tropospheric sulfur aerosols. Atmos. Environ. 12:39-53.

Dasgupta, P.K. 1981. Discussion, The reaction of nitrogen oxides with SO_2 in aqueous aerosols. Atmos. Environ. 15:875.

Dlugi, R., S. Jordan, and E. Lindeman. 1981. The heterogeneous formation of sulfate aerosols in the atmosphere. J. Aerosol Sci. 12:185-197.

Eatough, D.L., T. Major, J. Ryder, M. Hill, N.F. Mangelson, N.L. Eatough, and L.D. Hansen. 1978. The formation and stability of sulfite species in aerosols. Atmos. Environ. 12:263-271.

Goldsmith, J.R., and L.T. Friberg. 1977. Effects of air pollution on human health. In: Air Pollution, Volume II, Third Edition. A.C. Stern, ed. Academic Press: New York. pp. 457-610.

Hackney, J.D., and W.S. Linn. 1981. Experimental evaluation of air pollutants in humans. In: Measurements of Risks. G.G. Berg and H.D. Maillie, eds. Plenum Publishing Co.: New York. pp. 231-251.

Hackney, J.D., W.S. Linn, E.L. Avol., M.P. Jones, M.T. Kleinman, and R.M. Bailey. 1980. Human health effects of nitrogenous air pollutants: Recent findings from controlled-environment clinical studies. In: Nitrogen Oxides and Their Effects on Health. S.D. Lee, ed. Ann Arbor Science: Michigan. pp. 307-314.

Harrison, H., T.V. Larson, and C.S. Monkman. 1982. Aqueous phase oxidation of sulfites by ozone in the presence of iron and manganese. Atmos. Environ. 16:1039-1041.

Hitchcock, D.R., L.L. Spiller, and W.E. Wilson. 1980. Sulfuric acid aerosols and HCl release in coastal atmospheres: Evidence of rapid formation of sulfuric acid particulates. Atmos. Environ. 14:165-182.

Kagawa, J., K. Tsuru, T. Doi, T. Tsunoda, T. Toyama, and M. Nakaza. 1980. Lung function studies on intermittently exercising high school students exposed to air pollution. In: Nitrogen Oxides and Their Effects on Health. S.D. Lee, ed. Ann Arbor Science: Michigan. pp. 333-359.

Kaplan, D.J., D.M. Himmelblau, and C. Kanoka. 1981. Oxidation of sulfur dioxide in aqueous ammonium sulfate aerosols containing manganese as a catalyst. Atmos. Environ. 15:763-773.

Kleinman, M.T., W.J. Mautz, T. McClure, R.F. Phalen, and T.T. Crocker. 1984a. Enhancement of ozone-induced lung lesions by acidic aerosols. Presented at the 68th Annual Meeting of the Federation of American Societies for Experimental Biology, St. Louis, MO.

Kleinman, M.T., R.F. Phalen, R. Mannix, M. Azizian, and R. Walters. 1984b. Influence of Fe and Mn ions on the incorporation of radioactive $^{35}SO_2$ by sulfate aerosols. Atmos. Environ. (submitted).

Koenig, J.Q., W.E. Pierson, and R. Frank. 1980. Acute effects of inhaled SO_2 plus NaCl droplet aerosol on pulmonary function in asthmatic adolescents. Environ. Res. 22:145-153.

Koenig, J.Q., W.E. Pierson, M. Horike, and R. Frank. 1981. Effects of SO_2 plus NaCl aerosol combined with moderate exercise on pulmonary function in asthmatic adolescents. Environ. Res. 25:340-348.

Last, J.A., P.K. Dasgupta, and J.R. Etchison. 1980. Inhalation toxicology of sodium sulfate aerosol in rats. Toxicol. Appl. Pharmacol. 55:229-234.

Liebsch, E.J., and R.G. de Pena. 1982. Sulfate aerosol production in coal-fired power plant plumes. Atmos. Environ. 16:1323-1331.

McJilton, C.E., R. Frank, and R. Charlson. 1973. Role of relative humidity in the synergistic effect of a sulfur dioxide-aerosol mixture on the lung. Science 182:503-504.

McJilton, C.E., R. Frank, and R.J. Charlson. 1976. Influence of relative humidity on functional effects of an inhaled SO_2-aerosol mixture. Am. Rev. Respir. Dis. 113:163-169.

Mannix, R.C., R.F. Phalen, J.L. Kenoyer, and T.T. Crocker. 1982. Effect of sulfur dioxide-sulfate exposure on rat respiratory tract clearance. Am. Ind. Hyg. Assoc. J. 43:679-685.

Martin, L.R., D.E. Damschen, and H.S. Judeikis. 1981. The reactions of nitrogen oxides with SO_2 in aqueous aerosols. Atmos. Environ. 15:191-195.

Mazumdar, S., H. Schimmel, and I.T.T. Higgins. 1982. Relation of daily mortality to air pollution: An analysis of 14 London winters, 1958/59 - 1971/72. Arch. Environ. Health 37:213-218.

Oblath, S.B., S.S. Markovitz, T. Novakov, and S.G. Chang. 1981. Kinetics of the formation of hydroxylamine disulfonate by reaction of nitrite with sulfites. J. Phys. Chem. 85:1017-1021.

Phalen, R.F., J.L. Kenoyer, T.T. Crocker, and T.R. McClure. 1980. Effects of sulfate aerosols in combination with ozone on elimination of tracer particles inhaled by rats. J. Toxicol. Environ. Health 6:797-810.

Walters, R.B., and R.F. Phalen. 1982. Development of a high pressure air purification system for inhalation toxicology research. APHEL Report R-82-1. University of California, Irvine, Department of Community and Environmental Medicine. 10 pp.

Walters, R.B., R.F. Phalen, R.C. Mannix, and G.L. Smart. 1982. An aerosol and gas aging line suitable for use in inhalation toxicology research. Am. Ind. Hyg. Assoc. J. 43:218-222.

SESSION 2

INTEGRATED ASSESSMENT

NITRO COMPOUNDS IN ENVIRONMENTAL MIXTURES AND FOODS

Yoshinari Ohnishi,[1] Takemi Kinouchi,[1] Yoshiki Manabe,[1] Hideshi Tsutsui,[1] Hisashi Otsuka,[1] Hiroshi Tokiwa,[2] and Takeshi Otofuji[2]

[1]School of Medicine, The University of Tokushima, Tokushima 770, and [2]Fukuoka Environmental Research Center, Fukuoka 818-01, Japan

INTRODUCTION

Nitropyrenes (NPs) are the most potent mutagens that have been detected in environmental pollutants such as airborne particulates (Tokiwa et al., 1981b), car exhaust emissions (Lee et al., 1980; Gibson et al., 1981; Pederson and Siak, 1981; Schuetzle et al., 1981), photocopies (Löfroth et al., 1980; Rosenkranz et al., 1980), wastewater from gasoline stations (Ohnishi et al., 1983; Manabe et al., 1984), and used crankcase oil (Manabe et al., 1984). They are direct-acting mutagens in the Ames *Salmonella* mutation test; 1-NP, 1,3-diNP, 1,6-diNP, and 1,8-diNP produce 417; 65,500; 82,000; and 100,000 His^+ revertants/plate/nmol, respectively, from strain TA98 in the absence of rat liver S9 mix (unpublished data). To induce mutagenicity the NPs are converted to the activated forms by nitroreductases in the *Salmonella* tester strain (TA98). Since strains TA98NR and TA98/1,8-DNP_6 are defective in the specific activating enzymes, they show low mutagenicity by 1-NP and diNPs (Rosenkranz et al., 1981).

Four nitroreductases have been isolated from the intestinal anaerobic bacterium *Bacteroides fragilis* and found to be specific to the various mutagens containing NPs (Kinouchi and Ohnishi, 1983). By means of these specific nitroreductases, 1-NP and 1,6-diNP are finally converted to 1-aminopyrene (AP) and 1,6-diAP, respectively, *in vitro*, and these reduced chemicals are measured with high performance liquid chromatography (HPLC) to estimate the amounts of original NPs in the environmental mixtures (Ohnishi et al., 1983; Manabe et al., 1984).

Since people live indoors for more than 80% of a day, the effects of indoor air on human health are important. Moreover, the majority of cancer deaths are attributed to diet or life-style and cigarette smoking (Doll and Peto, 1981). We report here the mutagenicity and the amounts of 1-NP and 1,6-diNP in diesel exhaust particles, indoor particulates exhausted from a kerosene heater, and "yakitori" (grilled chicken). To evaluate the short-term assays of these chemicals in mixtures, carcinogenicity of 1-NP and 1,6-diNP was studied in BALB/c mice.

MATERIALS AND METHODS

Chemicals

1-Nitropyrene and 1,6-diNP were purified to 99.9% purity with 70% acetonitrile or 70% methanol by HPLC, as described previously (Kinouchi et al., 1982; Kinouchi and Ohnishi, 1983). Glucose-6-phosphate (G-6-P) was obtained from Sigma Chemical Co., and nicotinamide-adenine dinucleotide phosphate (NADPH) and G-6-P dehydrogenase were obtained from Oriental Yeast Co., Ltd. Diethyl ether and other reagent-grade chemicals were purchased from Wako Pure Chemical Industries, Ltd.

Samples and their Extraction

The particles exhausted from a small diesel engine (displacement 269 ml) idling at 2,000 revolutions per minute without a load were collected on a glass-fiber filter paper (Toyo GB-100R) for 31 min (Ohnishi et al., 1980, 1982). The particulate sample from a 1982 diesel-powered truck (displacement 5,785 ml) operated on a chassis dynamometer at the 10 mode cycle was collected on a Teflon-coated filter (Pallflex T60A20) by an Andersen high volume sampler for 23 min. The particles less than 1.1 μm in diameter were extracted by sonication with benzene-ethanol (4:1) for 30 min.

The particles in a 28-m^3 room warmed with a radiant kerosene heater were collected many times for 20-min periods on the Teflon-coated filters by a high volume sampler at a speed of 1.5 m^3/min and extracted by sonication with benzene-ethanol (4:1). The extracts were further fractionated with diethyl ether into neutral, acidic, and basic fractions by the procedure described previously (Manabe et al., 1984).

Pieces of raw chicken with or without a sauce (pH 4.9) were grilled above a city gas flame for various lengths of time and then cut into small pieces with scissors and extracted by sonication with benzene-ethanol. The chicken grilled for 5 min by this method was good for eating. The extracts were further fractionated, as described above.

Mutagenicity Assay

The bacterial strains used for the mutation test were *Salmonella typhimurium* *his*$^-$ strains TA98, TA100, TA98NR, and TA98/1,8-DNP_6. Each extract or fraction to be tested was dissolved in dimethyl sulfoxide (DMSO), and the mutagenic activity was measured by the Ames *Salmonella* mutagenicity test with preincubation at 37°C for 20 min, as described previously (Kinouchi and Ohnishi, 1983).

Measurement of Chemicals

The procedure for measuring 1-NP and 1,6-diNP was described previously (Manabe et al., 1984). In brief, extracted fractions to be analyzed were applied on a Zorbax ODS (4.6-mm inside diameter [i.d.] x 250 mm) or Merck RP-18 (7.9-mm i.d. x 250 mm) column. The fractions corresponding to 1-NP and 1,6-diNP eluted from the HPLC column were separately collected and incubated in a mixture containing the specific nitroreductase purified from *Bacteroides fragilis* (Kinouchi and Ohnishi, 1983) and NADPH to be reduced to the amino derivatives. The reduced materials were again applied separately on the Zorbax ODS column for HPLC, and the fluorescence of 1-AP or 1,6-diAP converted from 1-NP or 1,6-diNP was measured quantitatively using the fluorescence detector of the HPLC.

Carcinogenicity Test

Sixty male 6-week-old BALB/c mice (Charles River Co., Atsugi, Japan) were divided into three groups of 20 mice each. The mice in two of the groups were subcutaneously inoculated with 0.1 mg of 1-NP or 1,6-diNP dissolved in DMSO once a week for 20 wk. The remaining control group was similarly injected with a total of 4 ml of DMSO. All the animals were autopsied and histologically examined.

Portions of the tumors were homogenized and diluted by half with cold saline solution, and about 0.3-0.5 ml of the material was immediately transplanted subcutaneously into the backs of five male BALB/c mice (Tokiwa et al., 1984).

RESULTS

Mutagenicity of Diesel Particles

Methanol, dichloromethane, and benzene-ethanol (4:1) were used as solvents for sonication to extract mutagens in the particles exhausted from a small diesel engine. Benzene-ethanol was the best eluant among those used for extraction of mutagenic substances (Table 1) and NPs (Table 2). Diesel exhausts from a small engine and a truck showed the highest mutagenicity for strain TA98 in the absence of S9 mix. They contained 60-70 ng of 1-NP and about 1 ng of 1,6-diNP per milligram of extract, accounting for 4 and 9%, respectively, of the total mutagenicity of the extracts.

Mutagenicity of a Kerosene Heater in Indoor Air

Kerosene heaters, convective and radiant, are widely used in Japanese residences because of their efficiency and low cost. Room air without a heater shows very low mutagenicity, 1.7 His^+ revertants/plate/m^3 of air from strain TA98 in the absence of S9 mix (Table 3). However, the samples from a room containing emissions at the beginning of burning for 20 min show very high mutagenicity, 237 His^+ revertants/plate/m^3, and a considerable amount of 1-NP and 1,6-diNP. In contrast, emissions from a heater burning stably show low mutagenicity,

Table 1. Effects of Extract Solvents on Mutagenicity of Diesel Exhaust Particles

Diesel Engine	Extract Solvent	Revertants/Plate/m^3 of Exhaust Gas			
		TA98		TA100	
		(−)S9	(+)S9	(−)S9	(+)S9
Small engine (269 ml)	Methanol	10,400	16,600	16,200	10,800
	Dichloromethane	71,200	26,500	39,500	14,500
	Benzene-ethanol	135,000	37,000	45,500	22,700
Truck (5,785 ml)	Benzene-ethanol	69,700	27,400	66,400	14,400

9 His$^+$ revertants/plate/m^3. These results suggest that lighting the heater creates mutagenic substances such as NPs.

Mutagenicity of Yakitori

Yakitori is a popular food in Japan. Pieces of raw chicken with or without a sauce were grilled above a city gas flame for 7 min and extracted by sonication with benzene-ethanol. The basic fraction of the chicken grilled without the sauce was more mutagenic than the other fractions for *Salmonella* strain TA98 (Table 4). This may be due to the presence of amino acid pyrolysates such as Trp-P-1 and Trp-P-2. However, when the chicken was grilled with the sauce, the basic fraction of the extract showed lower mutagenicity in strain TA98 in the presence of S9 mix than did the same fraction without sauce. The neutral fraction showed high mutagenicity in strain TA98 in the absence of S9 mix. The mutagenicity of the neutral fraction for strains TA98NR and TA98/1,8-DNP$_6$ without S9 mix was low, suggesting that the fraction contains 1-NP and diNPs. In fact, the neutral fraction of the chicken grilled for 3, 5, and 7 min contained 3.8, 19, and 43 ng of 1-NP/g of grilled chicken, respectively, accounting for 3.0, 2.7, and 1.3%, respectively, of the total extract mutagenicity. 1,6-Dinitropyrene was not detected in this sample. The neutral fraction of the chicken grilled without the sauce for 7 min contained only 1.4 ng of 1-NP per gram of grilled chicken.

Table 2. Mutagenicity and Nitropyrene Concentration of Diesel Particulate Extracts

	Revertants/mg Extract					Mutagenicity (%)	
Sample	TA98	TA98NR	TA98/ 1,8-DNP$_6$	1-NP (ng/mg extract)	1,6-diNP (ng/mg extract)	1-NP	1,6-diNP
Small engine							
Methanol	260	156	840	17.0	0.03	11.1	3.2
Dichloromethane	1,570	850	370	37.5	0.97	4.0	17.3
Benzene-ethanol	3,240	1,700	496	71.3	1.03	3.7	8.9
Truck							
Benzene-ethanol	2,720	928	362	61.5	0.81	3.8	8.4

Carcinogenicity of 1-NP and 1,6-diNP

To investigate the possible carcinogenicity of 1-NP and 1,6-diNP, a total dose of 2 mg of each compound was subcutaneously inoculated into BALB/c mice. No tumors developed at the injection site in any of the mice injected with 1-NP or DMSO. In the group of mice injected with 1,6-diNP, the first tumor was found on day 112 after the first inoculation. Table 5 shows that 1,6-diNP induced tumors at the injection site in 10 of 20 mice; the tumors were first observed on days 112, 126, 202, 204, 245, 252, 259, 284, 301, and 315. The difference in the incidence of tumors between the 1,6-diNP group and the 1-NP or control group was statistically significant according to the chi-square test ($p < 0.002$). The tumors induced at the injection site in all mice were transplantable successively through five generations. All the transplanted tumors showed basically the same histological features as the original tumors.

Macroscopically, the original tumors were nodular, gray-white and solid, and weighed approximately 2 to 18 g. Central necrosis was observed in most tumors and some showed deep ulceration. Histologically, all the tumors showed a combination of storiform and pleomorphic areas and were diagnosed as malignant fibrous histiocytomas. Two tumors had a fascicular growth pattern and resembled fibrosarcoma, except for the scattered giant cells.

Table 3. Mutagenicity and Nitropyrene Concentration of Indoor Particulates Exhausted from a Kerosene Heater

	Revertants/m^3 of Air[a]					Mutagenicity[b] (%)	
Sample[c]	TA98	TA98NR	TA98/1,8-DNP_6	1-NP (ng/m^3)	1,6-diNP (ng/m^3)	1-NP	1,6-diNP
A	1.7	1.2	1.1	n.d.[d]	n.d.	n.d.	n.d.
B	34.9	23.7	7.0	0.147	0.025	0.71	20.1
C	237	133	17.8	1.62	0.149	1.16	17.6
D	8.9	5.7	4.0	0.044	0.001	0.84	4.4

[a]Particles were extracted with benzene-ethanol (4:1).
[b]1-Nitropyrene and 1,6-dinitropyrene induce 1.69 and 280 His^+ revertants/plate/ng, respectively.
[c]Sample A was collected in a room without a heater with a change of the filter every 20 min. Sample B was continuously collected for 8 h from the time of lighting the heater. Sample C was collected 21 times for 20 min at the beginning of burning after intervals of 20 min of ventilation. Sample D was continuously collected for 7 h after 60 min of ventilation with burning just after lighting the heater.
[d]Not determined.

DISCUSSION

Polycyclic aromatic hydrocarbons (PAHs) are produced by incomplete combustion of fuel. Since nitro-PAH derivatives are easily formed by nitration of the PAH with NO_2 under the acidic condition (Pitts et al., 1978; Tokiwa et al., 1981a), NPs in diesel exhausts and indoor air pollutants might be artifacts produced during filter sampling of the particles. Schuetzle (1983) reported that the nitro-PAH formation during filter sampling of diesel particles collected on Teflon filters from diluted exhaust containing less than 3 ppm NO_2 can be minimized by short sampling times of less than 23 min and at sampling temperature less than 43°C. Under these conditions, Schuetzle (1983) estimates that the maximum conversion of pyrene to NP would be 10% of the total NP, and under normal dilution tube sampling of diesel particulate, less than 10-20% of the NP would be converted during sampling. The particles collected from the kerosene heater and the diesel truck in our experiments were collected under these conditions,

Table 4. Mutagenicity of Yakitori Grilled in a Gas Range

Chicken Grilled	Fraction	Revertants/Plate/g of Grilled Chicken			
		TA98 (−)S9	TA98 (+)S9	TA98NR	TA98/ 1,8-DNP$_6$
With sauce	Crude extract	5,610	3,260	3,480	2,820
	Neutral	3,100	650	916	798
	Acidic	89	32	51	51
	Basic	795	2,170	596	464
Without sauce	Crude extract	2,280	10,900	1,260	1,230
	Neutral	600	252	323	568
	Acidic	195	45	85	21
	Basic	629	6,880	539	449

except for the concentration of NO_2 in the case of the diesel truck. Since the concentration of NPs during the first 20 min after lighting the kerosene heater was very high, rooms should be ventilated by opening windows and using a ventilator just after lighting the kerosene heater.

The sauce of yakitori increased the direct-acting mutagenicity. The sauce alone grilled with glass wool in the gas range did not produce highly active direct-acting mutagens and 1-NP. An electric range did not form 1-NP when chicken was grilled (unpublished data). A low concentration of NO_2 may be formed by grilling the chicken, although NO_2 is not produced by the electric range itself. Chicken immersed in acetate buffers (pH 3, 5, and 7) was grilled, and the neutral fraction extracted from the chicken showed high mutagenicity but not the proportional concentration of 1-NP. These results suggest that the acidity of the sauce might be very important for the mutagenicity of the grilled chicken and probably NP formation.

Of the NPs, 1,6-diNP was found to be carcinogenic when a total dose of 2 mg of the compound was subcutaneously injected into BALB/c mice; no tumors at the injection site were observed after injection of 1-NP. A total dose of 2 mg of 4-nitroquinoline 1-oxide (4-NQO) induced tumors at the injection site in 15 of 20 BALB/c mice by day 260

Table 5. Tumor Incidence in Nitropyrene-Treated Mice

Compound	Total Dose (mg/mouse)	Number of Effective Mice[a]	Number of Mice with Subcutaneous Tumors (%)
DMSO (control)		20	0 (0)
1-Nitropyrene	2.0	20	0 (0)
1,6-Dinitropyrene	2.0	20	10 (50)[b]

[a]Effective mice are those that survived beyond day 112, when the first tumor was found in a mouse treated with 1,6-diNP.
[b]Significantly different from 1-NP-treated and control group at $p < 0.002$ (by chi-square test).

(unpublished data). These results indicate that carcinogenicity of 1,6-diNP seems to be weaker than that of 4-NQO in these systems. However, yellowish granules of 1,6-diNP were formed in the subcutaneous tissue at the site of injection in about half of the mice treated with the chemical, suggesting that the smaller amount of 1,6-diNP was actually active at the site.

ACKNOWLEDGMENTS

This work was supported in part by funds from the Nissan Science Foundation and by grants-in-aid for scientific and cancer research from the Ministry of Education, Science and Culture and the Ministry of Health and Welfare of Japan.

REFERENCES

Doll, R., and R. Peto. 1981. The causes of cancer: Quantitative estimates of avoidable risks of cancer in the United States today. J. Natl. Cancer Inst. 66:1191-1308.

Gibson, T.L., A.I. Ricci, and R.L. Williams. 1981. Measurement of polynuclear aromatic hydrocarbons and their reactivity in diesel automobile exhaust. In: Polynuclear Aromatic Hydrocarbons: Chemical Analysis and Biological Fate. M. Cooke and A.J. Dennis, eds. Battelle Press: Columbus, Ohio. pp. 707-717.

Kinouchi, T., and Y. Ohnishi. 1983. Purification and characterization of 1-nitropyrene nitroreductases from *Bacteroides fragilis*. Appl. Environ. Microbiol. 46:596-604.

Kinouchi, T., Y. Manabe, K. Wakisaka, and Y. Ohnishi. 1982. Biotransformation of 1-nitropyrene in intestinal anaerobic bacteria. Microbiol. Immunol. 26:993-1005.

Lee, F.S.-C., T.M. Harvey, T.J. Prater, M.C. Paputa, and D. Schuetzle. 1980. Chemical analysis of diesel particulate matter and an evaluation of artifact formation. In: Sampling and Analysis of Toxic Organics in the Atmosphere. ASTM STP 721. American Society for Testing and Materials. pp. 92-110.

Löfroth, G., E. Hefner, I. Alfheim, and M. Møller. 1980. Mutagenic activity in photocopies. Science 209:1037-1039.

Manabe, Y., T. Kinouchi, K. Wakisaka, I. Tahara, and Y. Ohnishi. 1984. Mutagenic 1-nitropyrene in wastewater from oil-water separating tanks of gasoline stations and in used crankcase oil. Environ. Mutagen. 6:669-681.

Ohnishi, Y., K. Kachi, K. Sato, I. Tahara, H. Takeyoshi, and H. Tokiwa. 1980. Detection of mutagenic activity in automobile exhaust. Mutat. Res. 77:229-240.

Ohnishi, Y., H. Okazaki, K. Wakisaka, T. Kinouchi, T. Kikuchi, and K. Furuya. 1982. Mutagenicity of particulates in small engine exhaust. Mutat. Res. 103:251-256.

Ohnishi, Y., T. Kinouchi, Y. Manabe, and K. Wakisaka. 1983. Environmental aromatic nitro compounds and their bacterial detoxification. In: Short-Term Bioassays in the Analysis of Complex Environmental Mixtures III. M.D. Waters, S.S. Sandhu, J. Lewtas, L. Claxton, N. Chernoff, and S. Nesnow, eds. Plenum Press: New York. pp. 527-539.

Pederson, T.C., and J.-S. Siak. 1981. The role of nitroaromatic compounds in the direct-acting mutagenicity of diesel particle extracts. J. Appl. Toxicol. 1:54-60.

Pitts, J.N., Jr., K.A. Van Cauwenberghe, D. Grosjean, J.P. Schmid, D.R. Fitz, W.L. Belser, Jr., G.B. Knudson, and P.M. Hynds. 1978. Atmospheric reactions of polycyclic aromatic hydrocarbons: Facile formation of mutagenic nitro derivatives. Science 202:515-519.

Rosenkranz, H.S., E.C. McCoy, D.R. Sanders, M. Butler, D.K. Kiriazides, and R. Mermelstein. 1980. Nitropyrene: Isolation, identification, and reduction of mutagenic impurities in carbon black and toners. Science 209:1039-1043.

Rosenkranz, H.S., E.C. McCoy, R. Mermelstein, and W.T. Speck. 1981. A cautionary note on the use of nitroreductase-deficient strains of *Salmonella typhimurium* for the detection of nitroarenes as mutagens in complex mixtures including diesel exhausts. Mutat. Res. 91:103-105.

Schuetzle, D. 1983. Sampling of vehicle emissions for chemical analysis and biological testing. Environ. Health Perspect. 47:65-80.

Schuetzle, D., F.S.-C. Lee, T.J. Prater, and S.B. Tejada. 1981. The identification of polynuclear aromatic hydrocarbon (PAH) derivatives in mutagenic fractions of diesel particulate extracts. Int. J. Environ. Anal. Chem. 9:93-144.

Tokiwa, H., R. Nakagawa, K. Morita, and Y. Ohnishi. 1981a. Mutagenicity of nitro derivatives induced by exposure of aromatic compounds to nitrogen dioxide. Mutat. Res. 85:195-205.

Tokiwa, H., R. Nakagawa, and Y. Ohnishi. 1981b. Mutagenic assay of aromatic nitro compounds with *Salmonella typhimurium*. Mutat. Res. 91:321-325.

Tokiwa, H., T. Otofuji, K. Horikawa, S. Kitamori, H. Otsuka, Y. Manabe, T. Kinouchi, and Y. Ohnishi. 1984. 1,6-Dinitropyrene: Mutagenicity in *Salmonella* and carcinogenicity in BALB/c mice. J. Natl. Cancer Inst. 76.

THE EFFECT OF ATMOSPHERIC TRANSFORMATION UPON THE BACTERIAL MUTAGENICITY OF AIRBORNE ORGANICS: THE EPA EXPERIENCE

L.D. Claxton,[1] L. Cupitt,[2] R.Kamens,[3] and L. Stockberger[2]

[1]Health Effects Research Laboratory, U.S. Environmental Protection Agency, Research Triangle Park, North Carolina 27711, [2]Atmospheric Sciences Research Laboratory, U.S. Environmental Protection Agency, Research Triangle Park, North Carolina 27711, and [3]School of Public Health, University of North Carolina, Chapel Hill, North Carolina 27514

INTRODUCTION

Air pollution obviously can exist in almost any defined environment. For example, cigarette smoking contributes to pollutant levels in homes, businesses, and occupational settings. Automotive exhaust has some measurable impact upon most outdoor areas. Fugitive emissions are associated with many industrial processes. In addition, most defined environments contain pollutants from a mixture of sources, and these sources contribute proportionately different amounts of organic and inorganic materials. To complicate the picture even further, these atmospheric pollutants interact in various ways depending upon the environmental conditions (e.g., temperature, light, humidity, etc.). Individuals, therefore, are exposed to a very large number of organic and inorganic pollutants, each of which is present in relatively minute quantities and many of which are secondary rather than primary pollutants. In spite of these rather universally accepted observations, most health research has examined the toxicology of source compounds. Examples of source compounds include: (1) compounds that are the end products or major by-products of an industrial process, such as trichloroethylene; (2) compounds that are readily identified just prior to and immediately after release into the atmosphere, such as benzo(a)pyrene associated with coke oven emissions; and (3) compounds that are in common use within the environment and are likely to exhibit some form of toxicity, such as pesticides. Since these source compounds are subject to chemical reactions within the atmosphere, most populations of interest may not be exposed to these source compounds. Instead, they may be exposed to secondary pollutants (i.e., reaction by-products) of these compounds. (Occupational populations are likely to be an exception.) This phenomenon of testing primarily source chemicals is to be expected and is the outcome of logical thinking. Since ambient air contains thousands of chemicals--each in sub-ppm concentration--one tends to test those that can be most easily identified or are known to exist in ambient air. A variety of ambient studies have been conducted during recent years in an effort to identify the airborne organic toxicants to which the population at large is exposed. Most of these studies have been reported or reviewed previously (Hughes et al., 1980; Tice et al., 1980; Waters et al., 1978, 1981, 1983). In order to bridge a knowledge gap between those source compounds known to be emitted into the atmosphere and those compounds found in ambient air samples, EPA has been involved in a number of atmospheric transformation studies that involve the monitoring of genotoxic activity. Due to the complexity of these studies, several groups have been a part of the collaborative efforts. The purpose of this paper, therefore, is to provide a brief review of EPA's primary efforts to study the genotoxicity of atmospheric transformation products.

ATMOSPHERIC TRANSFORMATION OF AUTOMOTIVE EXHAUST ORGANICS

The initial effort concerning the genotoxicity of atmospheric transformation products involved the use of exhausts from a diesel automobile (Claxton and Barnes, 1981). In this study, diesel exhaust from a 4-cylinder diesel automobile was injected into the Calspan smog chamber under different conditions. After a period of exposure, the chamber was exhausted and filterable products were collected using Pallflex glass fiber filters. Organic material was extracted from the filters using methylene chloride. The organic material was then solvent-exchanged into dimethyl sulfoxide and tested in the *Salmonella typhimurium* plate incorporation assay.

The Calspan chamber is a cylindrical stainless steel chamber coated with a fluoroepoxy-type urethane developed by the Naval Research Laboratory, Washington, DC. It is 9.14 m in diameter and 9.14 m in height. Illumination, when used, was provided by forty-eight 40-W sun lamps, one hundred ninety-two 85-W black lamps, and forty-eight 215-W specially produced black lamps. The effects of a series of parameters (including presence or absence of UV irradiation, length of chamber exposure, presence of O_3, presence or absence of diesel aerosols, and addition of a reactive hydrocarbon such as propylene) were examined.

The first baseline study involved the collecting and testing of aerosols that were generated by the reaction of propylene, NO, and NO_2 without diesel exhaust. The collected products of this reaction did produce a mutagenic response. This background response could account for 10 to 35% of the mutagenic response seen in subsequent experiments with diesel exhaust. Two other primary lessons learned from the Calspan study were: (1) dark conditions and UV light conditions did not alter aerosol mutagenicity without other mitigating factor(s), such as increased levels of O_3, and (2) photochemical reactions have the potential to both increase and decrease the mutagenicity of airborne organic material. Due to constraints that had to be placed upon these experiments, one must be cautioned not to over-interpret the data and results. Cost constraints limited the study to independent individual experiments. The $NO_x/SO_2/O_3$ ratios were quite variable and could not be precisely duplicated. Experiments were done using a single automobile with a diesel engine. Technical difficulties also prevented the collecting of confirmatory and/or supporting chemical information. This study, however, did pave the way for future work in the genotoxicity of atmospherically transformed products.

PHOTOACTIVATION OF SINGLE COMPOUNDS IN A CONTINUOUSLY STIRRED TANK REACTOR (CSTR)

As seen from the Calspan study, atmospheric reactions are capable of generating organic mutagens from a single low-molecular-weight hydrocarbon. In order to attempt to understand these photochemical reactions and to identify the mutagenic components generated, studies were designed using a continuously stirred tank reactor (CSTR). This reactor, originally referred to as the Mobile Aerosol Reaction Chamber (MARC), is a large, cylindrical, Teflon bag held between two aluminum plates that are coated with Teflon. The Teflon bag is 8 m long and 2 m in diameter. This flow-through system provides a steady-state stream of reaction products to designated bioassay chambers. A complete description of the design of the system is available (Parks, 1982). Typically, four bioassay chambers connected to the CSTR by heated conduits are used. Each chamber is 3 ft 1 in. x 1 ft 6 in. x 1 ft 6 in. and is also Teflon coated. As a control, bacteria in one chamber are exposed to air from a clean air source. Another chamber is used to expose bacteria to the reactants before entry into the reaction chamber. The unfiltered products of the reaction chamber enter a third bioassay chamber, and products of the reaction chamber enter the other bioassay chamber after filtration using a Pallflex Teflon-coated glass fiber filter. The configuration of the CSTR system is shown in Figure 1. Fifty or fewer plates can be placed into each chamber and exposed for up to 20 h. In each biochamber, therefore, the organisms were exposed to a continuous, steady-state stream of components or products.

Initial efforts with the CSTR focused upon the hydrocarbon propylene. Propylene, NO, NO_2, and SO_2 were injected into the reactor at a concentration of less than 2 ppm and photoactivated. The residence time of these reactants in the CSTR for the initial experiments was 6.7 h. The bacteria were exposed within the biochambers for 20 h. Table 1 provides the results for two replicate experiments using *Salmonella typhimurium* TA100. The spontaneous controls (done within the laboratory and not within a biochamber) gave results within the normal range for this strain. After exposure to clean air or reactants before entry into the CSTR, the test plates also

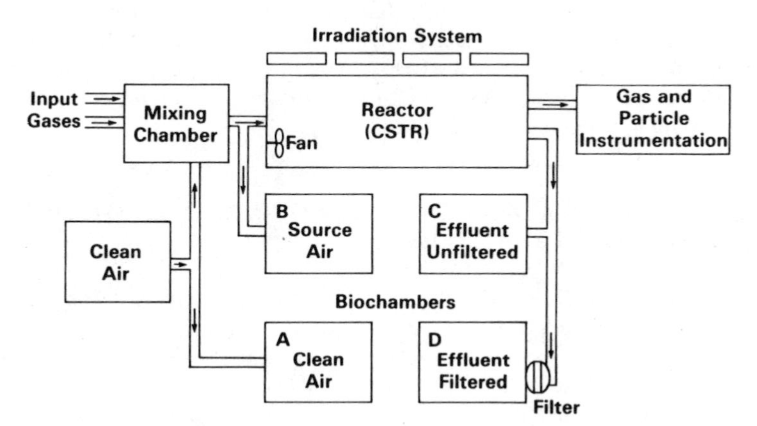

Figure 1. Schematic of the continuously stirred tank reactor (CSTR) bioassay system.

Table 1. *Salmonella* Mutagenicity Results for Continuously Stirred Tank Reactor (CSTR) Experiments Using Propylene

		Mean Number of Revertant Colonies per Plate per Experiment[b]				
Date	τ^a	Spontaneous	Clean Air	Reactants Before CSTR	Unfiltered	Filtered
5/24	6.7	150	147	142	551	581
6/08	6.7	122	171	174	485	572
Mean	6.7	136	159	158	518	577

[a]Mean residence time (h).
[b]Mean of 10 plates.

exhibited counts within the normal spontaneous range. Exposure to both the unfiltered and filtered products of the reaction chamber caused approximately a 4-fold increase in plate counts. Once again, as in the Calspan study, mutagenic products are the result of photochemical reactions involving a single low-molecular-weight hydrocarbon. Efforts have now begun seeking to identify these generated mutagenic products.

Once again--even with these more highly controlled experiments--one must be careful not to over-interpret the data. There are primarily three reasons for this caution. First, only a single condition can be tested during each exposure period. Secondly, artifacts may be formed at the air-agar interface of the bacterial plates. Lastly, the most active mutagens tend to contribute a minor proportion of the total organic mass and are the most difficult to identify.

ATMOSPHERIC TRANSFORMATION OF WOODSMOKE EMISSIONS USING DUAL OUTDOOR SMOG CHAMBERS

The University of North Carolina outdoor smog chambers located in Pittsboro, NC, have been used to study the effect of atmospheric interaction upon the mutagenicity of woodsmoke emission organics. Recent studies have been described in detail (Kamens et al., 1984). Dual outdoor chambers made of Teflon film were used for these studies. Each side of the octagonal base in 45 in. long. The walls on each side are

10 ft high, and the center of the coned roof is 15 ft high. All lighting is by sunlight. For generation of woodsmoke emissions, a commercially available Buck stove was used.

In these studies, the woodsmoke (1300-8000 $\mu g/m^3$) is allowed either to react with sub-ppm levels of NO_2, $NO_2 + O_3$, or O_3 or to age by itself. These experiments are done in the presence and/or absence of sunlight. Bacterial mutagenicity as indicated by the Ames assay (*S. typhimurium* TA98) did not increase after aging the dilute woodsmoke in the presence of low levels of combustion-generated NO_2. The addition of 0.3- to 0.9-ppm mixtures of $O_3 + NO_2$, however, did increase the mutagenicity of the particle extracts. A demonstratable increase in mutagenicity after $NO_2 + O_3$ injection occurs very rapidly, usually within 15 min. Although additions of NO_2 (above those of combustion-generated NO_2) alone increased the direct mutagenic activity of woodsmoke, the increase was not as great as from combined $NO_2 + O_3$ systems. Preliminary indications are that peat combustion gives results similar to woodsmoke. These types of simulated atmospheric studies do not precisely mimic real atmospheres; however, they do provide an understanding of atmospheric processes and the effect of these processes upon the mutagenicity of complex organic emissions.

CONCLUSIONS

Carcinogens and mutagens have been identified within ambient air samples (Hughes et al., 1980). Many of these toxic compounds, however, are the result of atmospheric reactions; therefore, ambient levels of toxicants do not directly correlate with the level of source compounds that are emitted into the ambient air (Yocom, 1982). The chamber studies presented demonstrate that atmospheric processes can both increase and decrease the mutagenicity of emitted organic compounds. The level of O_3 and NO_x gases, as well as light intensity, humidity, etc., appear to be very important in these processes. These studies also demonstrate that mutagenic products produced by ambient-like levels of pollutants can be monitored using bacterial bioassays.

REFERENCES

Claxton, L.D., and H.M. Barnes. 1981. The mutagenicity of diesel-exhaust particle extracts collected under smog-chamber conditions using the Salmonella typhimurium test system. Mutat. Res. 88:255-272.

Hughes, T.J., E. Pellizzari, L. Little, C. Sparacino, and A. Kolber. 1980. Ambient air pollutants: Collection, chemical characterization and mutagenicity testing. Mutat. Res. 76:51-83.

Kamens, R.M., G.D. Rives, J.M. Perry, D.A. Bell, R.F. Paylor, Jr., R.G. Goodman, and L.D. Claxton. 1984. Mutagenic changes in dilute wood smoke as it ages and reacts with ozone and nitrogen dioxide: An outdoor chamber study. Environ. Sci. Technol. 18(7):523-530.

Parks, R.M. 1982. HERL Biological Exposure Chamber Conceptual Design: Technical Note. EPA-600/S2-81-234, U.S. Environmental Protection Agency: Research Triangle Park, NC.

Tice, R.R., D.L. Costa, and K.M. Schaich. 1980. Genotoxic Effects of Airborne Agents. Vol. 25, Environmental Science Research, Plenum Press: New York. 658 pp.

Waters, M.D., S. Nesnow, J.L. Huisingh, S.S. Sandhu, and L. Claxton, eds. 1978. Application of Short-Term Bioassays in the Fractionation and Analysis of Complex Environmental Mixtures. Vol. 15, Environmental Science Research, Plenum Press: New York. 587 pp.

Waters, M.D., S.S. Sandhu, J.L. Huisingh, L. Claxton, and S. Nesnow, eds. 1981. Short-Term Bioassays in the Analysis of Complex Environmental Mixtures II. Vol. 22, Environmental Science Research, Plenum Press: New York. 524 pp.

Waters, M.D., S.S. Sandhu, J. Lewtas, L. Claxton, N. Chernoff, and S. Nesnow, eds. 1983. Short-Term Bioassays in the Analysis of Complex Environmental Mixtures III. Vol. 27, Environmental Science Research, Plenum Press: New York. 589 pp.

Yocom, J.E. 1982. Indoor-outdoor air quality relationships: A critical review. J. Air Pollut. Control Assoc. 32(5):500-520.

MUTAGENICITY STUDIES OF NEW JERSEY AMBIENT AIR PARTICULATE EXTRACTS

Thomas B. Atherholt,[1] Gerard J. McGarrity,[1] Judith B. Louis,[2] Leslie J. McGeorge,[2] Paul J. Lioy,[3] Joan M. Daisey,[3] Arthur Greenberg,[4] and Faye Darack[4]

[1]Department of Microbiology, Institute for Medical Research, Camden, New Jersey 08103, [2]Office of Science and Research, Department of Environmental Protection, Trenton, New Jersey 08625, [3]Institute of Environmental Medicine, New York University Medical Center, Tuxedo, New York 10987, and [4]Department of Chemical Engineering and Chemistry, New Jersey Institute of Technology, Newark, New Jersey 07102

INTRODUCTION

Mutagenic particulate matter has been detected in ambient air at many locations throughout the world (e.g., Hughes et al., 1980; Chrisp and Fisher, 1980; Tokiwa et al., 1983). Two studies identified locations with nondetectable levels of mutagenic material in the aerosol. These nonmutagenic sites were Hummelfjell, Norway (Alfheim and Møller, 1979) and Camp Paivika in the San Bernardino mountains of California (Pitts et al., 1977). Extracts of ambient air have been shown to be carcinogenic by injection into laboratory animals (Epstein et al., 1979). In addition, correlations have been made between the high levels of mutagenic activity of air particulate matter extracts and elevated lung cancer mortality rates (Walker et al., 1982). However, the actual risk that ambient levels of mutagenic particulate matter pose to human health is unknown. Epidemiological evidence indicates that only when levels of ambient particulate matter and other pollutants (e.g., sulfur oxides) are significantly elevated can acute detrimental health effects be quantitated (Holland et al., 1979). On the other hand, adverse health effects due to lower concentrations of air pollutants cannot be discounted due to the inherent insensitivity of epidemiological studies. Furthermore, recent studies have shown that extracts of diesel engine exhaust appear to contain comutagens in addition to direct- and indirect-acting mutagens (Li and Royer, 1982). Therefore, airborne pollutants such as those from wood and fossil fuel combustion, in addition to having mutagenic and/or carcinogenic properties, may act synergistically with mutagens and carcinogens to which people are exposed in other ways (e.g., eating and drinking).

New Jersey is a highly urban and industrialized state with high overall and lung cancer mortality rates compared to nationwide averages (Greenberg et al., 1981). The present report presents mutagenicity data from the Airborne Toxic Elements and Organic Substances (ATEOS) project (Lioy et al., 1983). The ATEOS project was a comprehensive investigation of ambient air at four New Jersey sites, focusing on substances that pose potential risks to human health. Parameters studied included levels of inhalable particulate matter (IPM) and the bacterial mutagenic activity of extractable organic matter (EOM) associated with IPM. Other parameters studied included levels of trace elements, volatile organic compounds, and polycyclic aromatic hydrocarbons (PAH). Sample locations included three urban sites and one rural site. Samples were collected for two 6-wk periods during the summer and winter of two successive years. By analyzing the levels of mutagenic activity over extended periods of time, "background" or "ambient" levels of mutagenicity at these sites were established. These results provide a data base that will serve as a reference point in future New Jersey studies.

MATERIALS AND METHODS

Sampling Procedures

Twenty-four hour samples of IPM ($d_{50} \leq 15\ \mu m$) were collected from 10 a.m. to 10 a.m. 7 d/wk for 6-wk periods in the cities of Newark, Elizabeth, and Camden, New Jersey and at a rural site, Ringwood State Park, near the New Jersey-New York border.

Four campaigns were conducted for the ATEOS project: two summer campaigns (July 6-August 14 of 1981 and 1982) and two winter campaigns (January 18-February 26, 1982, and January 17-February 25, 1983). This report summarizes data from both summer campaigns and the first of the two winter campaigns. Data analyses from the second winter campaign are not yet complete.

A size selective inlet Hi-Vol sampler (Andersen Model 7000, General Metal Works, Cleveland, OH) operated at 40 ft^3/min was used at each site. Inhalable particulate matter was collected on pre-ignited glass-fiber filters (Type AE, Gelman Instrument Co., Ann Arbor, MI). In addition, fine particulate matter (FP; $d_{50} \leq 2.5\ \mu m$) was also collected from these sites using Andersen samplers; each was equipped with an Aerotec #2 cyclone preselective inlet (25 ft^3/min). Details of the four sites and of the site selection process can be found elsewhere (Lioy et al., 1983).

Sample Processing

The filters containing particulate matter were stored in the dark in freezers or over dry ice during transport to the extraction laboratory and were then sequentially extracted in a Soxhlet extractor using cyclohexane (CX; 8 h), dichloromethane (DCM; 8 h), and acetone (ACE; 8 h). Blank control filters were also extracted using the same procedure. All filters were extracted within 4 d of collection. Details of the extraction procedure are described elsewhere (Daisey et al., 1979). Composite samples for each 39-d collection period ("whole period") were prepared for each site and each fraction to obtain sufficient extract for mutagenic analysis. Aliquots of extracts representing equivalent volumes of air for each day (each three days at Ringwood) were combined.

In addition, during the summer 1982 campaign, daily samples were combined according to a photochemical smog or "espisode" (EP) period (July 13-19) and the remaining "nonepisode" (NE) period (July 6-12 and July 20-August 14, 1982). The criteria used to choose the EP period were elevated concentrations of IPM and FP, sulfate, ozone and EOM, and meteorology.

Gravimetric analysis was performed on each composite using a Cahn electrobalance (Model 25, Cahn Instruments, Cerritos, CA) to determine the amount of nonvolatile organic material in each extract (Daisey et al., 1979). The CX and DCM composite samples were solvent exchanged to ACE using a small Kuderna-Danish evaporative concentrator with prepurified and filtered nitrogen gas ebullation (K-720002, K-569300, and K-569401, Kontes, Inc., Vineland, NJ). The nitrogen gas was introduced into each sample through glass capillary tubes.

Toxicity/Dose-Finding Assay

The toxicity test consisted of a standard Ames plate incorporation mutagenic assay (Maron and Ames, 1983). A small portion of each extract (0.15 ml) was serially diluted with ACE. To conserve extract, one plate per dose was used. *Salmonella* strain TA1537 without S9 metabolic activation was used for this test in the summer 1981 and winter 1982 campaign and strain TA98 without S9 was used in the summer 1982 campaign. Only selected extracts from the summer 1982 campaign were tested because of toxicity data available from the summer 1981 campaign. Toxicity was determined by visual and dissecting microscopic observations of the background lawn of cell growth on each plate. A reduction or absence of this lawn was evidence of sample toxicity. The results of toxicity tests were used to determine appropriate dose levels for the mutagenicity assay.

Mutagenicity Testing

Samples, in ACE, were reduced to a small volume (approximately 0.5 ml) in the small Kuderna-Danish evaporator described previously. Various amounts of sterile spectrophotometric-grade dimethylsulfoxide (DMSO; Schwarz/Mann, Orangeburg, NY) were added to each sample to achieve a predetermined solubles concentration. In this way, the organic extracts used in the Ames assay were contained in a solvent mixture composed primarily of DMSO and a small amount of ACE.

The standard Ames plate incorporation assay (Maron and Ames, 1983) was conducted using strains TA98, TA100, TA1535, TA1537, and TA1538 in the first two campaigns. For the summer 1982 campaign, strain TA97 was added (Levin et al., 1982) and strains TA1535 and TA1538 were omitted. Strain TA1535 was not used in the later campaigns because none of the sample extracts were mutagenic to this strain. Strain TA1538 was omitted because it provided no additional information beyond that provided by TA98. Most of the recommendations of de Serres and Shelby (1979) were also incorporated. Testing was conducted with and without S9 metabolic activation constituents in the form of Aroclor 1254-induced liver homogenate from male Nichols-Wistar rats (25 μl liver homogenate per plate [per 0.5 ml of S9 mix]). Mutagen controls (mean of two plates) were as follows: with metabolic activation, 2-aminoanthracene (all strains, 1 μg/plate); without metabolic activation, 2-nitrofluorene (TA98, TA1538, 50 μg/plate), N-methyl-N'-nitro-N-nitrosoguanidine (TA100, TA1535, 10 μg/plate), 9-aminoacridine (TA1537, 50 μg/plate, summer 1981 and winter 1982 campaigns), and ICR-191 (TA97 and TA1537, 1 μg/plate, summer 1982 campaign). Other assay controls (with and without S9) consisted of "blank" or spontaneous reversion rate controls (no sample or carrier solvent; mean of three plates); carrier solvent controls consisting of (a) DMSO (mean of six plates, summer 1981; mean of eight plates, winter 1982 campaign) or (b) DMSO plus a percentage of ACE equal to the highest sample ACE percentage (mean of three plates) and DMSO plus a percentage of ACE equal to the lowest sample percentage (mean of three plates) for the summer 1982 campaign samples. The viable count of each strain inoculum (dilution plating on nutrient agar plates) was also determined (mean of three plates). All plates were incubated for 48 h at 37°C. Bacterial colonies were counted on an automated colony counter (BioTran II, New Brunswick Scientific Co., Edison, NJ). To minimize intertest variation, all assays with a given strain for each solvent fraction (summer 1982) or all fractions (summer 1981 and winter 1982) were performed on the same day.

All urban sample extracts were tested at five or more doses in the nontoxic portion of the dose-response curve. Due to small sample quantitites, the rural site extract and filter blank extract were sometimes tested at only four dose levels. The doses of the filter blank extracts were chosen by using the same or a similar dilution scheme that was used to prepare the urban site fraction doses. All sample doses were assayed in duplicate. Due to small sample quantities, the rural sample extract from the summer 1982 campaign was tested only with strains TA98 and TA100 and the EP and NE extracts were tested only with TA98.

Data Analysis

Definition of mutagenicity was by standard criteria: (1) an increase of twofold or more in the background reversion rate for any strain, and (2) a dose response over at least three consecutive doses. The linear portion of each dose-response curve was analyzed using linear regression analysis to obtain a slope that indicated the number of revertant colonies formed per microgram of extract tested. Sample extract masses were not corrected for blank extract masses. Blank extract masses were negligible. The square of the correlation coefficient (r^2) and the 95% confidence intervals were also determined for each dose-response curve. Using the known volume of air filtered to obtain each composite extract, the amount of EOM (in micrograms) per cubic meter of air in each extract was calculated. The number of revertant colonies per microgram was then transformed to the number of revertant colonies per cubic meter of air using the following equation: $\text{rev}/\mu\text{g} \times \mu\text{g}/\text{m}^3 = \text{rev}/\text{m}^3$.

RESULTS

Sites to Site Comparisons

Data from each of the three solvent fractions from the four sites during the three campaigns indicated that in almost every case the three urban sites contained more mutagenic activity than the rural Ringwood site. This was true whether or not the data were analyzed in terms of revertants per microgram of extracted organic matter or revertants per cubic meter of air sampled. The data from the summer 1982 composite extracts tested with *Salmonella* strain TA98 are shown in Table 1. Data for the CX, DCM, and ACE fractions of each sample have been summed for only one purpose: to discern general trends among the various sites. We are aware, and examples have been documented (Hass et al., 1981; Hermann, 1981; Alfheim et al., 1984), that mutagenic as well as nonmutagenic chemicals can exert synergistic and/or antagonistic effects on the levels of activity of other mutagenic chemicals when such compounds are brought together as mixtures. For this reason, the actual number of revertants experimentally determined in a combined mixture of CX, DCM, and ACE fractions may not, in fact, be equal to the arithmetic sum of the mutagenic activities of the separate fractions (testing of combined fractions was not done in this study). The synergism or antagonism in the extracts from each of the various sites may be different due to differences in chemical composition.

The intersite patterns observed with the other *Salmonella* strains for the summer 1982 samples and with sample extracts from previous campaigns were for the most part similar to those observed with TA98. That is, of the three urban sites, Newark usually contained the most mutagenic activity and Camden usually contained the least mutagenic activity. Some solvent fraction extracts from Newark and Elizabeth (and less often all three urban sites) were statistically equivalent in terms of the number of revertants per microgram of extract. With other fraction extracts, Elizabeth and Camden had equivalent levels of activity, but these were less than that of Newark. During both summers, the mutagenic potency (revertants per microgram) of the CX and ACE extracts with TA98 in the absence of S9 metabolic activation of the rural Ringwood site was similar to that of some of the urban sites. This was not true for TA100 for the extracts from either of the summer campaigns or for either TA98 or TA100 for the extracts from the winter 1982 campaign.

Comparisons of the Summer 1981 and 1982 Campaigns

Table 2 shows the amount of EOM (in micrograms per cubic meter of air) for each solvent fraction from the urban sites (averaged values) and the rural site. The table also shows mutagenicity test data from strain TA98 without S9 metabolic activation. The amount of EOM in each fraction from the rural site was fairly constant from one summer to the next. Likewise, the number of revertants per cubic meter of air in the three fractions was similar from year to year. At the urban sites, the EOM levels in the CX and ACE fractions were similar for both summers, but a significantly greater amount of material per cubic meter of air was extracted in the DCM fraction in 1982 than in 1981. Conversely, the mutagenic potency in this fraction (revertants per microgram of extract) was lower in 1982 than in 1981. Since more material was extracted in 1982 than in 1981, the resultant mutagenic activity per cubic meter for both years in the DCM fraction was similar. Although different amounts of total extractable material were found in the DCM fractions between 1982 and 1981, similar amounts of mutagenic material were extracted during both years. Mutagenic activity per microgram and per cubic meter was similar from one summer to the next in the CX and ACE fractions at the urban sites.

Table 1. Comparison of Mutagenic Activities of Cyclohexane, Dichloromethane, and Acetone Whole Period Fractions from the Four ATEOS Sites During the Summer 1982 Campaign -- Strain TA98

		Urban						Rural	
	Fraction	Newark		Elizabeth		Camden		Ringwood	
Rev/μg (−S9)[a]	CX	0.83	(0.12)[b]	0.48	(0.09)	0.44	(0.10)	0.48	(0.13)
	DCM	0.95	(0.10)	0.92	(0.21)	0.69	(0.10)	0.25	(0.04)
	ACE	0.58	(0.09)	0.37	(0.03)	0.26	(0.03)	0.20	(0.04)
	Sum	2.36		1.77		1.39		0.93	
Rev/μg (+S9)	CX	1.57	(0.23)	1.10	(0.14)	1.04	(0.15)	0.71	(0.11)
	DCM	1.38	(0.10)	1.02	(0.09)	0.67	(0.09)	0.36	(0.09)
	ACE	0.46	(0.05)	0.36	(0.03)	0.25	(0.04)	0.14	(0.02)
	Sum	3.41		2.48		1.96		1.21	
Rev/m^3 (−S9)[c]	CX	2.72	(0.38)	1.22	(0.23)	0.81	(0.17)	0.45	(0.12)
	DCM	1.80	(0.18)	1.39	(0.33)	1.39	(0.20)	0.24	(0.04)
	ACE	4.01	(0.64)	2.26	(0.20)	1.80	(0.23)	0.80	(0.15)
	Sum	8.53		4.87		4.00		1.49	

(continued)

Table 1. Continued

		Urban						Rural	
	Fraction	Newark		Elizabeth		Camden		Ringwood	
Rev/m^3 (+S9)	CX	5.12	(0.75)	2.80	(0.35)	1.88	(0.26)	0.66	(0.10)
	DCM	2.64	(0.18)	1.55	(0.14)	1.34	(0.28)	0.35	(0.09)
	ACE	3.17	(0.35)	2.21	(0.20)	1.69	(0.28)	0.55	(0.09)
	Sum	10.93		6.56		4.91		1.56	

[a]Number of revertant colonies formed per microgram of extract tested.
[b]Numbers in parentheses are 95% confidence intervals.
[c]Number of revertant colonies per cubic meter of air.

Table 2. Comparison of Extractable Organic Matter and Its Mutagenic Activity (TA98, −S9) in the Summer 1981, Summer 1982, and Winter 1982 Campaigns

		Summer 1981			Summer 1982			Winter 1982		
		EOM	Mutagenicity		EOM	Mutagenicity		EOM	Mutagenicity	
Site	Fraction	μg/m³	Rev/μg	Rev/m³	μg/m³	Rev/μg	Rev/m³	μg/m³	Rev/μg	Rev/m³
Urbana[a]	CX	2.59	0.45	1.17	2.54	0.58	1.58	5.60	0.83	4.53
	DCM	0.80	2.73	2.18	1.81	0.85	1.53	1.80	3.28	5.81
	ACE	6.29	0.43	2.70	6.59	0.40	2.69	10.20	0.77	7.93
	Sum	9.68		6.05	10.94		5.80	17.60		18.27

(continued)

Table 2. Continued

Site	Fraction	Summer 1981			Summer 1982			Winter 1982		
		EOM	Mutagenicity		EOM	Mutagenicity		EOM	Mutagenicity	
		μg/m^3	Rev/μg	Rev/m^3	μg/m^3	Rev/μg	Rev/m^3	μg/m^3	Rev/μg	Rev/m^3
Rural	CX	1.06	0.28	0.30	0.93	0.48	0.45	1.66	0.42	0.69
	DCM	0.90	0.61	0.55	0.99	0.25	0.24	1.17	1.61	1.88
	ACE	3.24	0.19	0.62	4.01	0.20	0.80	5.76	0.29	1.65
	Sum	5.20		1.47	5.93		1.49	8.59		4.22

[a]Mean of data from the Newark, Elizabeth, and Camden sites.

Mutagenic activity with S9 metabolic activation was generally less in the summer of 1982 than 1981 (both number of revertant colonies per microgram of extract tested and number of revertant colonies per cubic meter of air). This may reflect lot to lot differences in S9 activation capabilities (a different lot was used for each campaign). However, each lot of S9 was tested in a dose-response experiment with 2-aminoanthracene and benzo(a)pyrene to determine the optimal concentration to use in the mutagenicity assay. The optimal activating concentration of both lots used in the three campaigns was 25 µl liver homogenate per plate (per 0.5 ml S9 mix).

Three strains, TA98, TA100, and TA1537, were used in both the summer 1981 and 1982 campaigns. Year-to-year differences similar to those of TA98 were found with TA100 (data not shown). One difference was higher levels of revertant colonies per cubic meter of air in the ACE extracts in the summer of 1982 compared to 1981 (the criteria for a mutagenic response were not fulfilled for this strain in 1981). Strain TA1537 showed a pattern of year-to-year differences at the urban sites (the rural site extracts were not tested with this strain in 1982) similar to that of TA98, with one exception. More material mutagenic to TA1537 (per microgram and per cubic meter of air) was recovered in the CX fraction in 1982 compared to 1981. Mean ($\bar{X}$) data for TA1537 for the urban sites were as follows: the number of revertant colonies per microgram of extract tested without S9 metabolic activation was 0.07 for 1981 and 0.31 for 1982. The number of revertant colonies per cubic meter of air without S9 metabolic activation was 0.18 for 1981 and 0.84 for 1982.

Comparisons of the Summer and Winter Campaigns

Daily levels of IPM and FP were measured (in micrograms per cubic meter of air) and geometrically averaged (Gm) for each of the four sites during each campaign. The Gm values from the three urban sites were then averaged and the resultant values from the two summer campaigns were averaged and compared with those of the winter campaign. No seasonal differences were observed between the levels of IPM or FP at the urban or rural sites (values that follow are expressed in micrograms per cubic meter): urban summer values were IPM, 48.1; FP, 34.9; urban winter values were IPM, 45.0; FP, 35.2; rural summer values were IPM, 26.5; FP, 21.0; and rural winter values were IPM, 24.5; FP, 23.6.

In contrast to seasonal levels of IPM and FP, levels of EOM at the urban sites were 61 to 82% higher in winter than in summer (see Table 2). A similar difference was observed at the rural site. Published data (Lioy et al., 1983; Harkov et al., 1984) on summer 1981 and winter 1982 levels of selected PAH at the four sites indicate that geometric mean and peak values were from five to seven times higher in winter than summer. This is in agreement with observed levels of CX-extractable organic compounds during summer versus winter (PAH are detected primarily in the CX extract). Winter levels of CX extractables were 57 to 120% higher (depending on whether urban or rural site data were analyzed and whether summer 1981 or summer 1982 data were analyzed) than summer levels (Table 2). Acetone solubles were 44 to 78% greater in winter than summer. Because of the large difference in the levels of DCM extractables between the summer of 1981 and 1982 at the urban sites, wintertime DCM increases were either large (125%) or nonexistent. At the rural site, summertime

DCM levels from year to year were similar and were 18 to 30% lower than the winter 1982 level.

Strains TA98, TA100, and TA1537 were used for mutagenicity testing in the summer and winter campaigns. Table 2 includes mutagenicity data for TA98 without S9 metabolic activation from the winter 1982 campaign in addition to that from the two summer campaigns. There are more mutagens per unit volume of air during winter than during summer. This is true at the rural as well as the urban sites. A similar pattern was observed in the presence of metabolic activation (data not shown). Winter increases occurred in all three solvent fractions at the urban sites; winter levels were 2.7 to 2.9 times greater (depending on fraction) than summer levels. Winter levels also increased by similar amounts in the DCM and ACE fractions at the rural site, but wintertime increases in mutagenic activity in the CX fraction were smaller at the rural site than at the urban sites.

The pattern of seasonal differences for TA100 was similar to that of TA98 (data not shown). Levels of mutagenic activity (number of revertant colonies per cubic meter of air without S9 metabolic activation) were higher in winter than summer for all three solvent fractions at the urban sites. Levels of mutagenic activity at the rural site were not high enough above experimental control values during the summer or winter to make meaningful comparisons.

With TA1537 (number of revertant colonies per cubic meter of air without S9 metabolic activation), levels of activity were higher in winter than summer at the urban sites only for the ACE fractions (data not shown). Seasonal differences in DCM-soluble mutagens were small and differences in CX solubles were either high or nonexistent, depending on whether summer 1981 or summer 1982 data were compared with the winter data. Again, low levels of mutagenic activity at the rural site made seasonal comparisons with TA1537 difficult, although winter levels did appear to be higher than summer 1981 levels in all solvent fractions (criteria for a mutagenic response were not fulfilled for some of the summer 1981 urban extracts). Rural site extracts were not tested with TA1537 during the summer 1982 campaign.

Effect of Metabolic Activation on Mutagenic Activity

Data analyzed in this section are from the summer 1982 campaign. The effect of metabolic activation on each sample was quantified by finding the percent increase in mutagenic activity in the presence versus the absence of S9. The results of analysis of the urban site data are shown in Table 3. Metabolic activation increased the level of mutagenic activity in the CX samples by an average of 153%. Of the four strains, the greatest effect was observed with TA100 and the smallest was observed with TA98. With TA98 only, the addition of S9 to the CX fractions resulted in a much smaller effect on the mutagenic activity (smaller increase) in the winter samples than in the summer samples (data not shown).

Metabolic activation increased the mutagenic activity in most, but not all, of the urban DCM samples. The average increase for all strains was 51%. Again, S9 had the least effect with TA98; the effect in each of the other strains was comparable.

Table 3. Effect of Metabolic Activation on Urban Site Mutagenic Activity of the Whole Period Composite Extracts -- Summer 1982[a]

	Percent Increase (Decrease)[b,c]		
Strain	CX	DCM	ACE
TA98	118	18	(10)
TA100	179	60	5
TA97	161	62	1
TA1537	154	63	28
$\overline{X}$[d]	153	51	6

[a]Metabolic activation performed using Aroclor 1254-induced rat liver homogenate.
[b]Each urban site percentage =

$$\frac{\text{rev}/\mu\text{g}(+\text{S9}) - \text{rev}/\mu\text{g}(-\text{S9})}{\text{rev}/\mu\text{g}(-\text{S9})} \times 100$$

[c]Each number is a mean of the three urban site percentages.
[d]Mean of data from the four tester strains.

The addition of S9 resulted in essentially no increase in mutagenic activity (net increase of 6%) in the polar ACE sample extracts. There was a decrease in activity of 10% with TA98 and an increase of 28% with TA1537. No effect was found with TA100 or TA97.

Comparison of *Salmonella* Strain Responses to Individual Solvent Fractions

Data analyzed are from the summer 1982 campaign. For comparison purposes only, the mutagenicity data for the three solvent fractions for each tester strain were summed and the percent that each fraction contributed to the total mutagenic activity in the three fractions was computed (data from the three urban sites were averaged). The percentages obtained are shown in Table 4. There were clear-cut strain specificities in terms of sensitivity to nonpolar (CX), moderately polar (DCM), and polar (ACE) mutagens. Strains TA98 and TA100 appeared to be more sensitive than TA97 and TA1537 to polar mutagens in the air sample extracts; the reverse was true for nonpolar mutagens.

Table 4. Comparison of the *Salmonella* Tester Strain Response to Mutagens in the Solvent Fractions -- Summer 1982[a]

	Urban Sites[b]				Rural Site
Fraction	TA98	TA100	TA97	TA1537	TA98
	Rev/µg (−S9)				
CX	32	41	44	46	52
DCM	46	38	43	42	27
ACE	22	21	13	12	21
	Rev/µg (+S9)				
CX	47	54	58	57	59
DCM	39	34	35	35	30
ACE	14	12	7	8	11
	Rev/m³ (−S9)				
CX	27	34	41	45	30
DCM	26	22	28	27	16
ACE	46	44	31	27	54
	Rev/m³ (+S9)				
CX	44	50	59	57	42
DCM	25	22	24	24	22
ACE	32	29	17	19	35

[a]Numbers are percent values. The three fractions combined equal 100%.
[b]Numbers are mean ($\overline{X}$) of data from three urban sites.

Data from only one strain were available for the rural site. In terms of potency (revertant colonies per microgram of extract tested without S9), the rural site contained less moderately polar and polar mutagens compared to the nonpolar mutagens than did the urban sites (TA98). In terms of revertants per cubic meter of air without S9, the rural site contained less moderately polar mutagens relative to the other fractions compared to the urban site.

Comparison of Mutagenic Responses of TA97 and TA1537

Strain TA97 is a relatively new tester strain that contains the same type of mutation (+1 frameshift) that TA1537 contains, although at a different gene locus (Levin et al., 1982). In addition, TA97 contains a mutational "hotspot" (a run of CG's) next to the +1 frameshift mutated area. It also contains the pKM101 plasmid. Thus, TA97 and TA1537 have the same spectrum of mutational sensitivity, but TA97 is more sensitive than TA1537 due to the adjacent mutational hotspot and the plasmid (Levin et al., 1982; Maron and Ames, 1983). We included both strains in this campaign with the hope of comparing the qualitative and quantitative responses of each strain to the mutagenic effects of these complex mixtures.

Qualitative comparisons can be made using the data shown in Table 4. When the relative responses of each strain to each of the three solvent fractions (containing mutagens of different polarities) were compared, it appeared that strains TA97 and TA1537 had a similar spectrum of mutational sensitivities. Note the differences in percentages between these strains and TA98 and TA100.

Comparison of the two strains in terms of sensitivity to the mutagens in these samples was not possible with the data available. Mutagenic sensitivity is defined here as the lowest dose in a dose-response slope of a given compound or mixture in which an increase over background in the number of revertants per plate is first observed. The slope of a dose-response curve has no bearing on sensitivity when comparing one strain with another unless the strains in question have identical spontaneous reversion rates. This is not the case with TA97 and TA1537. Rather, it is the point of intercept of the dose-response curve with the horizontal spontaneous reversion rate line that would occur in the absence of any mutagens. To determine this point, one needs to test low, nonmutagenic sample doses in addition to mutagenic doses. Nonmutagenic doses were not assayed in any of the CX, DCM, or ACE samples. Quantitative comparisons of sensitivity to toxic components in the extracts were possible by observing the number of total doses that fell on the linear portion of the dose-response curve for each strain on identical samples. A smaller number with one strain versus the other indicates that a plateau in the dose-response curve was achieved at a lower dose (fewer, lower doses fell on the linear portion of the curve) and hence that strain was the more sensitive of the two to toxic components in a given sample. The data (not shown) indicated that the two strains had similar levels of sensitivity to toxic components in the CX fractions. However, TA97 was more sensitive than TA1537 to toxic components present in the DCM and ACE sample extracts.

Comparison of Episode versus Nonepisode Sample Mutagenic Activities

Daily variations in the concentrations of total EOM (CX + DCM + ACE) at the four sites are shown in Figure 1. There appeared to be several days fulfilling EP criteria included in the NE composite samples at each of the sites. Therefore, the results should be interpreted cautiously.

The geometric mean of daily values of various parameters (IPM, FP, sulfate, ozone, and EOM) was slightly higher during the EP than during the NE period (data not shown). On the other hand, concentrations of various PAH in the EP and the NE composite sample extracts were actually slightly lower in EP composite samples than in the NE composite samples (e.g., benzo(a)pyrene: NE = 0.16 ng/m^3; EP = 0.09 ng/m^3; data are averages of individual values from the three urban sites).

The mutagenicity testing results are shown in Table 5. In general, there was no difference in mutagenic activity between the EP and NE samples. The exceptions to this generality are noted below.

There appeared to be more mutagens per cubic meter of air for the three solvent fractions in the EP sample compared to the NE sample at the rural (Ringwood) site. The differences were not significant, however. Dichloromethane-soluble mutagens appeared higher on a revertants per microgram basis during the NE period compared to the EP period for the urban sites (the Camden site displayed more revertants per

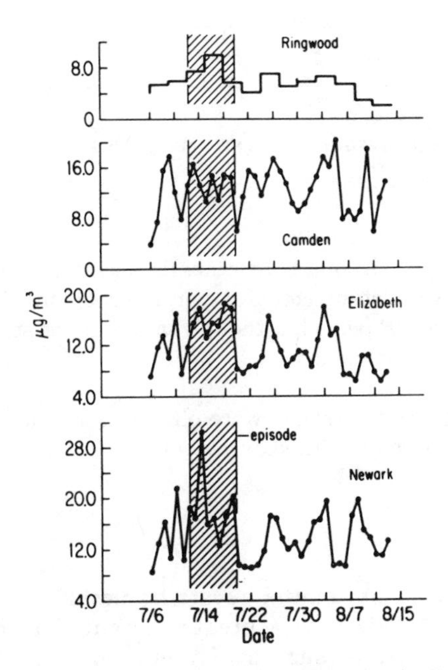

Figure 1. Variations in the concentrations of total EOM (CX + DCM + ACE) at four sites in New Jersey during the summer of 1982.

Table 5. Mutagenic Activity of Episode and Nonepisode Composite Samples -- Summer 1982[a]

		Rev/μg		Rev/m³	
Fraction	Sites	NE	EP	NE	EP
CX	Urban (Gm)[b]	0.65	0.55	1.49	1.33
	Rural	0.45	0.43	0.30	0.56
DCM	Urban (Gm)	1.31	0.80	1.74	1.30
	Rural	0.31	0.40	0.20	0.29
ACE	Urban (Gm)	0.52	0.31	2.80	2.26
	Rural	0.25	0.20	0.72	0.86

[a]Strain TA98 without S9 was used.
[b]Gm = geometric mean of data from three urban sites.

cubic meter of air as well without S9 metabolic activation). The urban sites also contained more potent (revertant colonies per microgram) ACE-soluble mutagens during the NE versus the EP period, although this was not statistically significant in all cases.

No statistically significant trends were observed for these two composite periods when the entire data base was analyzed as a whole.

DISCUSSION

Extracts of air particulate matter revert frameshift mutagen-sensitive tester strains (e.g., TA98 and TA1537) to a greater extent than base-pair substitution mutagen-sensitive tester strains such as TA100 and TA1535 (TA100 responds to frameshift mutagens as well). Strain TA98 was the most sensitive strain and this has been confirmed by others (e.g., Talcott and Wei, 1977; Tokiwa et al., 1980). Strain

TA1535 was not sensitive to mutagens present in any of the particulate extracts. This has also been observed by many other investigators (e.g., Dehnen et al., 1977; Madsen et al., 1982; Claxton, 1983). In fact, TA1535 is sensitive to a fairly small proportion of tested mutagens compared to some of the other tester strains (McMahon et al., 1979) and is no longer recommended for use in routine screening assays (Maron and Ames, 1983). From the lack of response of TA1535, one can infer what types of compounds are not present in these samples in significant amounts based on surveys of literature that list compounds that mutate this strain (e.g., de Flora, 1981; McCann et al., 1975).

Ambient air at the urban sites contained more mutagenic activity per cubic meter of air than the rural site. Similar urban effects on mutagenicity have also been noted by many investigators (e.g., de Raat, 1983; Tokiwa et al., 1980; Ohsawa et al., 1983; Pitts et al., 1977).

While levels of mutagenic activity at any given site were generally similar from the first summer to the next, substantial increases in activity were observed with every *Salmonella* tester strain for most fractions during the winter season. Explanations for these seasonal differences include one or more of the following: (1) the presence of wintertime mutagenic products resulting from the combustion of wood and fossil fuels for space heating (e.g., Daisey et al., 1980); (2) more rapid photochemical destruction of mutagens in the atmosphere during the summer; (3) increased volatilization of mutagens from the collection filters during the summer compared to winter (de Wiest et al., 1982; Chrisp and Fisher, 1980) and/or reduced deposition rates of vapor-phase mutagens onto airborne particulate matter during summer compared to winter due to higher summer temperatures; and (4) seasonal differences in atmospheric conditions. Seasonal differences in factors such as atmospheric inversion height levels may tend to increase wintertime levels of particulate matter. Atmospheric levels of Pb, caused mostly by auto emissions, should be fairly constant year round, but in fact, wintertime concentrations were higher than the summer (Lioy et al., 1983).

The CX fraction contains nonpolar compounds including PAH in addition to aliphatic hydrocarbons (Chrisp and Fisher, 1980; Daisey et al., 1979). Polycyclic aromatic hydrocarbons are promutagens in the Ames assay requiring metabolic activation for activity, and their presence would explain the effect of metabolic activation on mutagenic activity in the CX and possibly DCM fractions, although levels of PAH in DCM fractions are less than 10% of that found in CX fractions (unpublished data). Results of analysis of six PAH in the CX fractions from the summer and winter campaigns revealed that the winter samples contained about five to seven times the levels of these PAH as did the summer samples, but no corresponding increase in indirect mutagenic activity was observed in the winter samples (unpublished data). Several investigators have analyzed nonpolar solvent fractions of air extracts and concluded that unsubstituted PAH account for only a small portion of the total mutagenic activity (Dehnen et al., 1977; Talcott and Wei, 1977). Clearly, other biologically-active compounds were present in the CX fraction from both campaigns, but particularly the winter one, since most of the winter CX activity was of the direct-acting type. Nitrated PAH are possible candidates: (a) wintertime maxima of NO_2^- and NO_3^- in particles and of NO and NO_2 in the atmosphere have been observed (Fukino et al., 1982); (b) inorganic oxides of nitrogen have been shown by several investigators to react with PAH to create direct-acting mutagens (e.g., Pitts et al., 1978); and (c) nitro-PAH are responsible for a significant portion of the total mutagenicity of diesel particle extracts (Pederson and Siak, 1981; Salmeen et al., 1984) and extracts from other combustion sources (Kamens et al., 1984). In fact, nitrated PAH have recently been found in these CX extracts (J. Butler, NYU, personal communication).

Strain TA100 appeared to be the most sensitive of all the strains used to the effects of metabolic activation in the CX fraction; TA98 appeared to be the least sensitive of all the strains to these effects in the CX and DCM fractions.

Strains TA98 and TA100 appear to be more sensitive than TA97 and TA1537 to polar mutagens extracted from ambient air particulate matter; the reverse is true for nonpolar mutagens.

The mutagenic activity in the smog EP composite samples was no greater than that observed in the NE composite samples, although the compositing methodology used may have precluded a clear-cut separation between all EP and NE days. Therefore, the results should be interpreted cautiously.

Strains TA97 and TA1537 have a similar spectrum of mutational sensitivities to airborne mutagens, as expected. Quantitative comparisons in sensitivities could not be made with the data obtained in this study, but TA97 was more sensitive than TA1537 to toxic components in these air sample extracts.

ACKNOWLEDGMENTS

Our appreciation goes to the many laboratory technical personnel involved in this project. This project is supported by the Office of Science and Research, New Jersey Department of Environmental Protection.

REFERENCES

Alfheim, I., and M. Møller. 1979. Mutagenicity of long-range transported atmospheric aerosols. Sci. Total Environ. 13:275-278.

Alfheim, I., G. Becher, J.K. Hongslo, and T. Ramdahl. 1984. Mutagenicity testing of high performance liquid chromatography fractions from wood stove emission samples using a modified Salmonella assay requiring smaller sample volumes. Environ. Mutagen. 6:91-102.

Chrisp, C.E., and G.L. Fisher. 1980. Mutagenicity of airborne particles. Mutat. Res. 76:143-164.

Claxton, L.D. 1983. Characterization of automotive emissions by bacterial mutagenesis bioassay: a review. Environ. Mutagen. 5:609-631.

Daisey, J.M., M.A. Leyko, M.T. Kleinman, and E. Hoffman. 1979. The nature of the organic fraction of the New York City summer aerosol. Ann. N.Y. Acad. Sci. 322:125-142.

Daisey, J.M., T.J. Kneip, I. Hawryluk, and F. Mukai. 1980. Seasonal variations in the bacterial mutagenicity of airborne particulate organic matter in New York City. Environ. Sci. Technol. 14:1487-1490.

de Flora, S. 1981. Study of 106 organic and inorganic compounds in the Salmonella/microsome test. Carcinogenesis 2:283-298.

Dehnen, W., N. Pitz, and R. Tomingas. 1977. The mutagenicity of airborne particulate pollutants. Cancer Lett. 4:5-12.

de Raat, W.K. 1983. Genotoxicity of aerosol extracts: some methodological aspects and the contribution of urban and industrial locations. Mutat. Res. 116:47-63.

de Serres, F.J., and M.D. Shelby. 1979. The *Salmonella* mutagenicity assay: recommendations. Science 203:563-565.

de Wiest, F., D. Rondia, R. Gol-Winkler, and J. Gielen. 1982. Mutagenic activity of non-volatile organic matter associated with suspended matter in urban air. Mutat. Res. 104:201-207.

Epstein, S., K. Fuiji, and S. Asahina. 1979. Carcinogenicity of a composite organic extract of urban particulate atmospheric pollutants following subcutaneous injection in infant mice. Environ. Res. 19:163-176.

Fukino, H., S. Mimura, K. Inoue, and Y. Yamane. 1982. Mutagenicity of airborne particles. Mutat. Res. 102:237-247.

Greenberg, M., J. Caruana, J. Louis, and K. Steenland. 1981. Trends in cancer mortality in the New Jersey-New York-Philadelphia region 1950-1975. New Jersey Department of Environmental Protection: Trenton, NJ.

Harkov, R., A. Greenberg, F. Darack, J.M. Daisey, and P.J. Lioy. 1984. Summertime variations in polycyclic aromatic hydrocarbons at four sites in New Jersey. Environ. Sci. Technol. 18:287-291.

Hass, B.S., E.E. Brooks, K.E. Schumann, and S.S. Dornfeld. 1981. Synergistic, additive and antagonistic mutagenic responses to binary mixtures of benzo(a)pyrene and benzo(e)pyrene as detected by strains TA98 and TA100 in the Salmonella/microsome assay. Environ. Mutagen. 3:159-166.

Hermann, M. 1981. Synergistic effects of individual polycyclic aromatic hydrocarbons on the mutagenicity of their mixtures. Mutat. Res. 90:399-409.

Holland, W.W., A.E. Bennett, I.R. Cameron, C. du V. Florey, S.R. Leeder, R.S.F. Schilling, A.V. Swan, and R.E. Waller. 1979. Health effects of particulate pollution: reappraising the evidence. Am. J. Epidemiol. 110:533-659.

Hughes, T.J., E. Pellizzari, L. Little, C. Sparacino, and A. Kolber. 1980. Ambient air pollutants: collection, chemical characterization and mutagenicity testing. Mutat. Res. 76:51-83.

Kamens, R.M., G.D. Rives, J.M. Perry, D.A. Bell, R.F. Baylor, Jr., R.G. Goodman, and L.D. Claxton. 1984. Mutagenic changes in dilute wood smoke as it ages and reacts with ozone and nitrogen dioxide: an outdoor chamber study. Environ. Sci. Technol. 18:523-530.

Levin, D.E., E. Yamasaki, and B.N. Ames. 1982. A new Salmonella tester strain, TA97, for the detection of frameshift mutagens: a run of cytosines as a mutational hot-spot. Mutat. Res. 94:315-330.

Li, A.P., and R.E. Royer. 1982. Diesel-exhaust-particle extract enhancement of chemical-induced mutagenesis in cultured Chinese hamster ovary cells: possible interaction of diesel exhaust with environmental carcinogens. Mutat. Res. 103:349-355.

Lioy, P.J., J.M. Daisey, T. Atherholt, J. Bozzelli, F. Darack, R. Fisher, A. Greenberg, R. Harkov, B. Kebbekus, T.J. Kneip, J. Louis, G. McGarrity, L. McGeorge, and N.M. Reiss. 1983. The New Jersey project on airborne toxic elements and organic substances (ATEOS): a summary of the 1981 summer and 1982 winter studies. J. Air Pollut. Control. Assoc. 33:649-657.

Madsen, E.S., P.A. Nielsen, and J.C. Pedersen. 1982. The distribution and origin of mutagens in airborne particulates, detected by the Salmonella/microsome assay in relation to levels of lead, vanadium and PAH. Sci. Total Environ. 24:13-25.

Maron, D.M., and B.N. Ames. 1983. Revised methods for the Salmonella mutagenicity test. Mutat. Res. 113:173-215.

McCann, J., E. Choi, E. Yamasaki, and B.N. Ames. 1975. Detection of carcinogens as mutagens in the Salmonella/microsome test: assay of 300 chemicals. Proc. Natl. Acad. Sci. U.S.A. 72:5135-5139.

McMahon, R.E., J.C. Cline, and C.Z. Thompson. 1979. Assay of 855 test chemicals in ten tester strains using a new modification of the Ames test for bacterial mutagens. Cancer Res. 39:682-693.

Ohsawa, M., T. Ochi, and H. Hayashi. 1983. Mutagenicity in *Salmonella typhimurium* mutants of serum extracts from airborne particulates. Mutat. Res. 116:83-90.

Pederson, T.C., and J-S. Siak. 1981. The role of nitroaromatic compounds in the direct-acting mutagenicity of diesel particle extracts. J. Appl. Toxicol. 1:54-60.

Pitts, J.N., Jr., D. Grosjean, T.M. Mischke, V.F. Simmon, and D. Poole. 1977. Mutagenic activity of airborne particulate organic pollutants. Toxicol. Lett. 1:65-70.

Pitts, J.N., Jr., K.A. Van Cauwenberghe, D. Grosjean, J.P. Schmid, D.R. Fitz, W.L. Belser, Jr., G.B. Knudson, and P.M. Hynds. 1978. Atmospheric reactions of polycyclic aromatic hydrocarbons: facile formation of mutagenic nitro derivatives. Science 202:515-519.

Salmeen, I.T., A.M. Pero, R. Zator, D. Schuetzle, and T.L. Riley. 1984. Ames assay chromatograms and the identification of mutagens in diesel particle extracts. Environ. Sci. Technol. 18:375-382.

Talcott, R., and E. Wei. 1977. Airborne mutagens bioassayed in *Salmonella typhimurium*. J. Natl. Cancer Inst. 58:449-451.

Tokiwa, H., S. Kitamori, K. Takahashi, and Y. Ohnishi. 1980. Mutagenic and chemical assay of extracts of airborne particulates. Mutat. Res. 77:99-108.

Tokiwa, H., S. Kitamori, K. Horikawa, and R. Nakagawa. 1983. Some findings on mutagenicity in airborne particulate pollutants. Environ. Mutagen. 5:87-100.

Walker, R.D., T.H. Connor, E.J. MacDonald, N.M. Trieff, M.S. Legator, K.W. MacKenzie, Jr., and J.G. Dobbins. 1982. Correlation of mutagenic assessment of Houston air particulate extracts in relation to lung cancer mortality rates. Environ. Res. 28:303-312.

INTERURBAN VARIATIONS IN THE MUTAGENIC ACTIVITY OF THE AMBIENT AEROSOL AND THEIR RELATIONS TO FUEL USE PATTERNS

J.P. Butler, T.J. Kneip, F. Mukai, and J.M. Daisey

Institute of Environmental Medicine, New York University Medical Center, New York, New York 10016

INTRODUCTION

It has long been recognized that airborne pollutants may affect human health (National Research Council, 1978). The elevated incidence of respiratory disease (e.g., lung cancer) in urban areas has led to the investigation of airborne extractable organic matter (EOM) as a contributing factor (Task Group Report, 1978). Organic extracts of urban particulate matter have been shown to cause skin cancer when painted on mice (Heuper et al., 1962; Wynder and Hoffmann, 1965), and carcinogenic polycyclic aromatic hydrocarbons (PAHs), produced by the incomplete combustion of fossil fuels, have been identified in particulate organic matter extracts (Sawicki, 1967). In addition, short-term genetic bioassays have detected mutagenic compounds in EOM collected in a number of cities (Chrisp and Fisher, 1980; Hughes et al., 1980).

The major anthropogenic sources of EOM in urban areas are fuel combustion for space heating, power generation, and transportation; waste incineration; and industrial processes (Lamb et al., 1980; Simoneit and Mazurek, 1981). Depending upon the dominant combustion sources, the chemical composition of aerosols can differ among various cities (Daisey et al., 1983). Shifts in fuel use, including increased use of diesel-powered motor vehicles (National Academy of Sciences, 1981) plus the increasing utilization of coal and wood as a replacement for fuel oil, may also contribute to compositional differences of urban aerosols.

The purpose of the present study was to compare the bacterial mutagenicity of the ambient aerosol of five cities that differ in the nature of their combustion sources.

MATERIALS AND METHODS

Sample Collection

Particulate matter was collected in five cities during the past three years. Sampling periods and sample types are shown in Table 1, and a description of the sampling sites is presented in Table 2.

The particle sizes sampled ranged from fine particulate matter ($d_{50} \leq 2.5$ µm) to total suspended particulate matter ($d_{50} \leq 30$ µm). Although the samples differ in maximum particle size, this should have little effect on the mutagenicity of extractable organic matter concentrations, since most organic compounds in the ambient aerosol are in the fine particles (Van Vaeck and Van Cauwenberghe, 1978), particularly during winter (Daisey et al., 1982). In addition, mutagenic activity is detected almost entirely in this fraction (Talcott and Harger, 1980; Møller et al., 1982; Sorenson et al., 1982).

All of the samples were collected with high-volume samplers on pre-ignited Gelman AE glass fiber filters to reduce organic background. A size-selective inlet or

Table 1. Samples Collected and Composited for Testing in the Ames *Salmonella* Plate Incorporation Assay with Strain TA98

Site	Sampling Period	Sampling Duration	Type of Sample[a]	Particle Size (µm)
Philadelphia, PA	7/25-8/14/82	24 h	IPM	≤15
New York City	11/28/79-2/27/80	7 d	RSP	≤3.5
Elizabeth, NJ	1/18-2/22/82	24 h	FPM	≤2.5
Beijing, China	3/4-6, 4/12-14/81	8, 16, 24 h	TSP	~≤30
Mexico City	2/3-17/82	24 h	TSP	~≤30

[a]Definitions of abbreviations: FPM, fine particulate matter; IPM, inhalable particulate matter; RSP, respirable suspended particulate matter; TSP, total suspended particulate matter.

Table 2. Description of Sampling Sites

Site	Description
Philadelphia, PA	Samples were collected at the Fireboat Station, on the Delaware River, at a height of ~5 m. A residential area and Interstate Highway 95 are located several blocks to the west.
New York City	Samples were collected on the roof of the New York University Medical Center (60 m above the street), which is located in Manhattan between First Avenue and FDR Drive at 30th Street.
Beijing, China	Samples were collected at the Institute of Atmospheric Physics (~10 m above ground level), located at the northern edge of the city.
Mexico City	Samples were collected at the Universidad Autonoma Metropolitana-Azcapotzalco, at a height of ~7 m. The site is located to the north of the city and is surrounded by a number of industrial facilities and major traffic routes.
Elizabeth, NJ	Samples were collected on the roof of the two-story Board of Education Building. Oil refineries and the New Jersey Turnpike are located within 3 km of the site.

Aerotec No. 2 cyclone was used with the sampler to obtain the indicated 50% particle cut size. The samples were shipped (at ~0°C) to the New York University Laboratories and organic extractions were initiated within 24 h.

Sample Preparation

A Soxhlet apparatus was used to perform sequential extractions with 200 ml of cyclohexane (CX), followed by dichloromethane (DCM) and, finally, acetone (ACE) for 7.5 h with each solvent (Daisey et al., 1979). This procedure separates organic matter into nonpolar, moderately polar, and polar fractions, respectively. After the extracts were filtered, a rotary evaporator was used to reduce the volume to 10.0 ml. A Cahn Electrobalance was used to determine weights of duplicate 100-μl aliquots taken to dryness on a slide warmer at 40°C. The extracts were stored at –30°C. Samples were protected from light at all times.

Depending on the sampling period, extracts were stored for various lengths of time until mutagenicity testing was performed. Huisingh et al. (1978) saw only slight decreases in the mutagenicity of diesel extracts stored at refrigerator temperatures for 2 mo, although toxicity may have increased. The samples in the present study were stored at –30°C and there was little evidence of toxicity at the maximum doses used. Experiments in this laboratory have shown that in general, the mutagenicity per microgram of extracts that were stored at –30°C and retested 6-15 mo later was within ±40% of the original value.

Composites of extracts were prepared for mutagenicity testing by combining aliquots that represented equivalent volumes of sampled air. The composite extracts were then evaporated just to dryness under argon and redissolved in acetone. Sample concentrations were in the range of 2-4 mg/ml. For PAH analyses, aliquots of the cyclohexane-soluble fractions were taken to dryness, redissolved in acetonitrile, briefly sonicated, and then analyzed by high-pressure liquid chromatography (HPLC).

Mutagenicity Testing

The Ames *Salmonella* plate incorporation assay was performed as previously described (Ames et al., 1975; Maron and Ames, 1983), with the modification of adding histidine and biotin to the bottom rather than the top agar. The use of multiple tester strains was not possible due to the limited sample sizes; therefore, the extracts were tested with *Salmonella typhimurium* strain TA98, usually the most sensitive strain to airborne mutagens (McGarrity et al., 1982; Alink et al., 1983). Aroclor-1254-induced rat liver S9 fraction (Litton Bionetics) supplemented with the appropriate cofactors was used for metabolic activation. Most extracts were tested at five concentrations (15-664 μg/plate), in duplicate, with and without activation (50 μl S9/plate). Revertant colonies were counted after a 48-h incubation at 37°C. All extracts from all cities were tested first without S9; all were then retested 5 wk later with S9.

2-Nitrofluorene (5 μg/plate) was used as a positive control without metabolic activation (752 rev/plate), and 2-aminoanthracene (2.5 μg/plate) was tested in the presence of S9 (>2500 rev/plate). Composite extracts of filter blanks were also tested as controls.

Least squares analysis was used to calculate the slope of the linear portion of dose-response curves, including the 0 dose. Results are expressed as net revertant colonies per microgram of extract and per cubic meter of sampled air. The dose-response curves were highly linear, in general. Occasionally, the higher doses produced toxicity as seen by a leveling off of the curve as well as a slight reduction in the background bacterial lawn under microscopic examination. All extracts (except filter blanks) exhibited mutagenicity as defined by at least a 2-fold increase in reversion frequency and a dose-response relationship for at least three of the five doses tested.

Chemical Analyses: PAHs and Alkylating Agents

The cyclohexane-soluble fractions were analyzed for PAHs by HPLC as previously described by Daisey et al. (1983). The following PAH compounds were quantitated:

fluoranthene, pyrene, benzo(e)pyrene, benzo(k)fluoranthene, benz(a)anthracene, benzo(j)fluoranthene, benzo(b)fluoranthene, benzo(a)pyrene, benzo(ghi)perylene, and indeno(1,2,3-cd)pyrene.

Alkylating (and acylating) agents were detected in the ACE fraction by reaction with 4-(p-nitrobenzyl)pyridine (Agarwal et al., 1979). Alkylating agents include classes such as epoxides, lactones, and alkyl sulfates, many of which are known to be direct-acting mutagens (Hoffmann, 1980). Concentrations of alkylating agents were determined as equivalent micrograms of the positive control, styrene epoxide, per cubic meter.

RESULTS

The atmospheric concentrations of the three EOM fractions are presented in Table 3. Total EOM (CX + DCM + ACE) concentrations for the five cities, compared in Figure 1, ranged from 11.6 μg/m^3 for New York City to 39.1 μg/m^3 for Mexico City. The EOM concentrations were highest in Mexico City and Beijing, China. In general, the DCM fraction was the smallest percentage of total EOM while the ACE was the largest. Beijing was the exception, having nonpolar (CX) organic material constituting the largest proportion of EOM (55%).

Table 4 summarizes the mutagenic activities of the three EOM fractions in units of revertants per cubic meter and revertants per microgram. The sum of the mutagenic activities per cubic meter of the three fractions from each city are compared in Figure 2. Without S9, the highest activity per cubic meter, 22.1, was observed for

Table 3. Extractable Organic Matter Fraction Concentrations For All Sites

	EOM Concentration (μg/m^3)			
Site	CX	DCM	ACE	Total
Philadelphia, PA	2.7	2.1	8.1	12.9
New York City	3.5	1.7	6.4	11.6
Beijing, China	16.7	3.4	10.2	30.3
Mexico City	14.0	3.4	21.7	39.1
Elizabeth, NJ	7.5	1.9	8.4	17.8

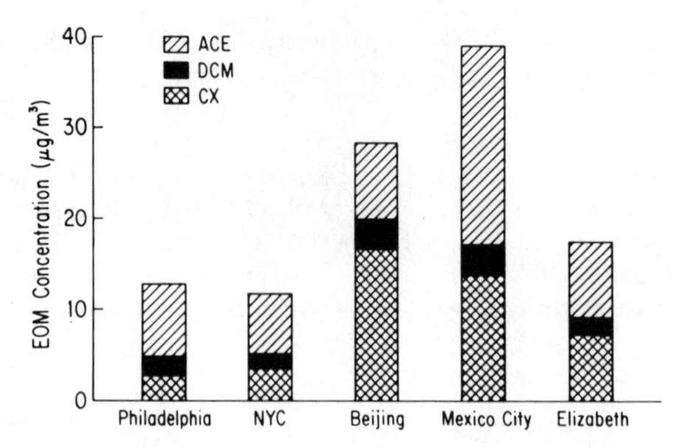

Figure 1. Extractable organic matter fraction concentrations as a proportion of the total at each site.

Table 4. Summary of Mutagenic Activity with Strain TA98

Site	Fraction	Net Revertants/μg		Net Revertants/m^3	
		Without S9	With S9	Without S9	With S9
Philadelphia, PA	CX	0.28	1.64	0.8	4.4
	DCM	0.52	1.15	1.1	2.4
	ACE	0.13	0.21	1.1	1.7
New York City	CX	0.98	2.00	3.4	7.0
	DCM	1.07	2.78	1.8	4.7
	ACE	0.74	0.55	4.7	3.5
Beijing, China	CX	0.19	2.96	3.2	49.4
	DCM	0.66	-[a]	2.2	-
	ACE	0.37	1.04	3.8	10.6
Mexico City	CX	0.67	2.34	9.4	32.8
	DCM	1.93	6.69	6.6	22.7
	ACE	0.28	0.22	6.1	4.8
Elizabeth, NJ	CX	0.42	1.38	3.2	10.4
	DCM	2.46	4.55	4.7	8.6
	ACE	0.26	0.36	2.2	3.0

[a]Not tested due to insufficient sample size.

Mexico City. Levels for New York, Beijing, and Elizabeth, NJ, were approximately half of that observed for Mexico City. The lowest activity per cubic meter was seen for the EOM from Philadelphia, which was from samples collected during the summer.

With microsomal activation, the total EOM activities per cubic meter were highest for Beijing and Mexico City; levels at Beijing were presumably higher than Mexico City since only the CX and ACE fractions from Beijing were available for testing with S9. The relative order of activities for the other three cities (+S9) was Elizabeth > New York City > Philadelphia.

The distribution of mutagenic activity between the three fractions differed considerably among the five cities (cf. Figure 2). For the Philadelphia and Mexico City samples, the direct-acting activity per cubic meter was distributed fairly evenly among the three fractions. For New York City and Beijing, the DCM fraction contributed the smallest proportion of total direct activity, while in Elizabeth, this fraction contributed almost half of the total activity.

The effects of S9 activation on the three fractions also varied among the cities. Metabolic activation increased the mutagenicity of all the CX extracts; Beijing showed a 16-fold increase in indirect-acting activity, the largest we have seen in this laboratory. The increases for the other cities ranged from 2 to 6 times higher. The tested DCM fractions also demonstrated an enhancement (2- to 3.5-fold) in mutagenic activity upon metabolic activation. The mutagenicity of ACE extracts with S9 added remained essentially unchanged with the exception of the ACE extract from Beijing. A 3-fold increase in activity was observed with S9 for this extract.

The concentrations of total EOM for the five cities did not parallel the total mutagenic activity per cubic meter, as can be seen by comparing Figures 1 and 2. For EOM concentrations, the relative order of the cities was Mexico City > Beijing > Elizabeth > Philadelphia ≃ New York City. For direct-acting mutagenic activity, the order was Mexico City > Elizabeth ≃ New York City ≃ Beijing > Philadelphia. With

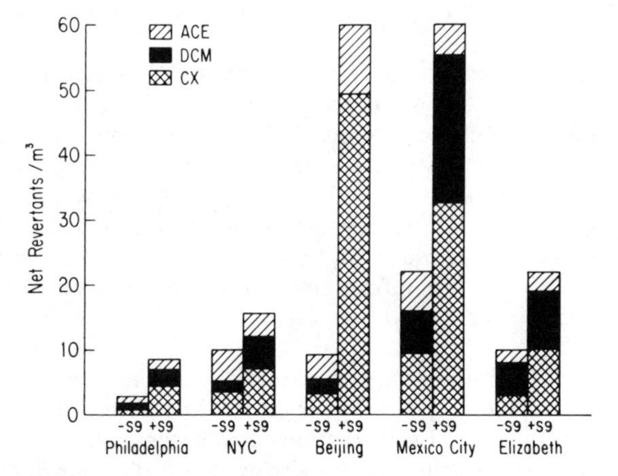

Figure 2. The sum and distribution of the mutagenic activities (±S9) per cubic meter of the 3 fractions from each site.

S9, the order was Beijing > Mexico City > Elizabeth > New York City > Philadelphia.

DISCUSSION

The five cities that are the subject of this study differ significantly in the nature of the combustion sources that contribute to atmospheric concentrations of EOM. The major anthropogenic sources of EOM in New York City are transportation and residual fuel oil (No. 4 and No. 6) combustion for space heating and power generation (Daisey and Kneip, 1981). In Elizabeth, NJ, the principal sources of EOM are motor vehicles and combustion of No. 2 fuel oil for space heating; this sampling site was within several kilometers of oil refineries (Lioy et al., 1982). The Philadelphia, PA, samples were collected in a largely residential area near a major traffic route (I-95) during a summer period in which contributions from secondary organic aerosol to EOM would be expected.

Mexico City is a densely populated (9.5 million people in 1500 km^2) urban center influenced by very heavy automotive traffic, metallurgical and petrochemical industries, and petroleum refining (Lioy et al., 1983). The automobiles are not fitted with emission control devices and use leaded gasoline. Beijing, China, is a heavily populated city that burns large quantities of coal for space heating, cooking, and power production but has little automotive traffic (Daisey et al., 1983). Bicycles are used as the principal mode of transportation.

The differences in the sources of EOM obviously affect both the total concentrations and the proportions of the three fractions in the ambient aerosols of these five cities. For New York City and Mexico City, the ACE fraction was 55% of total EOM and CX was one third of the total. The ACE fraction was 63% of EOM for the Philadelphia composite, while the CX and DCM fractions were present in similar proportions. In contrast, for Beijing and Elizabeth, the nonpolar CX fractions accounted for a larger proportion of total EOM: 55 and 42%, respectively. The ACE fractions were only 34 and 47% of the total for these two cities.

The relative EOM concentrations, seen in Figure 1, do not provide an adequate index of the mutagenic activities of the aerosols in these five cities as seen in Figure 2. For instance, although the concentration of the CX fraction was slightly higher in Beijing than Mexico City, the mutagenic activity per cubic meter (–S9) was 3 times higher in Mexico City than Beijing. Similarly, Beijing and New York City had almost the same CX mutagenicity (–S9), although the Beijing CX concentration was 5 times higher. Lack of correlation between EOM levels and mutagenicity per cubic meter (±S9) can be seen with the other fractions, as well, and from differences in the mutagenic potencies of the individual extracts.

The variations in the mutagenic activities of the extracts reflect the differences in the chemical compositions of the aerosols from the five cities. The differences in the relative proportions of the three EOM fractions among these cities are one indication of such compositional differences, but there is also evidence of chemical differences among fractions of a given type.

The CX fractions of all the cities showed an enhancement of mutagenic activity (rev/m^3) upon the addition of S9, indicating the presence of both direct- and indirect-acting mutagens. The Beijing sample, however, showed a much greater (16-fold) increase in indirect activity than did the CX extracts from the other cities. The activation-requiring class of PAHs was greatly elevated in Beijing (Daisey et al., 1983), as seen in other coal-burning cities (DeWiest et al., 1978). Emissions of PAHs per BTU have been shown to be one to two orders of magnitude higher for coal combustion than for oil or gas (Hangebrauck et al., 1967). Figure 3A shows that, with the exception of Beijing, there is a statistically significant correlation between the sum of the concentrations of the 10 measured PAHs in the CX fractions and the CX mutagenic activities (+S9) per cubic meter; in addition, the intercept of the line is close to 0. Beijing was omitted from the linear regression analysis since it deviates somewhat from this relationship. The sum of the 10 PAH compounds for the Beijing CX composite was extremely high (272.2 ng/m^3), yet compared to the other sites, indirect mutagenicity was not correspondingly elevated. This suggests aerosol compositional differences such as the presence of inhibitory compounds in the Beijing CX fraction or

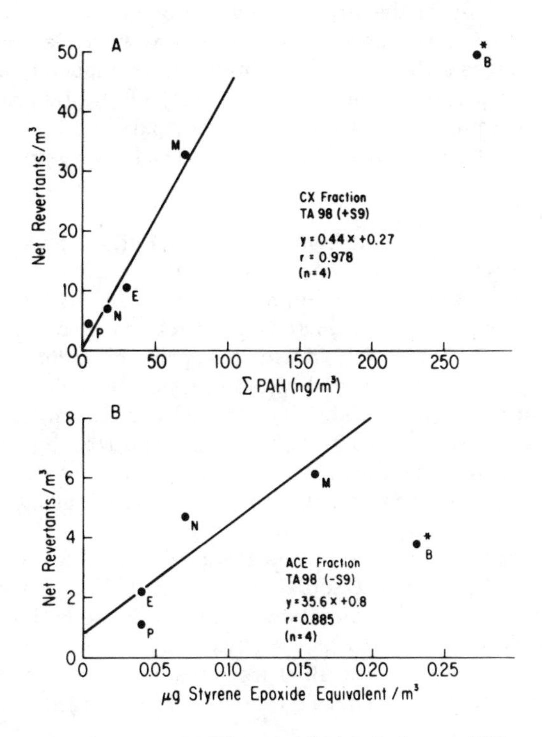

Figure 3. Linear regression curves: A, correlation between CX mutagenicity (+S9) per cubic meter and the sum of the concentrations of 10 PAHs in the CX fractions. B, correlation between ACE mutagenicity (−S9) per cubic meter and alkylating agent concentrations in the ACE fractions. E = Elizabeth, N = New York City, B = Beijing, M = Mexico City, P = Philadelphia. *Beijing values are not included in the regression analysis.

additional classes of mutagens in the other cities, such as nitro-PAH compounds. Nitro-PAHs could account for some of the indirect as well as direct mutagenic activity (Jäger, 1978; Wang et al., 1980; McCoy et al., 1983) in the CX extracts, and there is evidence for their presence in the CX extract (Butler, unpublished data).

The DCM fractions exhibited direct activity as well as a 2- to 3.5-fold enhancement with S9 activation. The Mexico City DCM appeared to include a greater proportion of indirect-acting compounds in this fraction than did the U.S. sites. Oxidized derivatives of PAHs are thought to be a major component of this fraction (Daisey, 1980); work in this laboratory has indicated that other classes may be contributing to a portion of the activity seen in this fraction (Butler, unpublished data).

The ACE fractions generally showed comparable mutagenic activities both with and without S9 activation, indicating the presence of primarily direct-acting mutagens. Beijing, with a 3-fold increase upon activation, was the exception. Alkylating and acylating agents, detected by reaction with 4-(p-nitrobenzyl)pyridine, are present in this fraction. As shown in Figure 3B, alkylating agents appear to be at least partially responsible for the direct-acting mutagenicity of the ACE fractions. The substantially decreased correlation ($r = 0.557$, $n = 5$) when Beijing is included in the linear regression suggests the presence of inhibitory compounds in this fraction. In addition, the greater proportion of indirect activity of the Beijing extract suggests significant chemical compositional differences, compared to the other sites, that are probably related to the large amounts of coal burned for space heating and power generation.

With only one site sampled during the summer (Philadelphia, PA), it is difficult to make seasonal comparisons. However, compared to the other two eastern U.S. cities with similar levels of EOM, Philadelphia consistently had the lowest levels of mutagenicity on both a per-microgram and a per-cubic-meter basis. Other studies have shown decreased mutagenic activities during the spring or summer compared to winter samples from the same city (Møller and Alfheim, 1980; Daisey et al., 1980), which were at least partially attributable to reductions in fuel combustion for space heating. In addition, there are greater proportions of polar, oxidized hydrocarbons produced photochemically in the aerosol during the summer (Daisey et al., 1984a). These have been found to be less active on a per-microgram basis (Daisey et al., 1984b).

The results reported here are consistent with a study of the comparative carcinogenic activity of the benzene-soluble organic (BSO) fraction of TSP collected in a number of U.S. cities (Heuper et al., 1962). This early study demonstrated that the same dose of BSO from different cities resulted in variable cancer rates in mice. Furthermore, although most of the BSO fractions exhibited carcinogenic activity, benzo(a)pyrene concentration was not a good index of carcinogenicity of the BSO.

CONCLUSIONS

This study has shown that interurban variations in concentrations of EOM do not necessarily reflect the relative biological activities of the urban aerosols as measured by the Ames assay. The types of fuels used for transportation, space heating, and power production, as well as the proportions of secondary organic aerosol in the EOM, can affect the chemical composition of the aerosols and, in turn, their mutagenic activity. The relationships between the sources of EOM, its chemical composition, and biological activity are not well understood at present. The results of this investigation suggest that a better understanding of such relations is important for assessing the potential impact of changing fuel use patterns in the U.S.

ACKNOWLEDGMENTS

The cooperation of the following investigators in providing samples is gratefully acknowledged: Prof. Yolanda Falcon, Universidad Autonoma Metropolitana-Azcapotzalco, Mexico; Prof. Wang Ming-xing, Institute of Atmospheric Physics, Beijing, China; Dr. Paul Lioy and his colleagues of the ATEOS project. We thank Dr. Thomas Atherholt of the Institute for Medical Research for providing the Ames tester strains.

This research was supported in part by the American Petroleum Institute and is part of a Center Program supported by the National Institute of Environmental Health Sciences, Grant No. ES00260, and the National Cancer Institute, Grant No. CA13343.

REFERENCES

Agarwal, S.C., B.L. VanDuuren, and T.J. Kneip. 1979. Detection of epoxides with 4-(p-nitrobenzyl)pyridine. Bull. Environ. Contam. Toxicol. 23:825-829.

Alink, G.M., H.A. Smit, J.J. van Houdt, J.R. Kolkman, and J.S.M. Boleij. 1983. Mutagenic activity of airborne particulate at non-industrial locations. Mutat. Res. 116:21-34.

Ames, B.N., J. McCann, and E. Yamasaki. 1975. Methods for detecting carcinogens and mutagens with the *Salmonella*/mammalian-microsome mutagenicity test. Mutat. Res. 31:347-364.

Chrisp, C.E., and G.L. Fisher. 1980. Mutagenicity of airborne particles. Mutat. Res. 76:143-164.

Daisey, J.M. 1980. Organic compounds in urban aerosols. Ann. N.Y. Acad. Sci. 338:50-69.

Daisey, J.M., and T.J. Kneip. 1981. Atmospheric particulate organic matter: Multivariate models for identifying sources and estimating their contributions to the ambient aerosol. In: Atmospheric Aerosol: Source/Air Quality Relationships. E.S. Macias and P.H. Hopke, eds. ACS Symposium Series, No. 167, American Chemical Society: Washington, DC. pp. 197-221.

Daisey, J.M., M.A. Leyko, M.T. Kleinman, and E. Hoffman. 1979. The nature of the organic fraction of the New York City summer aerosol. Ann. N.Y. Acad. Sci. 322:125-141.

Daisey, J.M., T.J. Kneip, I. Hawryluk, and F. Mukai. 1980. Seasonal variations in the bacterial mutagenicity of airborne particulate organic matter in New York City. Environ. Sci. Technol. 14:1487-1490.

Daisey, J.M., R.J. Hershman, and T.J. Kneip. 1982. Ambient levels of particulate organic matter in New York City in winter and summer. Atmos. Environ. 16:2161-2168.

Daisey, J.M., T.J. Kneip, W. Ming-xing, R. Li-xin, and L. Wei-xiu. 1983. Organic and elemental composition of airborne particulate matter in Beijing, spring, 1981. Aerosol Sci. Technol. 2:407-415.

Daisey, J.M., M. Morandi, G.T. Wolff, and P.J. Lioy. 1984a. Regional and local influences on the nature of airborne particulate organic matter at four sites in New Jersey during summer, 1981. Atmos. Environ. 18:1411-1419.

Daisey, J.M., P.J. Lioy, J.B. Louis, L.J. McGeorge, G.J. McGarrity, and T.B. Atherholt. 1984b. The mutagenic activity of the ambient aerosol during summertime: A comparison of photochemical pollution episodes with other periods. Paper 84-80.1 presented at the 77th Annual Meeting of the Air Pollution Control Association, San Francisco, CA.

DeWiest, F. 1978. Any factors influencing the dispersion and the transport of heavy hydrocarbons associated with airborne particles. Atmos. Environ. 12:1705-1711.

Hangebrauck, R.P., D.J. von Lehmden, and J.E. Meeker. 1967. Sources of Polynuclear Hydrocarbons in the Atmosphere. Publication 999-AP-33, U.S. Public Health Service: Cincinnati, OH. 44 pp.

Heuper, W.C., P. Kotin, E.C. Tabor, W.W. Payne, H. Falk, and E. Sawicki. 1962. Carcinogenic bioassays on air pollutants. Arch. Pathol. 74:89-116.

Hoffmann, G.R. 1980. Genetic effects of dimethyl sulfate, diethyl sulfate, and related compounds. Mutat. Res. 75:63-129.

Hughes, T.J., E. Pellizzari, L. Little, C. Sparacino, and A. Kolber. 1980. Ambient air pollutants: Collection, chemical characterization and mutagenicity testing. Mutat. Res. 76:51-83.

Huisingh, J., R. Bradow, R. Jungers, L. Claxton, R. Zweidinger, S. Tejada, J. Bumgarner, F. Duffield, M. Waters, V.F. Simmon, C. Hare, C. Rodriguez, and L. Snow. 1978. Application of bioassay to the characterization of diesel particle emissions. In: Application of Short-Term Bioassays in the Fractionation and Analysis of Complex Environmental Mixtures. M.D. Waters, S. Nesnow, J.L. Huisingh, S.S. Sandhu, and L. Claxton, eds. EPA-600/9-78-027, U.S. Environmental Protection Agency: Research Triangle Park, NC. pp. 382-418.

Jäger, J. 1978. Detection and characterization of nitro derivatives of some polycyclic aromatic hydrocarbons by fluorescence quenching after thin-layer chromatography: Application to air pollution analysis. J. Chromatogr. 152:575-578.

Lamb, S.I., C. Petrowski, I.R. Kaplan, and B.R.T. Simoneit. 1980. Organic compounds in urban atmospheres: A review of distribution, collection and analysis. J. Air Pollut. Control Assoc. 30:1098-1115.

Lioy, P.J., J.M. Daisey, A. Greenberg, B. Kebbekus, J. Bozzelli, G. McGarrity, and T. Atherholt. 1982. First Annual Report of the New Jersey Project on Airborne Toxic Elements and Organic Species (ATEOS). New York University Medical Center: New York.

Lioy, P.J., Y. Falcon, M.T. Morandi, and J.M. Daisey. 1983. Particulate matter pollution in Mexico City as measured during the winter of 1982. Aerosol Sci. Technol. 2:166 (extended abstract).

Maron, D.M., and B.N. Ames. 1983. Revised methods for the *Salmonella* mutagenicity test. Mutat. Res. 113:173-215.

McCoy, E.C., G. DeMarco, E.J. Rosenkranz, M. Anders, H.S. Rosenkranz, and R. Mermelstein. 1983. 5-Nitroacenaphthene: A newly recognized role for the nitro function in mutagenicity. Environ. Mutagen. 5:17-22.

McGarrity, G., T. Atherholt, J. Louis, and L. McGeorge. 1982. Mutagenicity of airborne particles at urban and rural sites in New Jersey. Paper 82-1.4 presented at the 75th Annual Meeting of the Air Pollution Control Association, New Orleans, LA.

Møller, M., and I. Alfheim. 1980. Mutagenicity and PAH-analysis of airborne particulate matter. Atmos. Environ. 14:83-88.

Møller, M., I. Alfheim, S. Larssen, and A. Mikalsen. 1982. Mutagenicity of airborne particles in relation to traffic and air pollution parameters. Environ. Sci. Technol. 16:221-225.

National Academy of Sciences. 1981. Health Effects of Exposure to Diesel Exhaust. National Academy Press: Washington, DC.

National Research Council. 1978. Airborne Particles. University Park Press: Baltimore, MD.

Sawicki, E. 1967. Airborne carcinogens and allied compounds. Arch. Environ. Health 14:46-53.

Simoneit, B.R.T., and M.A. Mazurek. 1981. Air pollution: The organic components. CRC Crit. Rev. Environ. Control 11:219-276.

Sorenson, W.G., W.-Z. Whong, J.P. Simpson, F.J. Hearl, and T.-m. Ong. 1982. Studies of the mutagenic response of *Salmonella typhimurium* TA98 to size-fractionated air particles: Comparison of the fluctuation and plate incorporation tests. Environ. Mutagen. 4:531-541.

Talcott, R., and W. Harger. 1980. Airborne mutagens from particles of respirable size. Mutat. Res. 79:177-180.

Task Group Report. 1978. Air pollution and cancer: Risk assessment methodology and epidemiological evidence. Environ. Health Perspect. 22:1-12.

Van Vaeck, L., and K. Van Cauwenberghe. 1978. Cascade impactor measurements of the size distribution of the major classes of organic pollutants in atmospheric particulate matter. Atmos. Environ. 12:2229-2239.

Wang, C.Y., M.-S. Lee, C.M. King, and P.O. Warner. 1980. Evidence for nitroaromatics as direct-acting mutagens of airborne particulates. Chemosphere 9:83-87.

Wynder, E.L., and D. Hoffmann. 1965. Some laboratory and epidemiological aspects of air pollution carcinogenesis. J. Air Pollut. Control Assoc. 15:155-159.

MUTAGENICITY ANALYSES OF INDUSTRIAL EFFLUENTS: RESULTS AND CONSIDERATIONS FOR INTEGRATION INTO WATER POLLUTION CONTROL PROGRAMS

Leslie J. McGeorge,[1] Judith B. Louis,[1] Thomas B. Atherholt,[2] and Gerard J. McGarrity[2]

[1]Office of Science and Research, New Jersey Department of Environmental Protection, Trenton, New Jersey 08625, and [2]Institute for Medical Research, Camden, New Jersey 08103

INTRODUCTION

The Ames *Salmonella*/microsomal assay has found extensive application in evaluations of the mutagenic activity of a variety of environmental sample types. A number of investigators have successfully utilized this test system with municipal (Rappaport, 1979; Ellis et al., 1982) and industrial wastewaters. Many of the industrial effluent studies have concentrated on specific manufacturing process types. Separate investigations have demonstrated that wastewaters from pulp mills (Douglas et al., 1980) and coal gasification processes (Epler, 1980) were mutagenic, while numerous textile mill effluents were reported to be nonmutagenic (Rawlings and Samfield, 1979). Two published projects have examined the mutagenicity of effluents discharged from a limited number of industrial process types. Stinett et al. (1981) utilized the Ames assay to evaluate the mutagenicity of 12 unconcentrated effluent samples from a number of types of organic chemical industries. These investigators reported two positive effluent results from organics and plastics facilities, although both of these results did not meet the conventional positive mutagenicity evaluation criterion of a dose response (Chu et al., 1981; Maron and Ames, 1983). The need for concentration of treated wastewater samples prior to mutagenicity testing is generally accepted (Epler, 1980). Somani and co-workers (1980) analyzed concentrates from 12 industrial effluents collected from 9 different process types. Only one sample, a foundry effluent, was found to be mutagenic.

In the present study, the Office of Science and Research (OSR), New Jersey Department of Environmental Protection (NJDEP), in conjunction with the Institute for Medical Research, expanded the scope of Ames analyses of industrial effluent concentrates to include a wide diversity of manufacturing processes. The project was partially funded by the U.S. Environmental Protection Agency (EPA) under Section 28 of the Toxic Substances Control Act (TSCA) (Public Law 94-469), and was designed to examine the practical utility of effluent mutagenicity analyses to state water pollution control agencies.

Presently, the tools commonly available to water pollution organizations for the control of the discharge of toxic substances include chemical analyses for the limited list of 126 priority pollutants and biomonitoring for pollutants that cause acute toxic effects (e.g., fish mortality). Interest in utilizing the Ames assay as a monitoring tool centers around its potential for nonspecific detection of toxics (i.e., mutagens) that have been associated with chronic disease (Hartman, 1983).

In addition to building an extensive database of Ames effluent results, issues that were addressed in this study included: presentation of data, ranking of sample mutagenic activities, reproducibility of results, and a limited evaluation of the effects of tertiary treatment technology. Priority pollutant analyses were performed concurrently with mutagenicity studies, and a discussion of these results in relation to the Ames data will be presented elsewhere.

MATERIALS AND METHODS

Site Selection

The site selection process conducted by OSR was designed to emphasize those New Jersey industries with the potential for discharging toxic (or, specifically, mutagenic) organic contaminants. Site selection was performed in consultation with the NJDEP Division of Water Resources. The following criteria were utilized:

1. Effluent type. With two exceptions, only direct, final discharges were sampled. All discharges contained some process wastewater.

2. Substantial flow rate. Discharge flow rates were generally greater than 50,000 gal/d.

3. Industrial process type. The EPA-designated "Primary Industry" list (Federal Register, 1980) was used as a guide to select a variety of manufacturing processes that were likely to discharge toxic substances. Selection was targeted toward processes involving organic materials.

4. Published mutagenicity data. Some industrial categories were selected because associated raw materials, products, or discharges had been reported to be mutagenic.

5. Evidence of organic contamination in the discharge from: (a) OSR New Jersey Industrial Survey for the use and/or discharge of 155 toxic substances; (b) monitoring results for the National Pollutant Discharge Elimination System (NPDES) permit parameters or EPA priority pollutants.

Sampling

Effluent collection was conducted between 1981 and 1983 at NPDES-designated sampling sites. Depending on the specific wastewater treatment processes utilized, such sampling may follow chlorination or cooling water dilution. Collection at the NPDES sites was performed to provide representative samples of the final effluent just prior to discharge.

All wastewater samples were composites collected with a portable, automatic sampler (Model 3000T, Manning Environmental Corporation, Santa Cruz, CA). All wetted parts of this sampler are composed of Teflon or glass. Prior to use, the sampler and borosilicate glass receiving bottle were rinsed with detergent (Alconox, New York, NY), tap water, reagent grade acetone, and distilled water. Two aliquots of the wastewater were collected through the sampler and used to rinse the receiving bottle.

The sampling method consisted of the collection of a 160-ml volume of wastewater every 15 min for a 24-h period. Of the resulting ~15-liter sample volume, 10 liters was generally utilized for Ames analyses, while chemical analyses were performed on the remaining volume. Intake waters were collected as grab samples. All samples were transported on ice and in the dark to the laboratory. For mutagenicity analyses, a 5-d holding time prior to extraction was followed. This holding time was shorter than that established by EPA for nonvolatile priority pollutants (Federal Register, 1979). Samples collected for total organic carbon (TOC) analyses were preserved with H_2SO_4.

Mutagenicity Analyses

Sample preparation. At approximately one third of the facilities sampled, chlorine was utilized in the waste treatment process for disinfection of sanitary waste. Wastewaters collected containing >0.5 mg/liter total residual chlorine were dechlorinated within 24 h of collection with a 2.8 mM solution of ferrous ammonium sulfate (FAS). For each liter of sample, 10 ml of FAS was used to neutralize 1 mg of chlorine. Dechlorination was conducted for two purposes: to prevent the formation of additional potentially mutagenic chlorination products during sample storage and processing, and to prevent the possible production of low-level artifactual mutagenic activity from chlorination of the resin material utilized for subsequent sample processing (Cheh et al., 1980b). Low concentrations of halogenated organic artifacts have been shown to be produced when 1-2 mg/liter active chlorine is passed through XAD-2 resin (Bean et al., 1978). FAS was used versus the more conventional dechlorinating agent, sodium thiosulfate, because the nucleophilic thiosulfate has been demonstrated to inactivate some direct-acting mutagens, while the nonnucleophilic FAS is largely ineffective in such inactivation (Cheh et al., 1980a). Use of FAS requires adjustment of the sample pH to ≤3.5 with H_2SO_4 to prevent oxidation of the dechlorinating agent and to hold the agent in solution. Nonchlorinated samples were processed at ambient pH (generally 6-8). Sample pH during extraction can influence the sorption efficiency of various organics (Junk et al., 1974). Dechlorinated samples are indicated clearly in the Results section.

Samples were filtered through solvent-extracted glass fiber filters (No. 934-AH, Whatman, Clifton, NJ) to remove particulate matter that would interfere with the resin extraction process. After air-drying, filters were sonicated in small volumes of Resi-analyzed acetone and dichloromethane (J.T. Baker, Phillipsburg, NJ), each for 10 min. These solvent extracts were combined with solvent eluates from the resin extraction procedure.

The styrene divinylbenzene resin Amberlite XAD-2 (Rohm and Haas, Philadelphia, PA) was utilized to adsorb relatively nonpolar organics from the wastewater samples according to a modification of the technique described by Van Rossum and Webb (1978). For the majority of samples, the XAD-2 resin was precleaned by the supplier (Applied Science, State College, PA) according to EPA methodology (EPA, 1978). Additional multiple washings of the resin with acetone and distilled water were performed in the extraction columns prior to use. For some initial samples, XAD-2 resin was obtained from Bio-Rad (Richmond, CA) or Brinkman (Westbury, NY) and prepared in the laboratory according to Atherholt et al. (1982). All resin material was used only once. Extractions were performed by passing ~5-liter sample volumes through 40-cm^3 bed volumes of resin. A flow rate of 15-20 ml/min was used with 2.5 cm inside diameter columns, resulting in a linear velocity (< 4 cm/min) that should optimize sorption of organics (Harris et al., 1981). Resin desorption was accomplished with successive washes of acetone (50 ml), dichloromethane (100 ml), and acetone (50 ml). All eluates and filter extracts were combined. Solvent volume reduction was performed with two successive Kuderna-Danish concentration steps. Extracts were evaporated to small solvent volumes (0.5-1.0 ml) and brought to final 2- to 5-ml acetone volumes for subsequent Ames analyses. An overall 2,000- to 5,000-fold sample concentration was achieved. Extracts were stored in amber glass vials at −20°C. The maximum holding time for all sample preparations prior to mutagenicity testing was 30 d.

Gravimetric analyses were performed on the final acetone extracts by drying two extract aliquots (40 and 80 μl) in aluminum weigh pans on a slide warmer at 45°C for 16-18 h. Dried extracts were weighed on an electrobalance (Model 25, Cahn, Cerritos, CA) with a precision of ±2%. Measured extract weights and volumes were utilized to calculate the extractable, nonvolatile organic concentration, referred to as residue concentration, in the original wastewater sample.

Laboratory blanks consisting of deionized, distilled water were extracted concurrently with approximately one half of the wastewater samples. Where appropriate, the blanks were chlorinated to the same concentration as the wastewaters and dechlorinated accordingly. Three field blank samples were collected through the sampling apparatus.

Ames assay procedures. Mutagenicity analyses were conducted with the *Salmonella typhimurium*/microsomal test using the plate incorporation method described by Ames et al. (1975) and modified by Maron and Ames (1983) and de Serres and Shelby (1979). All five original bacterial tester strains (TA98, TA100, TA1535, TA1537, and TA1538) were used whenever an adequate total weight of extractable sample material was available. For several samples, extractable organic residue quantities were low, and only the two generally more sensitive strains (TA98 and TA100) were employed (McCann et al., 1975).

All tester strains were routinely checked for confirmation of genotypes. Concurrent tests for spontaneous reversion, solvent control reversion, positive control reversion, and tester strain viability were performed with each experiment. Positive control compounds used in direct activity assays were: TA98 and TA1538, 2-nitrofluorene; TA100 and TA1535, N-methyl-N'-nitro-nitrosoguanidine (MNNG); TA1537, 9-aminoacridine. For indirect activity assays, 2-aminoanthracene was used for all strains. All positive control assays produced acceptable increased responses. For direct assays, the mean solvent control values (revertants/plate) for the period during which the assays were conducted were as follows: TA98, 40; TA100, 134; TA1535, 32; TA1537, 8; TA1538, 15. For indirect assays, the mean solvent control values were: TA98, 39; TA100, 123; TA1535, 26; TA1537, 8; TA1538, 18. Spontaneous control values were similar. Viability counts averaged 8.5×10^8 colony-forming units/ml.

Each experiment was performed both with and without metabolic activation. Activation was achieved with Aroclor 1254-induced rat liver homogenate (S9) prepared in the IMR laboratory from male Nichols-Wistar rats. The S9 volume utilized per plate was determined from TA98 and TA100 dose-response assays in which various doses of homogenate were tested with 2-aminoanthracene and benzo(a)pyrene (5 µg/plate). The S9 volume selected was the volume that demonstrated optimal activation of both of these indirect mutagens. For each S9 batch used in this study, a volume of 25 µl/plate (in 0.5 ml of S9 mix) was determined to be optimal. This S9 volume was comparable to the 20-µl volume recommended for general screening by Maron and Ames (1983). For one mutagenic sample of particular interest (number 7B), an S9 titration with the wastewater extract was performed, and the optimal S9 volume of 50 µl/plate was utilized.

All assays were conducted with duplicate plates per dose level. A minimum of five doses of sample extract were tested, generally up to a maximum dose of at least 1 mg/plate. Early experience with effluent samples demonstrated that positive responses were always observed for mutagenic samples at or below 1 mg, and responses were rarely linear above this dose. Where sample material was limiting, the number of strains was generally reduced before the maximum test dose was lowered below 1 mg/plate. Prior to mutagenicity assays, toxicity tests were performed using the standard Ames mutagenicity protocol, in which background lawns were examined for growth with a dissecting microscope. Toxicity test results were utilized to set maximum doses if a toxic response was observed below 1 mg/plate.

To assess the reproducibility of determinations of mutagenic activity levels, duplicate trials were performed on five positive sample extracts. When duplicate trials were conducted, results were reported for the trial that provided the most data points within the linear portion of the dose-response curve, generally the second trial.

Revertant colonies were counted on a automated colony counter (Bio-Tran II, New Brunswick Scientific Company, Edison, NJ). Machine counts were calibrated against manual counts for each assay and tester strain.

Analysis of results. A modified twofold rule concept was utilized to determine if a sample extract was mutagenic (Chu et al., 1981). A positive response was defined as a twofold or greater increase in revertants on sample plates over solvent control plates, and a clear dose response over three or more doses. A sample was classified as negative if neither criterion was satisfied, and marginal if only one criterion was met. Clearly, a negative result must be viewed as negative under the conditions of the test and may be more suitably described as nondetectable. Estimates of quantitative activity were derived from the initial linear portions of the dose-response curves (Maron and Ames, 1983). Linear regression analyses were used to determine the slopes of these lines. The r^2 values from these analyses were generally higher than 0.80 and frequently greater than 0.90. Mutagenicity results were reported in both revertants per microgram of extracted material and revertants per equivalent milliliter of wastewater. The residue concentration values were used to convert the activities per weight as follows: revertants/microgram x micrograms/equivalent milliliter = revertants/equivalent milliliter.

TOC Analysis

Nonvolatile TOC concentrations were measured for the last 18 of the 33 final discharge samples. Analyses were performed with the combustion infrared method (American Public Health Association, 1980) using a TOC analyzer (Model 915, Beckman Instruments, Fullerton, CA).

RESULTS AND DISCUSSION

Facilities Sampled

Utilizing the stated criteria, we selected 27 New Jersey industrial facilities for sampling. A categorized list of these facilities, including major products, is provided in Table 1. A number of primary industry categories were used in this table; also, several categories were formed specifically for the facilities in this study. At the 27 sites, a total of 33 final effluents, 2 other wastewaters, and 4 intake waters were sampled. The 2 other wastewaters were sampled prior to tertiary carbon treatment and collected concurrently with final effluent samples. The intake waters were sampled at industries with surface water sources that were known to have the potential for substantial pollutant contamination. One duplicate final effluent sample was collected. Repeat samples were taken at 5 facilities, 2 for the determination of reproducibility of results and 3 because facility conditions had changed.

Qualitative Mutagenicity Results

All laboratory and field blank samples produced negative mutagenicity results. Positive responses for wastewater samples were observed with each of the five tester strains, although rarely with TA1535. Mutagenic samples generally produced positive responses with several tester strains. Metabolic activation was required to maximize responses of some samples but decreased the activity of others.

The qualitative mutagenicity data provided in Table 1 reveal that of the 33 final effluent samples, 13 produced positive responses (39.4%), 6 produced marginal responses (18.2%), and 13 produced negative responses (39.4%). For 1 sample (3%),

Table 1. Qualitative Mutagenicity Results with Residue and TOC Concentrations Listed by Industrial Category

Industrial Category	Sample Number	Description	Ames Result[a]	Residue Concentration (mg/liter)	TOC Concentration (mg/liter)
Dye manufacture or use[b]	7A	Manufacture of dyes and epoxy resins	+	22.4	NA[c]
	7B	(Repeat)	+	21.0	190
	19	Textile dyeing and finishing	+	9.9	80
	21	Paper dyeing and manufacture	+	4.0	40
	22A	Manufacture of hair dyes, herbicide	+[d]	5.8	NA
	22B	(Repeat)	–[e]	0.3	NA
Resin manufacture	10	Rosin-derived	+[d]	57.0	NA
	14	Polyvinyl chloride	–	4.1	NA
	15	Ion exchange	–[f]	4.3	NA
	23	Polyvinyl chloride	–[d]	0.8	3
Misc. organic chemicals	2	Freon, fluoropolymers	–[f]	1.5	14
	3A	Industrial organics	–[g]	NA	NA
	3B	(Repeat)	+/–	0.9	13
	11A	Organic peroxides	+	18.7	NA
	11B	(Repeat)	+	13.5	26
	17A	Plasticizers	+/–[h]	3.1	NA
	17B	(Repeat)	+	6.8	98
	25A	Industrial organics	+	6.7	19
	25B	(Duplicate)	+	6.6	20

(continued)

Table 1. Continued

Industrial Category	Sample Number and Description		Ames Result[a]	Residue Concentration (mg/liter)	TOC Concentration (mg/liter)
Petroleum refining	6		–	6.5	NA
	8		–	5.4	NA
	16		+[d]	11.1	18
	24		–[d]	9.1	NA
Soaps and detergents	9	Surfactants	–	NA	NA
	18A	Cleaning compounds	+/–	7.8	21
	18B	(Second discharge)	+/–	1.9	6
Pharmaceuticals	5		–	2.2	15
	13		+[d]	37.3	300
Paper manufacture	20		+/–	4.0	NA
Inorganic chemicals	4	Specialty chemicals	–	0.9	4

(continued)

Table 1. Continued

Industrial Category	Sample Number and Description		Ames Result[a]	Residue Concentration (mg/liter)	TOC Concentration (mg/liter)
Electroplating	1		–[f]	1.1	5
Explosives	12		T[d]	NA	NA
Foundry	26	Ductile iron pipe	+	NA	21
Hazardous waste treatment	27	Incineration	+/–[f]	0.5	8

[a] +, positive; +/–, marginal; –, negative; T, cytotoxic.
[b] This category was formed due to the common Ames result. Note that facilities 7 and 21 could be placed under Resin Manufacture and Paper Manufacture, respectively.
[c] NA, not available.
[d] Dechlorinated sample.
[e] Repeat sampling was performed following installation of carbon treatment unit.
[f] Two strains (TA98 and TA100) were used in the assay.
[g] Result is for fractionated sample versus whole extract.
[h] Major product was not being manufactured during first sampling at this facility.

cytotoxicity masked any mutagenicity that may have been present. It must be assumed that the percentage of effluents determined to be mutagenic in this study was higher than if discharges from all types of industrial processes were included. This assumption is based on consideration of the toxics-oriented site selection criteria.

Table 2 presents the qualitative results for the wastewaters collected prior to granular activated carbon filtration and for their associated final effluents. Both wastewaters collected prior to carbon filtration gave positive results. Carbon treatment altered the classification of the mutagenic activity for facility number 3 from positive to marginal. The substantial effect that tertiary treatment had on reducing the mutagenic response per milliliter of wastewater at a dye facility (number 22) is shown in Figure 1.

Table 2 also provides the qualitative results for the intake water samples. The 4 surface waters produced 1 positive, 1 marginal, and 2 negative results. The positive result was not surprising considering that this particular surface water receives numerous industrial discharges in this area, and that other surface waters have been shown to be mutagenic (Kool et al., 1981). The quantitated activity of this intake water at a refinery facility (number 16) was lower than the associated discharge activity (see Tables 3 and 4).

Definitive conclusions cannot be drawn between the manufacture of specific products and the activity of effluent samples in the *Salmonella* mutagenicity assay for various reasons, such as the low number of samples in each category, product categories that are too broad to reflect specific raw materials, and variations in waste

Table 2. Comparison of Ames Results for Intake Waters, Wastewaters Prior to Carbon Treatment, and Final Effluents

	Ames Result[a]		
Effluent Sample Number	Intake Water	Wastewater Before Carbon Treatment[b]	Final Effluent
3B	−	+	+/−
22B	NA[b]	+	−
8	+/−	NA	−
9	−	NA	−
16	+	NA	+

[a] +, positive; +/−, marginal; −, negative.
[b] NA, not applicable.

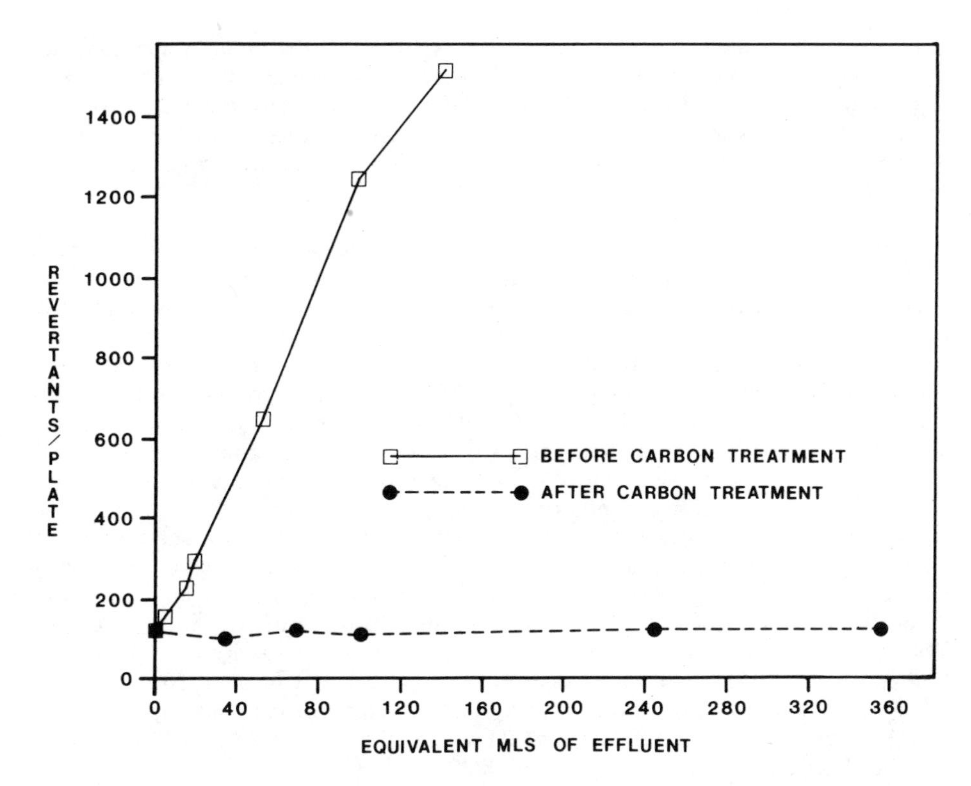

Figure 1. Effect of carbon treatment on mutagenic activity of secondary wastewater at dye facility (facility number 22). Each point represents the mean value of duplicate plate counts.

treatment processes. However, there was an apparent relationship between dye use and/or manufacture and positive results (see Table 1); five facilities in the dye category discharged mutagenic substances. Many dyes have been reported to produce positive Ames responses (Combes and Haveland-Smith, 1982). With one exception, the petroleum refinery effluents sampled produced negative results. Discharges from facilities utilizing primarily inorganic and/or volatile substances (sample numbers 1, 4, 14, and 23) gave negative results. As previously reported in the literature, a foundry effluent (sample number 26) was mutagenic (Somani et al., 1980).

Residue and TOC Concentrations: Relation to Mutagenicity Results

Table 1 includes the effluent residue concentrations and TOC concentrations, which had geometric mean values of 4.5 mg/liter and 22.7 mg/liter, respectively. These two measurements were linearly related ($r^2 = 0.85$), indicating that the magnitude of extractable organic material is dependent on the original concentration of total organics. The geometric mean percentage of the TOC recovered by the XAD concentration procedure was 18.2%. This relatively small fraction of recovered material demonstrates the selectivity of the XAD resin, presumably for nonpolar organics (Van Rossum and Webb, 1978). The percentage recovered here was somewhat higher than the $\leq$10% adsorption of dissolved organic carbon determined by Kool et al. (1981) for surface waters. These investigators demonstrated that the less-polar substances in surface waters were apparently responsible for the major portion of detected mutagenic activity.

Both residue and TOC concentrations were measurements of nonvolatile, organic material, and the *Salmonella* assay was used here to assess the activity of such substances. A visual comparison of either residue concentration or TOC concentration values with mutagenicity results (Table 1) reveals an apparent relationship between the magnitude of these values and the type of Ames response. Indeed, Kruskal-Wallis one-way analysis of variance by ranks (SAS, 1982) revealed significant differences in average residue concentrations ($p = 0.001$) and average TOC concentrations ($p = 0.003$) among the three Ames result types (positive, marginal, and negative). These findings for residue concentrations may be expected because of the selectivity of XAD-2 resin for the type of organics that has been associated with genotoxic activity in aqueous matrices. It was somewhat suprising, however, to note the significant difference in average TOC values among Ames result types, because of the broader nature of this measurement. TOC determinations are inexpensive and frequently conducted in environmental monitoring programs. In this data set, all samples with TOC concentrations of >25 mg/liter were positive in the Ames assay. In light of these limited results, it is suggested that TOC could be utilized to prioritize industrial effluents for mutagenicity testing. Clearly, such a suggestion does not imply that high TOC measurements in wastewaters would necessarily be associated with toxicity or, specifically, mutagenicity.

Negative results may have been obtained for samples with low residue concentrations because no organic nonvolatile mutagens were actually present. Another possibility to be considered is that the extractable organic concentrations were so low in these wastewaters that the total weight of available test material precluded the assay of sufficient doses (micrograms) of material to produce a measurable positive response. For instance, it was necessary to assay 5 effluents at maximum doses of <500 μg/plate; 2 of these samples were negative and 3 were marginal. These dose restrictions were present in this study despite the relatively large sample volumes extracted (10 liters) compared to those of an EPA interim recommended protocol (3 liters) (Williams and Preston, 1983). It is important to be aware that even if higher doses of extractable material would have produced positive results for some samples classified as negative, the impact on the environment of such discharges would most likely be substantially less than for the majority of positive samples. Any activity present per volume of such wastewaters would be low, demonstrated by the fact that negative samples generally were tested at maximum doses by volume (milliliters) that exceeded the maximum tested doses of positive samples. Consideration of the enviromental impact of potential toxic substances in an effluent must also be evaluated in light of discharge flow rate, and receiving stream uses and characteristics.

Quantitative Mutagenicity Results

OSR utilized the bar graph format to represent quantitative mutagenicity data for individual samples. Such a format, illustrated in Figure 2, provides rapid conveyance of information on all strain responses, the impact of S9, and the activities in both

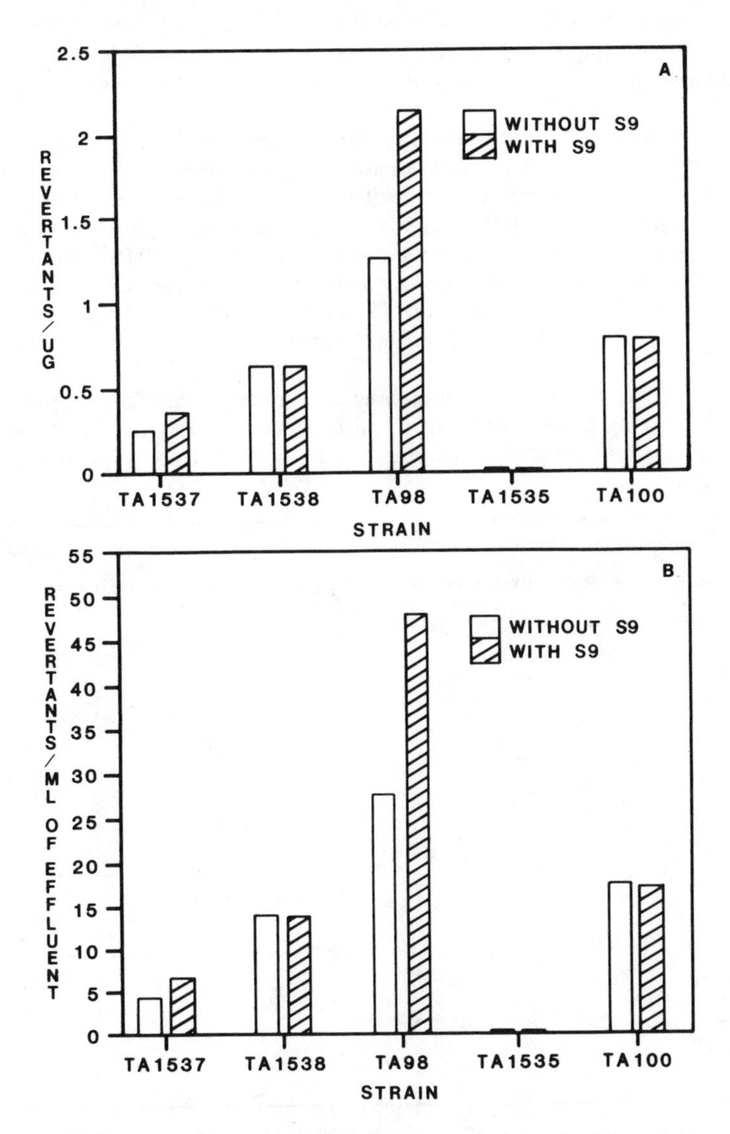

Figure 2. Summary presentation of the mutagenic activity of sample number 7A, effluent from a dye and epoxy resin facility. Activities per microgram (A) and per milliliter of effluent (B) are given. Bar heights represent the slopes of the linear portions of the dose-response curves.

revertants per microgram and revertants per milliliter. Figure 2 represents the mutagenic activities detected for one of the more potent samples, a dye and epoxy resin facility effluent (sample number 7A). OSR has also released a background and results document on Ames testing of effluents to provide a framework for interpretation of individual results (McGeorge et al., 1983).

Figure 3 provides a graphic example of the variation in the magnitudes of the positive responses detected. Strain TA100 dose-response curves are illustrated for five of the more active samples. It should be noted that for some of these samples, other strains and/or activation conditions produced the maximum responses.

Strains TA98 and TA100 were consistently more sensitive to individual sample extracts than the other three tester strains. Quantitative determinations of the responses of these two strains to the positive effluent samples and the one positive intake water sample are given in Tables 3 and 4. Estimates of potency (revertants per microgram) for all positive wastewater samples ranged over 1000 fold, from 0.02 to 27.0 (Table 3). Mutagenic activities expressed in revertants per equivalent milliliter ranged almost 500 fold, from 0.4 to 179 (Table 4). Levels of activity for the positive intake sample were lower than for all effluent samples.

In Tables 3 and 4, samples were ranked in a general order from higher responses to lower. Such a ranking is difficult to perform due to the variability in the most responsive strain and activation condition among this diverse group of samples.

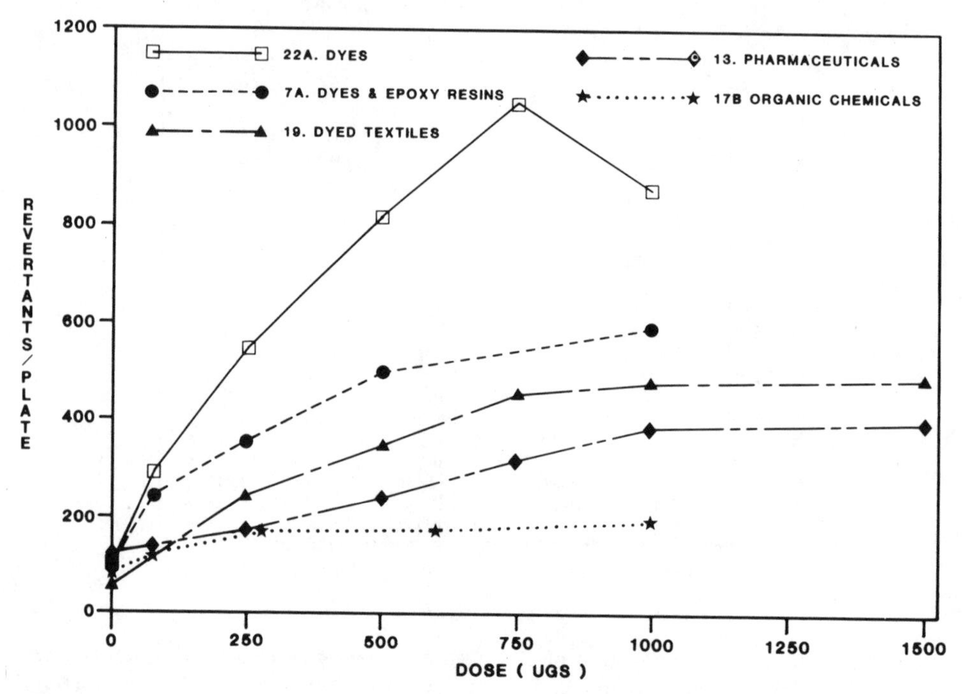

Figure 3. Dose-response curves for five positive samples. Assay results shown are for strain TA100 with S9. Each point represents the mean value of duplicate plate counts.

Table 3. Ranked Quantitative Mutagenicity Results by Weight for Positive Final Effluents and One Intake Water

Sample Number and Category	Revertants per Microgram[a]			
	TA98 (−S9)	TA98 (+S9)	TA100 (−S9)	TA100 (+S9)
25B Organic chemicals	3.49 ± 0.24	7.31 ± 0.64	9.46 ± 1.23	27.03 ± 3.18
25A (Duplicate)	3.52 ± 0.16	10.13 ± 0.54	11.65 ± 1.16	23.96 ± 3.53
7B Dyes	4.47 ± 0.28	5.27 ± 0.75	1.38 ± 0.19	0.90 ± 0.04[b]
7A (1st sample)	1.21 ± 0.10	2.13 ± 0.14	0.79 ± 0.19	0.77 ± 0.18
22 Dyes	0.58 ± 0.08	0.62 ± 0.07	0.80 ± 0.17	1.44 ± 0.16
17 Organic chemicals	0.56 ± 0.07	0.35 ± 0.04	–	0.26 ± 0.08
19 Dyes	0.44 ± 0.10	0.29 ± 0.04	0.27 ± 0.10	0.51 ± 0.08
11B Organic chemicals	+/–	0.12 ± 0.03	0.53 ± 0.15	0.46 ± 0.13
13 Pharmaceuticals	0.08 ± 0.02	+/–	0.49 ± 0.03	0.19 ± 0.01
11A Organic chemicals	0.04 ± 0.01	0.02 ± 0.01	0.32 ± 0.05	0.12 ± 0.02
16 Refinery	0.06 ± 0.02	0.09 ± 0.03	–	0.15 ± 0.09
21 Dyes	0.16 ± 0.06	–	–	–
26 Foundry	+/–	0.22 ± 0.01	–	+/–
10 Resins	0.03 ± 0.01	–	0.14 ± 0.03	0.04 ± 0.02
16 Refinery intake	0.01 ± 0.004	0.01 ± 0.004	–	–

[a]95% confidence intervals are provided. +/–, marginal; –, negative.

[b]Nonlinear slope.

Table 4. Ranked Quantitative Mutagenicity Results by Volume for Positive Final Effluents and One Intake Water

Sample Number and Category	Revertants per Milliliter[a]			
	TA98 (−S9)	TA98 (+S9)	TA100 (−S9)	TA100 (+S9)
25B Organic chemicals	23.03 ± 1.57	48.25 ± 4.20	62.46 ± 8.11	178.42 ± 20.97
25A (Duplicate)	23.47 ± 1.07	67.57 ± 3.60	77.67 ± 7.73	159.79 ± 23.55
7B Dyes	93.73 ± 5.81	110.61 ± 15.71	28.91 ± 4.05	18.78 ± 0.80[b]
7A (1st sample)	27.01 ± 2.15	47.65 ± 3.09	17.67 ± 4.18	17.31 ± 3.96
13 Pharmaceuticals	3.08 ± 0.89	+/−	18.41 ± 1.27	6.89 ± 0.52
22 Dyes	3.36 ± 0.47	3.59 ± 0.43	4.59 ± 0.98	8.32 ± 0.92
19 Dyes	4.34 ± 1.01	2.90 ± 0.41	2.64 ± 1.03	5.06 ± 0.81
17 Organic chemicals	3.78 ± 0.47	2.40 ± 0.27	−	1.79 ± 0.54
10 Resins	1.71 ± 0.51	−	7.74 ± 1.94	2.44 ± 1.05
11B Organic chemicals	+/−	1.57 ± 0.43	7.12 ± 2.07	6.12 ± 1.72
11A (1st sample)	0.80 ± 0.21	0.38 ± 0.11	6.00 ± 0.87	2.17 ± 0.37
16 Refinery	0.71 ± 0.21	0.97 ± 0.36	−	1.65 ± 0.96
26 Foundry	+/−	1.01 ± 0.44	−	+/−
21 Dyes	0.64 ± 0.23	−	−	−
16 Refinery intake	0.23 ± 0.08	0.17 ± 0.08	−	−

[a]95% confidence intervals are provided. +/−, marginal; −, negative.
[b]Nonlinear slope.

Although samples could be narrowly ordered within one strain and activation condition, for any practical application, environmental managers require a general overall ranking to prioritize further examinations of certain discharges. The ranking here was performed by considering the activity level within each test condition as well as the number of strains responding. The magnitudes of the other three tester strain responses were also considered. Table 3 shows that two discharges (sample numbers 25 and 7) had clearly higher potencies (revertants per microgram) than the other wastewaters. In general, industries manufacturing organic chemicals or using or producing dyes had more potent discharges than some other categories. When the ranked order of samples on an activity-per-milliliter basis (Table 4) is compared to the order on an activity-per-microgram basis, the influence that a high concentration of extractable organic matter can have is illustrated by sample number 13. This pharmaceutical effluent was ranked ninth by activity per weight, but fifth by activity per volume due to its high residue concentration (37.3 mg/liter).

Reproducibility of Results

In this study, two comparable final effluent samples were collected on separate occasions at the same facilities. In both cases, the samples produced the same positive qualitative mutagenicity results (sample numbers 7A and 7B and 11A and 11B). Strain sensitivity was also consistent: TA98 was the most responsive both times for facility 7, and TA100 was most responsive for facility 11. In order to compare the OSR *Salmonella* mutagenicity data to other qualitative results for the same discharges, Ames data were obtained from EPA's Region II New Jersey NPDES Industrial Screening Program (Fikslin, 1982). Despite the fact that this EPA laboratory utilized a dichloromethane solvent extraction sample preparation and generally collected the discharge samples at least one year prior to OSR, 7 out of 8 mutually sampled effluents produced the same qualitative Ames responses. EPA obtained results similar to OSR's for sample numbers 6, 7, 8, 11, 14, 16, and 22. Only sample number 24 produced a positive result when tested by EPA and a negative result one year later when assayed by OSR. These data may indicate that on a qualitative basis, Ames responses to industrial effluent samples may have some consistency over time.

Duplicate trials conducted in the IMR laboratory provide an indication of the assay reproducibility for effluent extracts on a quantitative basis. The mean percent difference of the 5 such duplicate trials with strains TA98 and TA100 (data not shown) was only 24.5%. The percent difference for each of these individual trials ranged from 0 to 56%.

Reproducibility considerations for the Ames analyses of effluent samples include not only the assay itself but also the extraction procedure. The determinations of extractable organic concentrations were relatively consistent in the two repeated, comparable samplings. Sample numbers 7A and 7B had residue concentrations of 22.4 and 21.0 mg/liter, and sample numbers 11A and 11B had concentrations of 18.7 and 13.5 mg/liter. The duplicate samples collected simultaneously at the industrial organics facility (numbers 25A and 25B) provide the best assessment of overall testing reproducibility. Residue concentrations of these samples were almost identical (6.7 and 6.6 mg/liter), and the percent difference in the mutagenic activity (revertants per microgram and revertants per milliliter) was only 16% for both strains and activation conditions (Tables 3 and 4). A graphical illustration of the encouraging similarity in mutagenic activities of the two duplicate samples is provided in Figure 4. This figure presents the dose-response curves for the most sensitive strain and activation condition. Additional duplicate analyses should be performed to determine if such reproducibility is to be routinely expected.

CONCLUSIONS AND IMPLICATIONS

The Ames assay was shown to have broad application for the assessment of the mutagenic activity of a diversity of industrial effluent types. Only one sample had sufficient cytotoxic effects to mask mutagenicity. A wide range of positive responses was quantified and ranked, providing environmental managers with some guidance as to what may be considered a "high" or "low" Ames response to effluent extracts. A concise format for data presentation was developed.

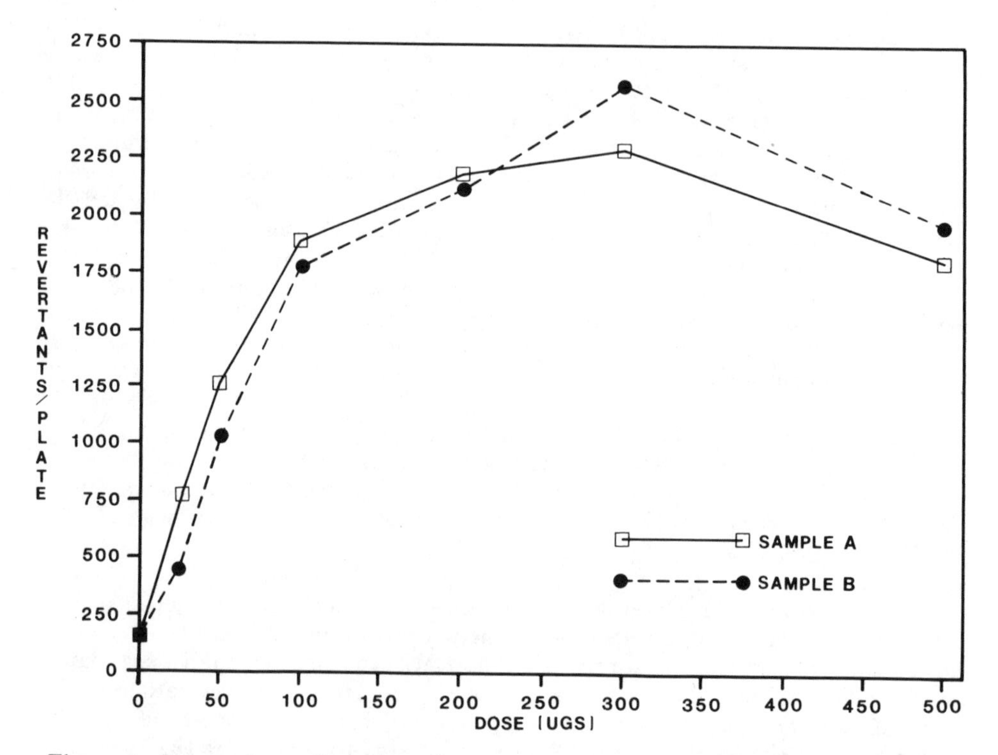

Figure 4. Comparison of mutagenic responses produced by duplicate samples collected at facility number 25, an organic chemicals industry. Assay results are for TA100 with S9. Each point represents the mean value of duplicate plate counts.

Some requirements of environmental monitoring quality assurance/quality control procedures have been reviewed and met. Laboratory and field blanks were demonstrated to be negative. Results of multiple assay trials and duplicate sample analyses revealed adequate reproducibility. Accuracy is an area that needs to be investigated through the testing of performance evaluation samples of complex mixtures by numerous laboratories. A pilot EPA-sponsored program to do just that is now in progress (Preston, 1984).

Water pollution control administrators may choose to utilize the Ames assay in a number of ways. Its primary function is foreseen as a screening tool to alert administrators to a potential toxics discharge problem, sometimes where none was previously suspected. Such screening could be done periodically for certain industrial categories by the environmental agency, or requested as a monitoring requirement in selected discharge permits. Positive Ames results for a specific wastewater could lead the agency to strongly encourage that particular industrial facility to further characterize its wastewater. Such further characterization might entail priority and nonpriority pollutant chemical analyses, or mutagenicity testing of specific effluent fractions or process streams to determine the compounds or processes responsible for the activity. OSR has already conducted work on chemical fractionation procedures and chemical group identification (e.g., aromatic amines) of positive effluent extracts. Another approach would be to evaluate, in concert with other pollutant measurements, the effects of control technologies and/or waste segregation techniques. In this study, carbon treatment was shown to be particularly effective in reducing the level of mutagenic substances from one secondary wastewater discharged from a dye facility. Such additional treatment may be effective for other mutagenic effluents.

The detection of mutagens in industrial effluents indicates that some constituent(s) of these wastewaters discharged to surface waters have the capability to cause an adverse biological effect. Such an effect could be deleterious from a public health standpoint, due to potential human exposure through contamination of potable water supplies, recreational waters, or edible aquatic biota. The Ames assay provides one mechanism for monitoring, and possibly controlling, the introduction of mutagens into surface waters.

ACKNOWLEDGMENTS

The authors would like to express their appreciation for the expert sampling and data analysis assistance provided by T. Fields, G. Harker, E. Stevenson, R. Lauer, and D. Keck, and the skilled technical laboratory assistance of V. Kwiatkowski and T. Schuck. This research was partially supported by the U.S. Environmental Protection Agency through TSCA Cooperative Agreement CS806854-01.

REFERENCES

American Public Health Association. 1980. Standard Methods for the Examination of Water and Wastewater, 15th ed. American Public Health Association: Washington, DC.

Ames, B., J. McCann, and E. Yamasaki. 1975. Methods for detecting carcinogens and mutagens with the *Salmonella*/mammalian-microsome mutagenicity test. Mutat. Res. 31:347-364.

Atherholt, T., G. McGarrity, and J. Louis. 1982. Mutagenic activity of sterilization indicators. Environ. Mutagen. 4:93-101.

Bean, R., R. Riley, and P. Ryan. 1978. Investigation of halogenated components formed from chlorination of marine water. In: Water Chlorination: Environmental Impact and Health Effects, Vol. 2. R. Jolley, H. Gorcheu, and D. Hamilton, eds. Ann Arbor Science Publishers: Ann Arbor, MI. pp. 223-233.

Cheh, A., J. Skochdopole, C. Heileg, P. Koski, and L. Cole. 1980a. Destruction of direct-acting mutagens in drinking water by nucleophiles: implications for mutagen identification and mutagen elimination from drinking water. In: Water Chlorination: Environmental Impact and Health Effects, Vol. 3. R. Jolley, W. Brungs, and R. Cumming, eds. Ann Arbor Science Publishers: Ann Arbor, MI. pp. 803-815.

Cheh, A., J. Skochdopole, P. Koski, and L. Cole. 1980b. Nonvolatile mutagens in drinking water: production by chlorination and destruction by sulfite. Science 207:90-92.

Chu, K., K. Patel, A. Lin, R. Tarone, M. Linhart, and V. Dunkel. 1981. Evaluating statistical analyses and reproducibility of microbial mutagenicity assays. Mutat. Res. 85:119-132.

Combes, R., and R. Haveland-Smith. 1982. A review of the genotoxicity of food, drug and cosmetic colours and other azo, triphenylmethane and xanthene dyes. Mutat. Res. 98:101-248.

Douglas, G., E. Nestman, J. Betts, J. Mueller, E. Lee, H. Stitch, R. San, R. Bruozes, A. Chmelaukas, H. Paavila, and C. Walden. 1980. Mutagenic activity of pulp mill effluents. In: Water Chlorination: Environmental Impact and Health Effects, Vol. 3. R. Jolley, W. Brungs, and R. Cumming, eds. Ann Arbor Science Publishers: Ann Arbor, MI. pp. 865-880.

Ellis, D., C. Jone, R. Larson, and D. Schaeffer. 1982. Organic constituents of mutagenic secondary effluents from waste treatment plants. Arch. Environ. Contam. Toxicol. 11:373-382.

EPA. 1978. Preparation of XAD-2 Adsorbent Resin. 600/7-78-201, U.S. Environmental Protection Agency.

Epler, J. 1980. The use of short-term tests in the isolation and identification of chemical mutagens in complex mixtures. In: Chemical Mutagens, Principles and Methods for Their Detection, Vol. 6. F. de Serres and A. Hollaender, eds. Plenum Press: New York. pp. 239-270.

Federal Register. 1979. Guidelines establishing test procedures for the analysis of pollutants. Fed. Regist. 44(233) (December 3).

Federal Register. 1980. Consolidated permit application forms for EPA programs. Fed. Regist. 45(98) (May 19).

Fikslin, T. 1982. Personal communication. EPA Region II Laboratory, Edison, NJ.

Harris, J., M. Cohen, Z. Gresser, and M. Hayes. 1981. Characterization of Sorbent Resins for Use in Environmental Sampling. 600/S2-80-193, U.S. Environmental Protection Agency: Research Triangle Park, NC.

Hartman, P. 1983. Mutagens: some possible health impacts beyond carcinogenesis. Environ. Mutagen. 5:139-152.

Junk, G., J. Richard, M. Grierer, D. Witiak, J. Witiak, M. Arguello, R. Vick, H. Svec, J. Fritz, and G. Calder. 1974. Use of macroreticular resins in the analysis of water for trace organic contaminants. J. Chromatogr. 99:745-762.

Kool, H., C. van Kreijl, H. Van Kranen, and E. de Greef. 1981. The use of XAD-resins for the detection of mutagenic activity in water I. Studies with surface water. Chemosphere 10:85-98.

Maron, D., and B. Ames. 1983. Revised methods for the *Salmonella* mutagenicity test. Mutat. Res. 113:173-215.

McCann, J., N. Spingarn, J. Kobori, and B. Ames. 1975. Detection of carcinogens as mutagens: bacterial tester strains with R factor plasmids. Proc. Natl. Acad. Sci. U.S.A. 72:979-983.

McGeorge, L., J. Louis, T. Atherholt, and G. McGarrity. 1983. Mutagenicity Analyses of Industrial Effluents: Background and Results to Date. New Jersey Department of Environmental Protection.

Preston, J. 1984. Personal communication. U.S. Environmental Protection Agency, NEIC, Denver, CO.

Rappaport, S. 1979. Mutagenic activity in organic wastewater concentrates. Environ. Sci. Technol. 13:957-961.

Rawlings, D., and M. Samfield. 1979. Textile plant wastewater toxicity. Environ. Sci. Technol. 13:160-164.

SAS (Statistical Analyses System). 1982. Version 82.3. SAS: Cary, NC.

de Serres, F., and M. Shelby. 1979. The *Salmonella* mutagenicity assay: recommendations. Science 203:563-565.

Somani, S., R. Teece, and D. Schaeffer. 1980. Identification of carcinogens and promoters in industrial discharges into and in the Illinois River. J. Toxicol. Environ. Health 6:315-331.

Stinnett, S., D. Nobel, E. Brown, and H. Love. 1981. Mutagenic Testing of Industrial Wastes from Representative Organic Chemical Industries. 600/2-81-007, U.S. Environmental Protection Agency.

Van Rossum, P., and R. Webb. 1978. Isolation of organic water pollutants by XAD resins and carbon. J. Chromatogr. 150:381-392.

Williams, L., and J. Preston. 1983. Interim Procedures for Conducting the *Salmonella*/Microsomal Mutagenicity Assay (Ames Test). U.S. Environmental Protection Agency: Las Vegas, NV.

ISOLATION OF MUTAGENIC COMPOUNDS FROM SLUDGES AND WASTEWATERS

M. Wilson Tabor,[1] John C. Loper,[1,2] Betty Lu Myers,[1] Laura Rosenblum,[1] and F. Bernard Daniel[3]

[1]Department of Environmental Health and [2]Department of Microbiology and Molecular Genetics, University of Cincinnati Medical Center, Cincinnati, Ohio 45267, and [3]Health Effects Research Laboratory, U.S. Environmental Protection Agency, Cincinnati, Ohio 45268

INTRODUCTION

Extracts of sludges and effluent wastewaters from municipal sewage treatment plants have been shown to be complex mixtures of organic chemicals (e.g., Baird et al., 1980; Bedding et al., 1982; Ellis et al., 1982; Johnston et al., 1982; Naylor and Loehr, 1982a, b; Clevenger et al., 1983; Hrubec et al., 1983; Strachan et al., 1983). During the past few years, these residue organics have been characterized as toxic and/or mutagenic in a variety of bacterial, animal, and plant test systems (e.g., Rappaport et al., 1979; Baird et al., 1980; Neal et al., 1980; Babish et al., 1982; Hopke et al., 1982; Johnston et al., 1982; Babish et al., 1983; Clevenger et al., 1983; Hopke and Plewa, 1983; Hrubec et al., 1983; Maciorowski et al., 1983; Hopke et al., in press; Meier and Bishop, 1984). The major mutagens appeared to be in the nonvolatile residue organic materials from sewages that included industrial sources. Although many volatile compounds, including priority pollutants, have been identified and quantified (e.g., Bedding et al., 1982; Harrold et al., 1982; Johnston et al., 1982; Naylor and Loehr, 1982a, b; Petrasek et al., 1983; Strachan et al., 1983), the chemical identity and source of the vast majority of the mutagens is unknown (Nellor et al., 1982).

Chemical characterization of these mutagens is necessary to assess the importance of the compounds as potential hazards to human health and to develop methods to reduce their concentrations in wastewaters and sludges. Progress toward this goal has been slow, however, primarily due to the lack of reliable methods for the isolation of the compounds from wastewaters and sludges for chemical and biological characterization. By evaluating or adapting existing methods and by developing new ones, our research is directed toward the identification of the major mutagens in wastewaters and sludges.

MATERIALS AND METHODS

Samples

Wastewater and sludge samples were taken at the U.S. Environmental Protection Agency Municipal Environmental Research Laboratory-Cincinnati (EPA/MERL-CIN) Technical and Evaluation Facility at the Mill Creek Sewage Treatment Plant of the Metropolitan Sewer District in Cincinnati, OH. The sample set consisted of influent wastewater screened prior to acquisition to remove debris, 20 L; primary sludge, 9 L; secondary sludge, 9 L; and secondary effluent, 17 L. Wastewater samples were collected in 20-L stainless steel pressure reservoirs (Amicon, Lexington, MA), and the sludge samples were collected in 4-kg amber glass jars fitted with Teflon-lined caps. The effluent wastewater was stored at 2°C for processing within 2 d. The remainder of the samples were divided into smaller aliquots and then frozen at -20°C for processing at a later time.

Bacterial Strains and Bioassays

Salmonella typhimurium strains TA98 and TA100 were obtained from B.N. Ames. The properties of the strains and their uses with standard positive and negative controls were as previously described (Loper et al., 1978). Microsomal activation of experimental samples and control compounds was accomplished using polychlorinated biphenyl (Aroclor 1254)-induced rat liver microsomes from Litton Bionetics (Charleston, SC). Detection of mutagenic activity in experimental samples was based on a dose-dependent response that exceeded the control (zero dose) value by at least 2-fold (total revertant colonies per plate/control colonies per plate ≥ 2). Mutagenic activity was determined from initial rates of dose-response curves obtained in replicate experiments consisting of a minimum of four doses per strain per experiment.

Preparation of Residue Organics from Wastewaters

Nonvolatile organics were extracted from wastewaters by chromatography using either XAD-2 columns alone or XAD-2 and XAD-7 columns in series (Loper et al., 1982; Loper and Tabor, 1983). Influent wastewater was passed through 200 cm^3 of tightly packed silanized glass wool and then through a 1-cm x 25-cm column of prewashed Celite 545 (Fisher Scientific, Pittsburgh, PA). (NOTE: The Celite 545 was prewashed by the following procedure: slurry the filter aid in 500 ml of American Society for Testing Materials (ASTM) Type IV water by swirling, settle briefly and decant the supernatant fluid containing the fine particles, and repeat the process with a second aliquot of ASTM Type IV water; the aqueous slurry is poured, according to standard procedures, into a glass chromatography column fitted with a Teflon draw-off valve. Following removal of the water to the top of the Celite, the column is washed successively with two 100-ml aliquots of nanograde acetone, 100 ml of nanograde hexane, 100 ml of nanograde acetone, and is then rinsed with five column volumes of ASTM Type I water.) The Celite 545 and the biological filters were soxhlet extracted 48 h with 250 ml methylene chloride. The XAD columns were eluted with hexane:acetone (85:15 by volume), and eluates of residue organics were concentrated, processed, and assayed for mutagenic activity as previously described (Loper et al., 1982; Loper and Tabor, 1983). For control Celite 545 processed in a similar manner, extracted residues showed no mutagenicity to TA98 or TA100 tested in the absence and presence of microsomal activation.

Effluent wastewaters were processed via the XAD column procedure without Celite 545.

Preparation of Residue Organics from Sludges

Sludge samples were thawed, allowed to settle overnight at 4°C, and the supernatant fluids were decanted. The supernatant solution comprised approximately 75% of the volume of the parent sludge sample. The slurry remaining after decanting was centrifuged for 20 min at 8000 x g at 6°C. This supernatant fluid was combined with the above decanted solution and the pool was stored for processing via the XAD procedure previously described. The remaining moist pellet was processed for the preparation of residue organics extracts via the alternative methods described later. A weighed portion of each sludge pellet was oven dried at 125°C for 24 h under a negative pressure (at 30 mm Hg) and then reweighed for the calculation of moisture loss. These values were used in expressing the bioassay results in terms of net revertant colonies per gram dry weight of sludge sample.

Soxhlet extraction of sludge samples. A 50-g portion of moist pellet obtained from either the primary or secondary sludge was extracted via the following modified version of the method described by Hites (Elder et al., 1981; Jungclaus et al., 1978). The moist pellet was added to a borosilicate glass thimble fitted with a fritted disc of 40- to 60-μm pore size (Fisher Scientific). The thimble was placed in a standard soxhlet extraction apparatus (e.g., a Pyrex 09-557 series, Fisher Scientific) and the sample was extracted with 300 ml of nanograde isopropanol for 7 h. This extraction was followed by a 7-h extraction with 300 ml of nanograde benzene. These extracts were passed through separate colunns, 1 cm x 25 cm, of anhydrous sodium sulfate to remove traces of water. (NOTE: sodium sulfate, reagent grade, low in nitrogen and suitable for Kjeldahl analysis was muffled at 500°C for 6 h prior to use.) Each drying column was washed with 100 ml of the respective solvent followed by 50 ml of the 85:15 hexane:acetone solvent system. These wash eluates were combined with the respective extracts and the two samples were concentrated 100 times via rotary evaporation under a negative pressure (at 30 mm Hg) at a temperature of 50°C. Samples then were concentrated to volumes of a few milliliters via evaporation at 50°C under dry nitrogen. Immediately prior to bioassay, an aliquot of this residue organics solution was concentrated to dryness via gentle evaporation, and the residue was dissolved in dimethyl sulfoxide for mutagenesis testing.

Methylene chloride extraction of sludge samples. A 50-g sample of moist pellet obtained from either the primary or secondary sludge was extracted via the following modified version of the U.S. Environmental Protection Agency Environmental Monitoring and Support Laboratory-Cincinnati (USEPA/EMSL-CIN) Method 624S/625S (Billets and Lichtenburg, 1983). The sample was homogenized for 1 min in 300 ml of nanograde methylene chloride using a Sorvall Omnimixer (Du Pont Co., Newtown, CT) at the high speed setting. Overheating during homogenization was avoided by immersing the sample container in an ice bath. The homogenate was adjusted to pH ≥ 11 by the addition of 1.0 N sodium hydroxide, homogenization was repeated, and the sample was centrifuged at 1500 x g for 10 min to separate the phases. The methylene chloride layer was removed and the extraction/centrifugation procedure was repeated twice. The residual aqueous layer was retained for further extraction. The methylene chloride solutions of the base/neutral extracts were combined and dried by passage through a column of anhydrous sodium sulfate, as described earlier. The column was washed with an additional 100 ml of methylene chloride followed by 50 ml of the 85:15 hexane:acetone solvent system. The wash eluates were combined with the sample eluates and the resultant was concentrated via Kuderna-Danish evaporation to a volume of 10 ml. Immediately prior to bioassay, an aliquot of this residue organics solution was concentrated to dryness via gentle evaporation, and the residue was dissolved in dimethyl sulfoxide for mutagenesis testing.

An additional 300 ml of methylene chloride was added to the residual aqueous layer, the sample was homogenized as before, and the pH of the homogenate was then adjusted to ≤ 2 by the addition of 6.0 N hydrochloric acid. Homogenization was repeated and the suspension was centrifuged at 1500 x g for 10 min to separate the phases. The methylene chloride layer was removed and the extraction/centrifugation procedure was repeated twice. The methylene chloride solutions of the acid extracts were combined and dried as before. The sample and wash eluates were combined, concentrated, and bioassayed as described for the base/neutral extract.

Milling/extraction of sludge samples. A 100-g portion of moist pellet obtained from either the primary or secondary sludge was mixed with 600 g of anhydrous sodium sulfate, and the mixture was added to a 1-L ball mill (U.S. Stoneware, Akron, OH). Eight 1-in. diameter stainless steel bearings and 60 $\frac{1}{2}$-in. stainless steel bearings were added, and the sample was milled for 5 to 9 h. Milling was complete when the sample had a consistency of flour. Milled samples were stored at -20°C until further extraction.

Eight 4-g aliquots of the milled sample were added to a glass extraction thimble for soxhlet extraction, as described above. (NOTE: 12 g of moist sludge yields 84 g of milled sample.) The sample was first extracted with 300 ml of high performance liquid chromatography (HPLC)-grade 1,1,2-trichloro-1,2,2-trifluoroethane (i.e., Freon 113) for 2 h. Following Freon extraction, the sample was first extracted for 2 h with 300 ml of methylene chloride and was then extracted for 2 h with isopropanol. Each of the three extracts was concentrated one hundred times via rotary evaporation under a negative pressure (at 35 mm Hg) at a temperature of either 30°C for the Freon and methylene chloride extracts or 50°C for the isopropanol extracts. The extracts were further concentrated to a small volume via evaporation at 40°C under dry nitrogen. Immediately prior to bioassay, an aliquot of a residue organics concentrate was concentrated to dryness via gentle evaporation, and the residue was dissolved in dimethyl sulfoxide for mutagenesis testing.

HPLC Separation of Residue Organics

High performance liquid chromatography analytical-scale separations were performed on a Waters Associates (Milford, MA) ALC/GPC 205 system fitted with a 254-nm absorbance detector and a 3.9-mm x 30-cm prepacked column of 10-μm silica particles bonded with octadecylsilane, as previously described (Tabor and Loper, 1980; Loper and Tabor, 1981). Analytical-scale HPLC separations were accomplished using a water to acetonitrile linear gradient. Solvents were degassed by sonication under reduced pressure for 15 min immediately prior to use.

Other Chemicals

American Society for Testing Materials Type I water and Type IV water were generated in our laboratory using a Continental-Millipore Water Conditioning System, as previously described (Tabor and Loper, 1980). Nanograde solvents (Mallinckrodt, St. Louis) and HPLC-grade solvents (Fisher Scientific) were used without further purification. All other chemicals were reagent grade and were used as obtained.

RESULTS AND DISCUSSION

Sampling Strategy for Municipal Treatment Plants

The sampling strategy for this study was to sample both municipal treatment plant raw material (influent water) and plant products (effluent water/sludges) (Figure 1). The samples were taken as composite grab samples in accordance with recommendations on sampling for the analysis of priority pollutants in municipal treatment plant sludges (Billets and Lichtenburg, 1983).

At the treatment plant studied, approximately 1% of the total influent water mass leaves the plant as sludges. The primary sludge represents a removal of 35-40% of the suspended solids and biochemical oxygen demand from the influent wastewater; the secondary sludge represents a removal of 85-90% of the remaining suspended solids

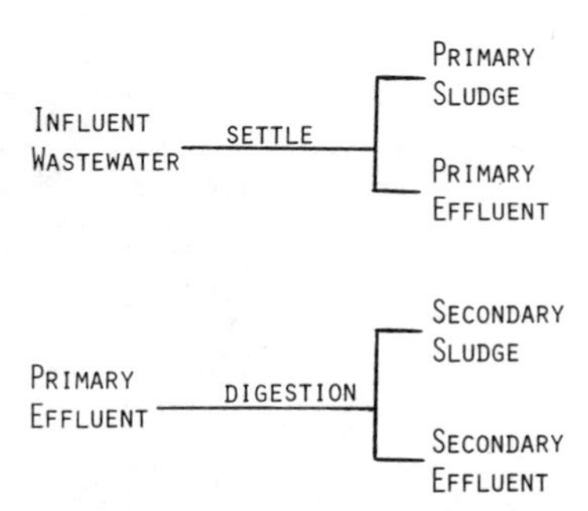

Figure 1. Schematic of municipal sewage treatment at the EPA/MERL-CIN Technical and Evaluation Facility, Cincinnati, OH.

and biochemical oxygen demand. It was recently reported (Petrasek et al., 1983) that the activated sludge treatment process at the plant where we performed our studies was not a highly effective system for controlling the discharge of toxic compounds as measured by the removal of priority pollutant organics spiked into the process stream.

Extraction of Residue Organics from Wastewater Samples

Residue organics were concentrated from the wastewater samples, both influent and effluent waters, according to the system outlined in Figure 2. For the influent water, it was necessary to clarify the sample. Centrifugation was tried but was inefficient, and the filtration procedure involving silanized glass wool Celite and bacterial filters was adopted. All of the TA98 direct-acting mutagenesis for the influent water sample was associated with the residue organics extracted from the Celite. No microsomal-dependent TA98 or TA100 mutagenesis was observed for this residue organics extract.

Further extraction of residue organics from the influent water was accomplished via XAD chromatography (Figure 2), according to our previously described method (Loper and Tabor, 1982; Loper et al., 1982). Bioassay of the residue organics extracted from each component of the system showed microsomal-dependent TA98 mutagenicity to be associated only with the bacterial filters and the XAD-2. No other mutagenic activity, either TA100 or direct-acting TA98, was observed for these residue organics.

The effluent wastewater sample was processed directly via the XAD column procedure (Figure 2). Of the TA98 direct-acting mutagenicity recovered, approximately 80% was eluated from the glass wool column. Neither TA98 microsomal-dependent mutagenesis nor TA100 mutagenesis was observed in the glass wool eluate for effluent wastewater. The remainder of the TA98 direct-acting mutagenicity, and all of the TA98 microsomal-dependent mutagenicity, observed for the effluent wastewater

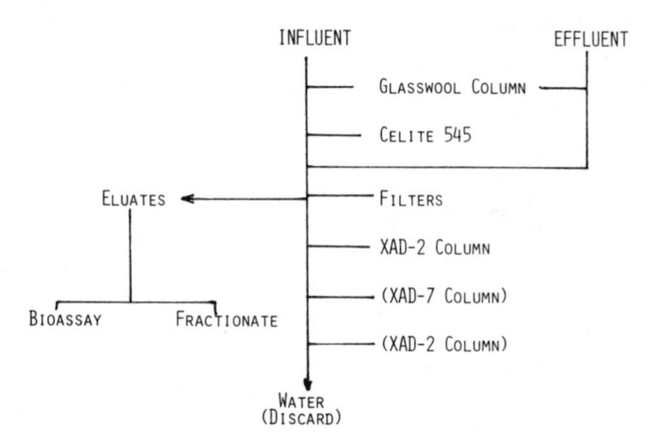

Figure 2. Flow diagram for concentrating residue organics from influent and effluent wastewater samples.

residue organics was in the XAD-2 eluate. None of the effluent wastewater eluates had mutagenic effects for TA100.

These results for the secondary effluent wastewater are in general agreement with those reported by Baird et al. (1980, 1981) and Nellor et al. (1984). Moreover, these authors did observe mutagenesis to both TA98 and TA100 for some tertiary effluent wastewaters and chlorinated effluent wastewaters. In other studies of mutagenicity of effluent wastewaters from municipal sewage plants that treated influent wastewaters primarily from industrial sources (e.g., Rappaport et al., 1979; Johnston et al., 1982; Neal et al., 1980), mutagenic activity to both TA98 and TA100 was observed.

Extraction of Residue Organics from Sludge Samples

One of the overall objectives of our research was to compare existing methods for the extraction of residue organics from sludge to our method of sludge extraction. The approach employed with the primary and secondary sludge methods is outlined in Figure 3. Because the water content of sludge samples varies, all samples were settled, decanted, and centrifuged prior to extraction. An aliquot of the moist pellet was dried to constant weight in order to reference mutagenesis assays to activity per gram of dry weight.

The three methods used for extraction of the residue organics from the sludges were (1) a modification of the USEPA/EMSL-CIN Method 624S/625S (Figure 4) (Billets and Lichtenburg, 1983); (2) a modification of the soxhlet extraction method of Hites (Figure 5) (Elder et al., 1981; Jungclaus et al., 1978); and (3) our ball milling method (Figure 6). The resultant residue organics from each extraction procedure were assayed for TA98 and TA100 mutagenicity. None of the residue organic extracts were mutagenic to TA100, but both direct-acting and microsomal-dependent TA98 mutagenicity was found for the primary and secondary sludge samples. These results are summarized in Table 1 for each of the three sludge extraction methods; Table 1 also

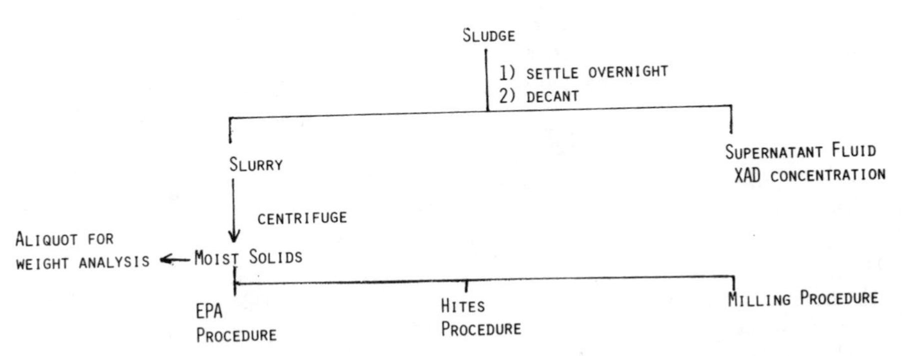

Figure 3. Flow diagram of strategy for processing primary and secondary sludge samples.

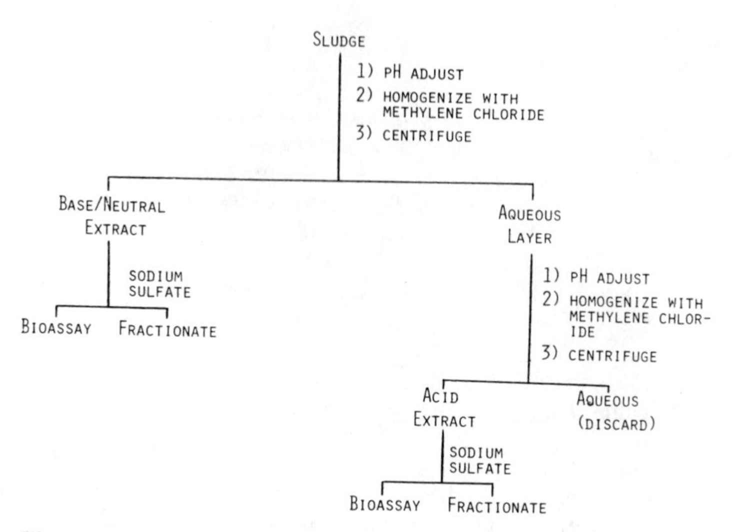

Figure 4. Flow diagram for EPA method for the preparation of residue. Organic extracts from moist pellet sludge samples. (Billets and Lichtenburg, 1983).

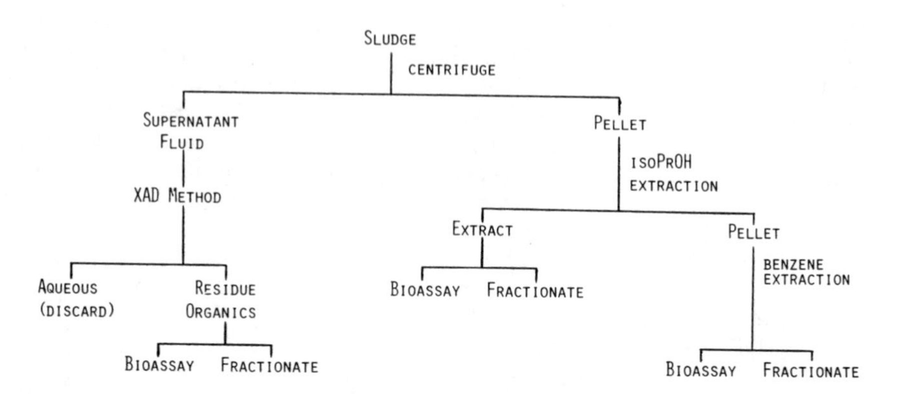

Figure 5. Flow diagram of Hites method for the preparation of residue organic extracts from moist pellet sludge samples (Jungclaus et al., 1978).

contains a summary of the TA98 mutagenic activities found in the influent and effluent waste waters.

The amount of mutagenicity found in the residue organics prepared by each method varied by as much as a factor of 3.5. Nevertheless, the primary sludge yielded more TA98 direct-acting mutagenicity than the secondary sludge; the secondary sludge appeared to contain more TA98 microsomal-dependent mutagenic activity than direct-acting mutagenicity. Part of the variation in recoveries of mutagenic activity by the individual methods for a particular sludge sample may be artifacts of the methods. In

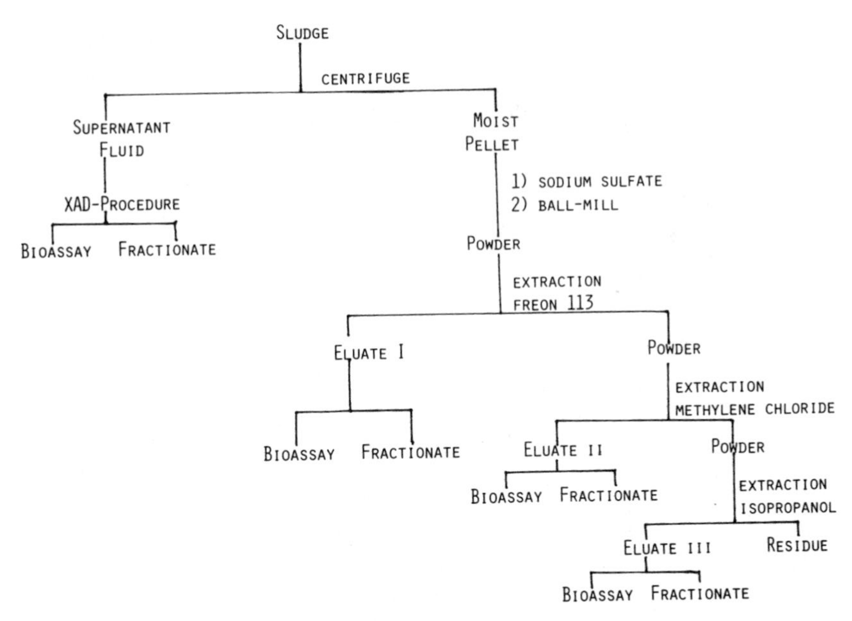

Figure 6. Flow diagram for the preparation of residue organic extracts from moist pellet sludge samples by the milling procedure.

particular, it is probable that the EPA and the Hites methods are altering the constituents in the sample. The EPA method involves extractions under extremes of pH, both basic and acidic, conditions that favor many commonly known organic reactions. Under acidic conditions, for example, secondary and tertiary alcohols are dehydrated to form alkenes. Under basic conditions, not only do secondary and tertiary alkyl halides readily undergo nucleophilic substitution reactions by hydroxide ions to form alcohols, but also many aldehydes and ketones with methylene hydrogens alpha to the carbonyl group undergo base catalyzed condensation reactions (e.g., the aldol condensation reaction). Furthermore, both acidic and basic conditions readily promote hydrolysis of compound classes such as amides and esters. From these few examples of chemical reactions possible under conditions of high or low pH, it is evident that for either primary or secondary sludge, the EPA sludge extraction method may either destroy mutagenic components in the original sample or cause nonmutagenic components to be converted to mutagenic constituents. These possible problems with the EPA method are under investigation.

The Hites procedure has been shown to be applicable to the extraction of a wide variety of classes of organic compounds from moist solid samples such as river sediments (Jungclaus et al., 1978; Elder et al., 1981). For the primary sludge, the Hites extraction procedure gave significantly lower mutagenicity recoveries compared to the EPA and ball mill procedures (Table 1). These results indicate that either the Hites procedure is not effective in extracting the mutagenic constituents from this sludge or that mutagens in the sample are being deactivated via some chemical process involving the sample and the method. As to the latter possibility, both the moderately

Table 1. TA98 Mutagenic Activity for Wastewater Treatment Plant Influent Water, Effluent Water, and Sludges[a]

Microsomal Activation	Influent Water Net Revertants/L	Primary Sludge Net Revertants/g Dry Wt			Secondary Sludge Net Revertants/g Dry Wt			Secondary Effluent Water Net Revertants/L
		EPA	Hites	Mill	EPA	Hites	Mill	
−	.09	218.0	0	161.0	24.0	97.0	68.0	2.8
+	12.0	251.0	34.0	75.0	227.0	107.0	65.0	3.8

[a]Mutagenic activity in TA98 expressed as net revertants in thousands of colonies.

high temperatures required for the solvents for an extended period of time in the operation of the soxhlet extraction process of the Hites procedure and the chemical constitution of the primary sludge could contribute to the low recovery of mutagenic activity. Primary sludges are complex mixtures of chemicals that usually contain high concentrations of metals (Naylor and Loehr, 1982a; Clevenger et al., 1983). These constituents can catalyze many organic reactions, including elimination, oxidation/reduction, and condensation reactions, particularly under conditions where heat is applied over an extended period of time.

Likewise, the disparity in mutagenicity via the Hites versus the mill procedure for the secondary sludge sample (Table 1) may be attributable to the nature of the compounds present in the sample. Activated sludge contains many new chemicals in the secondary sludge as the result of biochemical reactions accomplishd by microorganisms (Bedding et al., 1983). Many metabolic intermediates and products are heat labile and thereby have a potential for conversion to different compounds, i.e., artifacts of the real sample, under the moderately high temperature conditions employed in the Hites procedure. These possible problems with the Hites method are under investigation.

The ball mill procedure developed in this study for extracting sludges avoids many of the aforementioned problems with the EPA and Hites procedures. Sludges are milled with anhydrous sodium sulfate, producing a dehydrated sample of uniform consistency that is easily manipulated. The sample is soxhlet extracted with Freon 113 (nonpolar) and then methylene chloride (moderately polar), both of which are solvents with boiling points $<45°C$, although these extractions are followed by a higher boiling polar solvent (isopropanol). The soxhlet extraction time for each solvent is only 2 h and, thus, compares favorably with the Hites procedure, which requires a time of 7 h or more (Jungclaus et al., 1978). Longer times for soxhlet extractions using other solvents have been employed to extract residue organics from sludges (e.g., Babish et al., 1982; Strachan et al., 1983). In a separate study (data not shown), multiple extractions of the sample with each sequential solvent did not significantly increase or decrease the yield of mutagenic activity. However, the sequential extraction procedure employed (i.e., Freon 113, then methylene chloride, etc.) yielded residue organics with the greatest amounts of mutagenic constituents in the nonpolar solvents (Table 2). The use of alternative solvents, e.g., acetone, and alternative salts for dehydration/milling, e.g., a basic salt like sodium carbonate, in the ball mill procedure are under investigation.

HPLC Separation of Residue Organics from Sludges

Analytical-scale HPLC separations of residue organics from sludges were run as a preliminary guide to preparative-scale HPLC separations for mutagen isolation (Tabor and Loper, 1980). Both primary and secondary sludge residue organics prepared by the three methods (Figures 3-6) were examined. Representative chromatograms for the residue organics from the primary sludge prepared via our ball mill procedure are shown in Figure 7. In these chromatograms of the reverse phase HPLC separation, the nonpolar solvent (Freon 113) extract of residue organics is composed mainly of nonpolar components, as detected by absorbance at 254 nm. The second residue organics extract of the sequential process (the moderately polar methylene chloride sample) contains a broad distribution of compounds, ranging from

Table 2. TA98 Mutagenic Activity of Residue Organics from Sludge Sample Extracts Prepared by the Ball Mill Procedure[a]

Microsomal Activation	Residue Organics Extract					
	Primary Sludge			Secondary Sludge		
	Freon 113	Methylene Chloride	Isopropanol	Freon 113	Methylene Chloride	Isopropanol
−	91.0	71.0	0	31.0	35.0	1.0
+	21.0	42.0	13.0	39.0	14.0	11.0

[a]Mutagenic activity in TA98 expressed as net revertants in thousands of colonies.

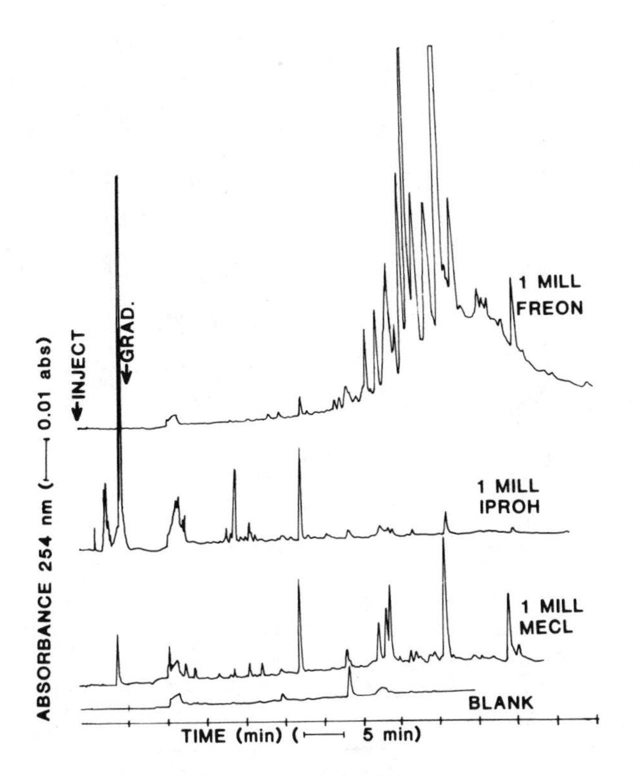

Figure 7. Reverse phase HPLC separation of residue organics extracted from primary sludge by the milling procedure (1 Mill). Samples representing 200-μg equivalents of sludge were fractionated by injecting the sample and eluting with water, followed by a linear gradient from 100% water to 100% acetonitrile. Chromatograms shown (top to bottom) are Freon 113-extracted residue organics, isopropanol-extracted residue organics, methylene chloride-extracted residue organics, and a solvent blank.

polar to nonpolar. The residue organics from the polar solvent extraction (isopropanol) are composed almost totally of polar constituents. Similar HPLC results were observed for the secondary sludge ball mill procedure extracts (data not shown). This distribution of compounds agrees with the mutagenesis data (Table 2) and the data of others (Maciorowski et al., 1983; Hopke et al., in press) in that the bulk of the mutagenic constituents in wastewater treatment plant products appears to be from relatively nonpolar compounds. Using the preliminary data from the analytical HPLC separations as a guide, current research efforts are directed to mutagen isolation from the ball mill procedure residue organics, using our previously developed approach (Tabor and Loper, 1980; Tabor, 1983).

Analytical HPLC separations of the residue organics extracted from the primary sludge via the EPA and Hites procedures are shown in Figures 8 and 9, respectively.

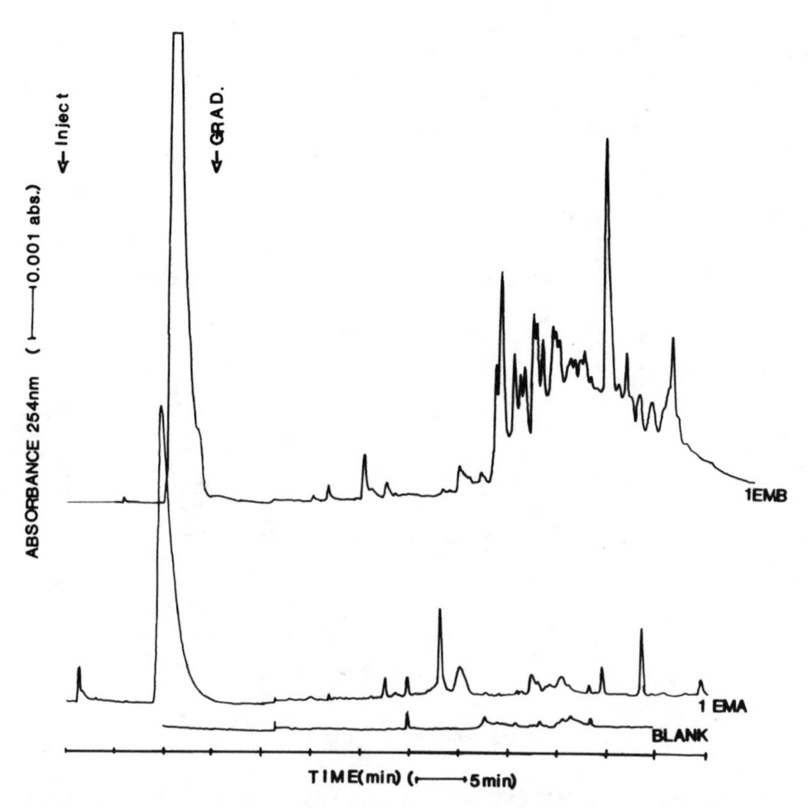

Figure 8. Reverse phase HPLC separation of residue organics extracted from primary sludge by the modified EPA 624S-625S procedure (IE). Samples representing 200-µg equivalents of sludge were fractionated by injecting the sample and eluting with water, followed by a linear gradient from 100% water to 100% acetonitrile. Chromatograms shown are the base/neutral-extracted residue organics, 1EMB; the acid-extracted residue organics, 1EMA; and a solvent blank.

For the residue organics obtained via each procedure, the first extraction step (see Figures 4 and 5) yielded the bulk of the 254-nm absorbing material separated via reverse phase HPLC. For the second EPA extraction step (acid extracts), very few constituents were found (Figure 8). However, the second extraction step in the Hites procedure yielded at least one major group of semipolar compounds (Figure 9). The mutagenesis data for each of these extracts (Table 3) is in agreement with the HPLC results in that all of the TA98 mutagenic activity extracted via the EPA procedure was in the base/neutral residue organics, whereas the Hites procedure showed TA98 mutagenicity in both the isopropanol- and benzene-extracted residue organics. Considering the possibility that both of these procedures may be producing mutagenic artifacts of extraction, we are pursuing the isolation of these mutagens for compound identification.

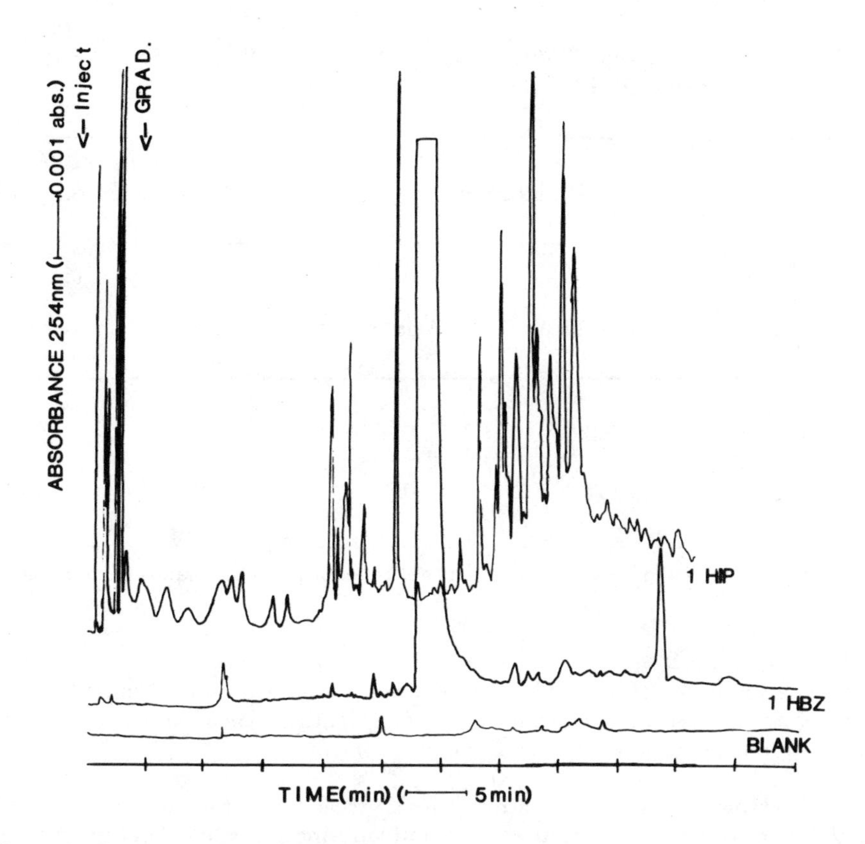

Figure 9. Reverse phase HPLC separation of residue organics extracted from primary sludge by the modified Hites procedure (IH). Samples representing 200-µg equivalents of sludge were fractionated by injecting the sample and eluting with water, followed by a linear gradient from 100% water to 100% acetonitrile. Chromatograms shown (top to bottom) are the isopropanol-extracted residue organics, 1 HP; the benzene-extracted residue organics, 1 HBZ; and a solvent blank.

CONCLUSIONS

The overall objective of our work is the isolation of nonvolatile organic mutagens of municipal sewage treatment plant influent/effluent wastewaters and sludges to characterize the mutagens both chemically and biologically. A biological approach to the chemical fractionation of residue organics is the method of choice for the isolation of hazardous compounds from such complex mixtures. We have previously described a general analytical fractionation and bioassay procedure for the isolation of mutagens from residue organics extracted from environmental samples (Tabor and Loper, 1980; Loper and Tabor, 1981; Loper and Tabor, 1982; Loper et al., 1982). To focus the analytical fractionation of residue organics to mutagen isolation, the *Salmonella* mutagenesis assay is well suited as a bioassay of subfractions (Loper, 1980a, b).

Table 3. Residue Organics TA98 Mutagenic Activity for Primary Sludge Extracts Prepared by the EPA and Hites Procedures[a]

	EPA Extracts		Hites Extracts	
Microsomal Activation	Base/Neutral	Acid	Isopropanol	Benzene
–	218.0	0	0	0
+	251.0	0	21.0	13.0

[a]Mutagenic activity in TA98 expressed as net revertants in thousands of colonies.

However, a methodological problem evident when we began our studies to isolate mutagens from wastewaters and sludges was the identification of methods appropriate for the extraction of residue organics from these samples.

In this report, we have described a general procedure for the extraction of residue organics from two types of wastewaters: influent and effluent. This procedure is similar to that described by others in their studies of wastewaters in which various XAD resins were employed for residue organic concentration (e.g., Rappaport et al., 1979; Baird et al., 1980; Hopke et al., 1982; Johnston et al., 1982). Also, our method is based on our XAD procedure originally described for the concentration of residue organics from drinking water (Loper and Tabor, 1982; Loper et al., 1982). However, the isolation of residue organics from sludges presented a different problem. Our results suggest that two previously described general approaches, the EPA procedure (Billets and Lichtenburg, 1983) and the Hites procedure (Jungclaus et al., 1978; Elder et al., 1981) appear to be producing mutagenic artifacts of isolation or destroying labile mutagens during isolation. Therefore, we have developed a procedure for isolating mutagenic residue organics from sludges that features not only gentle conditions but also produces a homogeneous sample that is easily manipulated.

In a broader sense, our results show that the wastewater and sludges of a major municipal sewage treatment plant impacted heavily by industrial effluents contain TA98 mutagens either in high concentrations or of a high potency. The level of mutagenic activity was considerably higher on a per mass basis in the treatment plant products, sludges and effluents, than in the treatment plant influent. This is of significance to individuals responsible for the treatment of municipal sewage in their overall concern for the presence of toxic/biohazardous organics in the treatment plant products (personal communication, Fred Bishop, USEPA/MERL-CIN, Cincinnati, OH; Hrudey and Smith, 1983). This concern has raised questions such as (1) does the treatment process remove toxic organics from the influent wastewater and (2) does the treatment process result in an increase in the amount of toxic organics in the treatment plant products?

By providing more information as to the identity of mutagenic compounds, their frequency of occurrence, their possible source, and their levels in treatment plant products, the results of our research effort will assist in the definition of waste management procedures.

ACKNOWLEDGMENTS

Our appreciation goes to Mr. Richard Hutchenson and Mr. Michael Niemi for their expert laboratory assistance and to Mrs. Nancy Knapp and Mrs. Mary Jo Frost for skillful and unstinting labors. This research was supported by USEPA CR810792-01.

REFERENCES

Babish, J.G., B.E. Johnson, B.O. Brooks, and D.J. Lisk. 1982. Acute toxicity of organic extracts of municipal sewage sludge in mice. Bull. Environ. Contam. Toxicol. 29:379-384.

Babish, J.G., B.E. Johnson, and D.J. Lisk. 1983. Mutagenicity of municipal sewage sludges of American cities. Environ. Sci. Technol. 17:272-277.

Baird, R., J. Gute, C. Jacks, R. Jenkins, L. Neisess, B. Scheybeler, R. Van Sluis, and W. Yanko. 1980. Health effects of water reuse: A combination of toxicological and chemical methods for assessment. In: Water Chlorination: Environmental Impact and Health Effects, Vol. 4, Book 2. R.L. Jolley et al., eds. Ann Arbor Science Publishers Inc.: Ann Arbor, MI. pp. 925-935.

Baird, R.B., C.A. Jacks, R.L. Jenkins, J.P. Gute, L. Neisess, and B. Scheybeler. 1981. A high performance macroporous resin concentration system for trace organic residues in water. In: Chemistry in Water Reuse, Vol. 2. W.J. Cooper, ed. Ann Arbor Science Publishers Inc.: Ann Arbor, MI. pp. 149-169.

Bedding, N.D., A.E. McIntyre, R. Perry, and J.N. Lester. 1982. Organic contaminants in the aquatic environment I. Sources and occurrence. Sci. Total Environ. 25:143-167.

Bedding, N.D., A.E. McIntyre, R. Perry, and J.N. Lester. 1983. Organic contaminants in the aquatic environment II. Behavior and fate in the hydrological cycle. Sci. Total Environ. 26:255-312.

Billets, S., and J.J. Lichtenburg. 1983. Interim Methods for the Measurement of Organic Priority Pollutants in Sludges. Physical and Chemical Methods Branch. Environmental Monitoring and Support Laboratory, U.S. Environmental Protection Agency: Cincinnati, Ohio, 45268. pp. 1-70.

Clevenger, T.E., D.D. Hemphill, K. Roberts, and W.A. Mullins. 1983. Chemical composition and possible mutagenicity of municipal sludges. J. Water Pollut. Control Fed. 55(12):1470-1475.

Elder, V.A., B.L. Proctor, and R.A. Hites. 1981. Organic compounds near dumpsites in Niagara Falls, New York. Biomed. Mass Spectrom. 8:409-415.

Ellis, D.D., C.M. Jone, R.A. Larson, and D.J. Schaeffer. 1982. Organic constituents of mutagenic secondary effluents from wastewater treatment plants. Arch. Environ. Contam. Toxicol. 11:373-382.

Harrold, D.E., J.C. Young, and M. Asce. 1982. Extraction of priority pollutants from solids. American Society of Civil Engineers 108:EE6. pp. 1211-1227.

Hopke, P.K., M.J. Plewa, J.B. Johnston, D. Weaver, S.G. Wood, R.A. Larson, and T. Hinesly. 1982. Multitechnique screening of Chicago municipal sewage sludge for mutagenic activity. Environ. Sci. Technol. 16:140-147.

Hopke, P.K., and M.J. Plewa. 1983. The Evaluation of the Mutagenicity of Municipal Sewage Sludge. EPA-600/1-83-016.

Hopke, P.K., M.J. Plewa, and D.L. Weaver. (in press). Comparison of the mutagenicity of sewage sludges. Environ. Sci. Technol.

Hrubec, J., C.F. van Kreijl, C.F.H. Morra, and W. Slooff. 1983. Treatment of municipal wastewater by reverse osmosis and activated-carbon-removal of organic micropollutants and reduction of toxicity. Sci. Total Environ. 27:71-88.

Hrudey, S.E., and D.W. Smith. 1983. Water -- reclamation and reuse. J. Water Pollut. Control Fed. 55:662-674.

Johnston, J.B., R.A. Larson, J.A. Grunau, D. Ellis, and C. Jones. 1982. Identification of Organic Pollutants and Mutagens in Industrial and Municipal Effluents. Final report submitted to the Illinois Environmental Protection Agency, Project FW-38. Institute for Environmental Studies, University of Illinois: Urbana, IL. 103 pp.

Jungclaus, G.A., V. Lopez-Avila, and R.A. Hites. 1978. Organic compounds in an industrial wastewater: A case study of their environmental impact. Environ. Sci. Technol. 12:88-96.

Loper, J.C. 1980a. Overview of the use of short-term biological tests in the assessment of the health effects of water chlorination. In: Water Chlorination: Environmental Impact and Health Effects, Vol. 3. R.L. Jolley, W.A. Brungs, and R.L. Cumming, eds. Ann Arbor Science Publishers Inc.: Ann Arbor, MI. pp. 937-945.

Loper, J.C. 1980b. Mutagenic effects of organic compounds in drinking water. Mutat. Res. 76:241-268.

Loper, J.C., and M.W. Tabor. 1981. Detection of organic mutagens in water residues. In: Short-Term Bioassays in the Analysis of Complex Environmental Mixtures II. M.D. Waters, S.S. Sandhu, J.L. Huisingh, L. Claxton, and S. Nesnow, eds. Plenum Press: New York. pp. 155-165.

Loper, J.C., and M.W. Tabor. 1983. Isolation of mutagens from drinking water: something old, something new. In: Short-Term Bioassays in the Analysis of Complex Environmental Mixtures III. M.D. Waters, S.S. Sandhu, J. Lewtas, L. Claxton, N. Chernoff, and S. Nesnow, eds. Plenum Press: New York. pp. 165-181.

Loper, J.C., D.R. Lang, R.S. Schoeny, B.B. Richmond, P.M. Gallagher, and C.C. Smith. 1978. Residue organic mixtures from drinking water show *in vitro* mutagenic and transforming activity. J. Toxicol. Environ. Health 4:919-938.

Loper, J.C., M.W. Tabor, and S.K. Miles. 1982. Mutagenic subfractions from nonvolatile organics of drinking water. In: Water Chlorination: Environmental Impact and Health Effects, Vol. 4. R.L. Jolley, W.A. Brungs, J.A. Cotruvo, R.B. Cumming, J.S. Mattice, and V.A. Jacobs, eds. Ann Arbor Science Publishers Inc.: Ann Arbor, MI. pp. 1199-1210.

Maciorowski, A.F., L.W. Little, L.F. Raynor, R.C. Sims, and J.L. Sims. 1983. Bioassays -- procedures and results. J. Water Pollut. Control Fed. 55:801-816.

Meier, J.R., and D.F. Bishop. 1984. Effectiveness of conventional treatment processes for removal of mutagenic activity from municipal wastewaters. Presented at the 15th Annual Meeting of the Environmental Mutagen Society, Montreal, Canada.

Naylor, L.M., and R.C. Loehr. 1982a. Priority pollutants in municipal sewage sludge. Part 1. Biocycle 23(4):18-22.

Naylor, L.M., and R.C. Loehr. 1982b. Priority pollutants in municipal sewage sludge. Part 2. Biocycle 23(6):37-41.

Neal, M.W., L. Mason, Jr., D.J. Schwartz, and J. Saxena. 1980. Assessment of Mutagenic Potential of Mixtures of Organic Substances in Renovated Water. EPA 600/1-81-016.

Nellor, M.H., R.B. Baird, and J.R. Smyth. 1984. Health Effects Study Final Report on Groundwater Recharge. County Sanitation Districts of Los Angeles County: Whitter, CA. 588 pp.

Petrasek, A.C., I.J. Kugelman, B.M. Austern, T.A. Pressley, L.A. Winslow, and R.H. Wise. 1983. Fate of toxic organic compounds in wastewater treatment plants. J. Water Pollut. Control Fed. 55:1286-1296.

Rappaport, S.M., M.G. Richard, M.C. Hollstein, and R.E. Talcott. 1979. Mutagenic activity in organic wastewater concentrates. Environ. Sci. Technol. 13(8):957-961.

Strachan, S.D., D.W. Nelson, and L.E. Sommers. 1983. Sewage sludge components extractable with nonaqueous solvents. J. Environ. Qual. 12:69-74.

Tabor, M.W. 1983. Structure elucidation of 3-(2-chloroethoxy)-1,2-dichloropropene. A new promutagen from an old drinking water residue. Environ. Sci. Technol. 17:324-329.

Tabor, M.W., and J.C. Loper. 1980. Separation of mutagens from drinking water using coupled bioassay/analytical fractionation. Int. J. Environ. Anal. Chem. 8:197-215.

SESSION 3

RISK ASSESSMENT

APPLICATION OF SHORT-TERM TESTS IN MONITORING OCCUPATIONAL EXPOSURE TO COMPLEX MIXTURES

Harri Vainio[1] and Marja Sorsa[2]

[1]International Agency for Research on Cancer, 69372 Lyon Cedex 08, France, and
[2]Institute of Occupational Health, SF-00290 Helsinki 29, Finland

INTRODUCTION

Over recent years, a great deal of effort has been put into developing sensitive experimental methods applicable in the identification and evaluation of mutagenic and carcinogenic compounds in the environment. The quantitative evaluation of impacts on health, however, is made extremely difficult by the fact that, in real life, no one is exposed solely to a single chemical but, instead, simultaneously to a multitude of chemicals and complex mixtures that interact differently in the human body.

The carcinogenic and mutagenic activities of a substance may be modified by the presence of another compound in a variety of ways (Boutwell et al., 1982). The implications of the synergistic, antagonistic, additive, and inhibitory interactions of chemical exposures are especially important to take into consideration in interpreting the potential health hazards of environmental samples. A further complication related to the evaluation of complex environmental mixtures is the difficulty in establishing techniques that will provide samples free of artifacts that can be introduced during collection, storage, concentration, fractioning, and handling of the complex sample during the testing procedure.

The use of short-term tests for identifying genetic effects has special significance in evaluating environmental samples, the complexity of which cannot allow the detection of a single hazardous compound. The relative inexpensiveness of the short-term tests permits testing of a large number of different fractions and combinations--a necessity in the sensible use of short-term tests.

The most successful approach to studying complex environmental samples, therefore, is a combination of chemical fractionation and bioassay techniques that should allow the identification of the genotoxic component(s), its primary source or origin, and the possible spontaneous or artificially induced changes in the nature of the chemical. Further, it should lead to indications of the most suitable control technology.

The present paper discusses applications of the available short-term test methods to the control of complex mixtures and multi-exposures typical of the occupational environment. We also present some ideas about a general strategy for health surveillance of people occupationally exposed to complex mixtures or to a multitude of chemicals, using the hospital personnel handling anticancer drugs as an example.

EVALUATION OF COMPLEX MIXTURES IN THE OCCUPATIONAL ENVIRONMENT

Exposures encountered in the occupational environment are usually much more extreme than those occurring in the general environment via ambient air, drinking water, etc., and are also usually more easily controllable. In some occupational situations, exposures are well characterized and a hazard can be focused on a single chemical entity: e.g., on benzene in the chemical industry or manufacturing plants. In such cases, preventive and hygienic measures can be selected readily, and environmental and biological monitoring should be straightforward. However, in most occupational environments, the spectrum of chemicals to which the population is exposed is wide, and although there may be epidemiological evidence of an increased frequency of disease, no firm association with any single causal factor can be established.

Complex mixtures (of industrial origin and those that occur as a result of pyrolysis or combustion) may be divided into three general classes:

1. "commercial mixtures," such as mineral oils, carbon blacks, and creosotes, the composition of which varies from sample to sample, depending on the source, method of preparation, etc., but which nevertheless will meet certain physical and (sometimes) chemical specifications;

2. "experimentally-generated mixtures," which might include such mixtures as cigarette tar condensate and diesel exhaust, generated and collected under laboratory conditions;

3. "environmental samples," such as water samples or airborne materials collected from, for example, the workroom air or as stationary or personal breathing zone samples.

On-site methods for monitoring ambient air in complex exposure situations at worksites are still in their infancy. Almost the only eukaryotic monitoring systems applied to the study of atmospheric mutagens are plants; e.g., the *Tradescantia* staminal hair cell and the maize marker locus (waxy, yellow-green) have been applied routinely (Constantin, 1982; Plewa, 1982). A preliminary study has been made of the use of populations of *Drosophila melanogaster*, placed at specific sites in a rubber factory, to monitor for possible exposures to mutagens in the ambient air, using sex-linked recessive lethal frequencies as the end point (Donner et al., 1983b).

Off-site methods are being used increasingly to study potential contamination of occupational environments with mutagenic agents. Bacterial mutagenicity tests have been the method used most frequently, e.g., in testing ambient air samples from vulcanizing sites in a rubber factory (Hedenstedt et al., 1981) and in testing ambient air and personal filter samples from workers in foundries (Skyttä et al., 1980; Schimberg et al., 1981). Fumes collected from stainless steel welding were mutagenic in the *Salmonella*/microsome test (Hedenstedt et al., 1977; Maxild et al., 1978). Fumes produced during the curing of different rubber polymers and mixtures of rubber in experimental vulcanizing conditions were mutagenic in various short-term bioassays (Donner et al., 1983a). Different fractions of aerosols produced by thermal degradation of polyurethane foam were also mutagenic in bacterial assay systems (Zitting et al., 1980).

It is essential that proper sampling procedures be used when taking such complex ambient mixtures, in order to guarantee the representativity of the sample with regard to worker exposure. Respirable particles or vapors, although highly critical in human exposure, may easily escape the sampler and cause underestimation of the real hazard (Chrisp and Fisher, 1980). The extraction and fractionation methods may also be overly selective: the mutagenicity may be spread over several fractions or it may be extinguished by the toxicity of the sample (Claxton, 1982).

Using the approach of continuous chemical fractionation and short-term assays, the investigator can accumulate information on the actual compounds responsible for the biological effects. Even in view of the difficulties outlined above, therefore, short-term tests provide a possibility for monitoring complex ambient exposure and are thus of importance in following changes in occupational conditions. Short-term tests can also aid in identifying the specific hazardous compounds involved and in establishing priorities for more definitive chemical analysis and monitoring, and for further testing in comparative systems, including whole animals, for mutagenesis and carcinogenesis.

A number of precautions should be observed in interpreting experimental data on complex mixtures. Along with the obvious bias that could accompany the choice of nonrepresentative samples, methods of storage and sample treatment may influence the results of any assay. In addition, biological activity in a crude material collected by environmental sampling may be difficult to detect because of the overall toxicity of the mixture or because of very low concentrations of the active components, or both. Materials are therefore often concentrated and/or fractionated to allow more accurate testing; the possible disadvantages entrained, however, include loss or modification of specific components or loss of possible synergistic effects.

For all three classes of complex mixtures, therefore, short-term tests may be helpful in identifying the active components. They can also function as a measure of biological activity, allowing comparison of baseline data with changes in the environmental conditions or in the processes.

MONITORING OF WORKERS EXPOSED TO COMPLEX MIXTURES

Environmental and biological monitoring are complementary approaches for evaluating human exposure to complex mixtures. The main goal of biological monitoring in health surveillance programs is either to ensure that the current or past exposure is "safe" or to detect potential excessive exposure before the occurrence of detectable adverse health effects. Biological monitoring is essentially a preventive medical activity. The information obtained by these methods is also essential in obtaining reliable data for epidemiological studies. Quantitative risk calculations in epidemiological studies are irrelevant without reliable information on the exposures. Unfortunately, data on past exposures are almost always unsatisfactory; it is to be hoped that future developments in biological monitoring will also help epidemiologists to make more reliable dose-response evaluations.

Biological monitoring is performed by analyzing biological specimens obtained from the exposed individual--mostly urine or blood. Dose estimation is achieved mainly by analyzing the *uptake*, i.e., the amount absorbed by the body. In a few instances, biological monitoring can be performed by monitoring a specific *effect* of the chemical on the body. Presently, however, this approach is rarely applicable to routine situations. It may still be an ideal method of monitoring, since it allows individual differences in susceptibility to be taken into consideration. This type of monitoring may prove especially useful in monitoring complex multiple exposures, a situation which is generally the case in the occupational environment. Short-term tests presently available for biological monitoring of occupational exposure to complex mixtures were summarized in a recent international seminar (Berlin et al., in press).

Methods for the analysis of body fluids for specific chemicals as well as bioassays for testing the ability of the body fluid sample to induce mutations in the indicator organism (e.g., bacteria) are available. Established analytical chemical assays are most appropriate for demonstrating the exposure when specific substances are known to occur in the environment. Biological assays, such as mutagenic activity in mice, are most useful when exposures are suspected but the specific chemicals (or possible interactions between chemicals) are unknown, or when analytical techniques are not available or cannot be used because of the complexity of the exposure. These situations are not infrequent in workplaces. In cases where workers are exposed to complex mixtures comprising hundreds of chemical entities, the possibilities for monitoring a single chemical are limited. In such situations, indicators that are nonspecific as to the identity of chemicals but specific with respect to the nature of the health hazards are advantageous (cf. Vainio et al., 1981).

Mutagenic Activity in Urine

Since recognition of the close association between mutagenicity and carcinogenicity, assays of mutagenicity in biological fluids have been brought into use in order to detect, and to some extent also quantitate, exposure to mutagenic and carcinogenic chemicals. Urine has been by far the most widely used body fluid for the detection of mutagenicity (Legator et al., 1982).

There are obvious reasons for the popularity of urine: sampling is easy, chemicals are excreted into urine (although usually only after metabolic transformation), and urine can be manipulated (e.g., concentrated). In most cases, mutagenic activity is detectable only after the urine samples have been concentrated and treated with deconjugating enzymes, such as b-glucuronidase and arylsulfatase.

Detection of mutagenic activity in urine is used mainly as a nonspecific indicator of exposure to mutagenic compounds, and these assays appear to be promising tools for identifying possible hazardous working situations. The advantages of this method have been summarized recently (Vainio et al., in press). Since the assay is nonspecific--i.e., it is specific only as to the nature of the exposure (mutagenicity) but not to the individual chemicals--special attention must be paid, when using this assay to monitor occupational exposures, to confounding exposures, such as to drugs and living habits. One advantage of the urine assay is that the mutagenicity data can be related to qualitative and quantitative chemical analyses of the urine in order to elucidate further the nature of the exposure.

The urinary mutagenicity assay is thus readily applicable to the monitoring of human exposures in certain situations. It can be used to identify hazards in workplaces, such as in chemical production, and can also be used to provide guidance in selecting appropriate hygienic measures.

Since quantitative risk assessment can be conceived as being composed of two aspects--the exposure and the hazard--exposure assessment is of critical importance. A presumption of hazard can be made if the exposure can be demonstrated, e.g., by showing that the individuals suspected for exposure do excrete mutagens in their urine. However, as most chemicals are excreted in conjugated, "detoxified" form, the presence of mutagenic activity in concentrated urine does not imply that there will necessarily be a further biological effect. Thus, data from the urinary mutagenicity assay should be considered as an exposure indicator and not as an effect indicator.

Measurement of Effects

Cytogenetic analyses of peripheral lymphocytes have been advocated as a biological indicator of genetic effects on somatic cells (cf. Sorsa et al., 1982a; Vainio and Sorsa, 1983). The appearance of induced cytogenetic changes in lymphocytes of exposed persons has already shown itself to be a sensitive indicator of low-level exposures (e.g., occupational exposures to ethylene oxide or vinyl chloride).

It should be noted that even though induced chromosome damage in somatic cells is deemed undesirable, the presence of cytogenetic changes cannot be used to predict specific adverse health effects in an individual. However, they do give an estimate of the magnitude of an exposure that could increase the risk of disease in a group of people. Consequently, cytogenetic study of chromosomal aberrations, micronuclei, and sister chromatid exchanges is a relevant method of choice in the evaluation of a hazard in a group of individuals exposed to known or suspected mutagens and carcinogens. Such studies should, however, be carefully designed, with appropriate controls and with proper methods for quality control and for statistical analysis of the data. Being a laborious task, a cytogenetic study should never be used without some previous knowledge about the clastogenicity of the exposing chemicals.

Cytogenetic data can, in principle, be used for qualitative and quantitative risk estimation. The cytogenetic studies that provide the bulk of the available data on human effect monitoring are limited as to their usefulness in risk identification procedures. Large-scale studies are time-consuming and expensive, and analysis is subjective, even when carried out by experienced investigators. A positive result exhibiting a dose response that is biologically plausible is generally regarded as a warning sign for a genotoxic hazard. Care should be exercised in interpreting negative studies. The nondividing peripheral lymphocytes are insensitive tools for monitoring effects produced by chemicals that act at the S phase of growth. Since many chemical mutagens, in fact, act at the S phase, their effects may thus go undetected by cytogenetic analysis of peripheral lymphocytes.

OCCUPATIONAL HEALTH HAZARDS WITH ANTICANCER DRUGS

The increasing importance of chemotherapy in the treatment of malignancies has made it necessary to investigate not only the side effects of anticancer drugs in patients but also the possible health hazards for personnel handling them. The first concerns about risks to health care personnel were expressed over a decade ago (Ng, 1970). Since then, it has become evident that many of the anticancer drugs are carcinogenic in animals (IARC, 1981) and that some have *sufficient evidence* for carcinogenicity in humans (IARC, 1982). In most instances, a combination chemotherapy is given to the patients, and therefore it is seldom possible to pinpoint a single drug as responsible for the carcinogenic effect observed.

Anticancer Drugs in the Workroom Air

Release of anticancer drugs in the workroom during their handling can occur in several ways. Opening of the ampullae and withdrawal of needles from vials may result in spillage. With rubber-stopper vials, there is a possibility of pressure buildup causing spray-out of the material. Release can result also from sloppy handling of excess materials and waste. A recent occupational hygiene survey showed relatively small and episodic amounts of anticancer drugs in the workroom air (de Werk Neal et al., 1983). Fluorouracil (0.1-82.3 ng/m^3) was detected in air during 200 of 320 h monitored. Cyclophosphamide (370 ng/m^3) was present during 80 h. Under these conditions, fluorouracil and cyclophosphamide were the most commonly used drugs.

The penetration through intact human skin of cyclophosphamide, one of the most commonly used alkylating cytostatics, has been shown by chemical analysis among nurses in a controlled handling situation (Hirst et al., 1984).

The limited results available suggest that personnel handling anticancer drugs may be exposed to them by inhaling the aerosols generated or through skin contact with the drug or urine of patients (Hirst et al., 1984, Venitt et al., 1984). Furthermore, these drugs may also be absorbed through the skin.

Mutagenic Activity in Urine

The first findings about possible exposure of hospital personnel to cancer chemotherapeutic drugs came from a study of Falck et al. (1979) where association was made between increased mutagenicity in urine and contact with cytostatic drugs. Urinary mutagenicity measured by the bacterial fluctuation test was significantly higher among the group of oncology nurses (samples taken after the working shift) as compared with office clerks but still considerably lower than patients receiving chemotherapy. After improvements in the handling procedures, a clear decrease of urine mutagenicity was observed in the same nurses (Falck et al., 1981).

Both positive and negative results have later been reported about urinary mutagenicity in relation to occupational handling of cytostatic drugs (see Table 1). The inconsistencies are not necessarily contradictory; negative responses may be due to insensitivity of the assay to detect minimal exposures or to safe handling of the drugs. Furthermore, optional timing of the sample for detection of mutagenicity may not have been considered, or the fact that alkaloid- and antimetabolic-type (for example) cytostatics do not respond in bacterial assays (IARC, 1981) may have been omitted. The possibility of false positive results due to remnant marker amino acids in the urine

Table 1. Results from Urine Mutagenicity Assays Among Hospital Personnel Handling Cancer Chemotherapeutics

Type of Test	Indicator Strains	Metabolic Activation	Result	Reference
Fluctuation assay (macroscale)	*S. typhimurium* TA100, TA98 *E. coli* WP2uvrA	S-9	+	Falck et al. (1979)
Ames test (plate incorporation)	*S. typhimurium* TA98, TA100, TA1535	-	+	Anderson et al. (1982)
Ames test (plate incorporation)	*S. typhimurium* TA100	S-9	+	Bos et al. (1982)
Fluctuation assay (macroscale)	*E. coli* WPZuvrA	S-9	–	Kolmodin-Hedman et al. (1983)
Ames test (treat-and-plate)	*S. typhimurium* TA98, TA1535	S-9	–	Gibson et al. (1984)
Fluctuation assay (microscale)	*S. typhimurium* TA98, TA100	S-9	–	Gibson et al. (1983)
Ames test (plate incorporation)	*S. typhimurium* TA98, TA100	S-9	–	Staiano et al. (1981)

concentrates has been recently brought up (Gibson et al., 1984; Venitt et al., 1984). This may indeed be a problem in microscale fluctuation assays, but not in macroscale-type tests.

Cytogenetic Studies

Many anticancer drugs, such as alkylating cytostatic drugs and some antibiotic-type chemotherapeutics, have been shown to induce both chromosomal aberrations and sister chromatid exchanges (SCEs) *in vitro* in the human and rodent and *in vivo* in cancer patients (Dobos et al., 1974; Gebhart et al., 1980a, b; Au et al., 1980; Aronson et al., 1982). On the basis of this information, it thus appears feasible to use cytogenetic methods also in monitoring possible occupational exposures to these agents. A few cytogenetic studies have been performed showing that low-level exposures can be detected at the group level.

As compared with office workers, the group of oncology nurses showed a significantly increased frequency of SCEs in their lymphocytes (Norppa et al., 1980).

Even at the individual level, there was a tendency for a correlation between the SCE frequency and the amount of alkylating cytostatics handled (Sorsa et al., 1982b).

Waksvik et al. (1981) also reported an increase in SCEs and in chromosome type gaps among nurses handling cytostatic drugs. However, breaks were not significantly increased among the oncology nurses.

In a recent study, the frequency of chromosomally aberrant lymphocytes was found to be significantly higher in the group of oncology nurses than in the control groups (laboratory workers, hospital clerks) (Nikula et al., 1984). In this study, the aberrations were typically chromosome type breaks, which were significantly increased among the nurses as compared to the reference groups. The authors concluded that the observed increase in chromosome type aberrations may have been due to long-term occupational exposure to cytostatic agents.

No effects on the level of SCEs from different cytostatic-handling duties was seen in a small group of nurses (Kolmodin-Hedman et al., 1983).

HANDLING OF CYTOSTATIC DRUGS

Extensive variation in exposures and possibilities for exposure exist depending on the frequency, amounts, and procedures of handling. Furthermore, only certain types of drugs, primarily the alkylating cytostatics, can theoretically and on the basis of experimental and patient studies be expected to cause chromosomal effects or increased urinary mutagenicity (IARC, 1981).

The available hygienic measurements as well as studies on mutagenic activity in urine and cytogenetic studies all point to the possibility that hospital personnel handling cancer chemotherapeutics may be exposed to low levels of these drugs. Many recommendations have been published for the handling of anticancer drugs (e.g.:

National Institutes of Health, 1982). The occupational exposure to anticancer drugs is a typical situation where the level of exposure is usually low but may simultaneously occur to a multitude of chemicals and may continue for decades. The health risks have been clearly demonstrated among patients receiving these drugs, but the scientific data concerning the risk to health care personnel are almost totally lacking. Still, even in the absence of any epidemiological end point studies (such as cancer or reproductive failures) among health care personnel, the evidence from biological monitoring studies has led to considerable improvements in working conditions. The concomitant decrease of unnecessary exposure to anticancer drugs has obviously also decreased the possibility of undesirable ill-health manifestations.

ACKNOWLEDGMENTS

The authors are grateful to Ms. L. Haroun for her helpful comments, to Ms. E. Heseltine for editorial assistance, and to Ms. J. Cazeaux for secretarial help.

REFERENCES

Anderson, R.W., W.H. Puckett, W.J. Dana, T.V. Nguyen, J.C. Theiss, and T.S. Matney. 1982. Risk of handling injectable antineoplastic agents. Am. J. Hosp. Pharm. 39:1881-1887.

Aronson, M.M., R.C. Miller, R.B. Hill, W.W. Nichols, and A.T. Meadows. 1982. Acute and long-term cytogenetic effects of treatment of childhood cancer: Sister-chromatid exchanges and chromosome aberrations. Mutat. Res. 92:291-307.

Au, W., O.I. Sokova, B. Kopnin, and F.F. Arrighi. 1980. Cytogenetic toxicity of cyclophosphamide and its metabolites in vitro. Cytogenet. Cell Genet. 26:108-116.

Berlin, A., M. Draper, K. Hemminki, and H. Vainio, eds. (in press). Monitoring of Human Exposure to Carcinogenic and Mutagenic Agents. IARC Scientific Publication No. 59, International Agency for Research on Cancer: Lyon, France.

Bos, R.P., A.O. Leenars, J.L.G. Theuws, and P.Th. Henderson. 1982. Mutagenicity of urine from nurses handling cytostatic drugs, influence of smoking. Int. Arch. Occup. Environ. Health 50:359-369.

Boutwell, R.K., A.K. Verma, C.L. Ashendel, and E. Astrup. 1982. Mouse skin: A useful model for studying the mechanism of chemical carcinogenesis. Carcinogenesis 7:1-12.

Chrisp, C.E., and G.L. Fisher. 1980. Mutagenicity of airborne particles. Mutat. Res. 76:143-164.

Claxton, L. 1982. Review of fractionation and bioassay characterization techniques for the evaluation of organics associated with ambient air particles. In: Genotoxic Effects of Airborne Agents. R.R. Tice, D.L. Costa, and K.M. Schaich, eds. Environmental and Scientific Research, Vol. 25, Plenum Press: New York. pp. 19-34.

Constantin, M.J. 1982. Plant genetic systems with potential for the detection of atmospheric mutagens. In: Genotoxic Effects of Airborne Agents. R.R. Tice, D.L. Costa, and K.M. Schaich, eds. Environmental and Scientific Research, Vol. 25, Plenum Press: New York. p. 159.

de Werk Neal, A., R.A. Wadden, and N.L. Chiou. 1983. Exposure of hospital workers to airborne antineoplastic agents. Am. J. Hosp. Pharm. 40:597-601.

Dobos, M., D. Schuler, and G. Fekete. 1974. Cyclophosphamide-induced chromosomal aberrations in nontumorous patients. Hum. Genet. 22:221-227.

Donner, M., K. Husgafvel-Pursiainen, D. Jenssen, and A. Rannug. 1983a. Mutagenicity of rubber additives and curing fumes. Scand. J. Work Environ. Health 9, Suppl. 2:27-37.

Donner, M., S. Hytönen, and M. Sorsa. 1983b. Application of the sex-linked recessive lethal test in Drosophila melanogaster for monitoring the work environment of a rubber factory. Hereditas 99:7-10.

Falck, K., P. Grohn, M. Sorsa, H. Vainio, E. Heinonen, and L.R. Holsti. 1979. Mutagenicity in urine of nurses handling cytostatic drugs. Lancet i:1250-1251.

Falck, K., M. Sorsa, and H. Vainio. 1981. Use of the bacterial fluctuation test to detect mutagenicity in urine of nurses handling cytostatic drugs. Mutat. Res. 85:236-237.

Gebhart, E., J. Lösing, and F. Wopfner. 1980a. Chromosome studies on lymphocytes of patients under cytostatic therapy: I. Conventional chromosome studies in cytostatic interval therapy. Hum. Genet. 55:53-63.

Gebhart, E., B. Windolph, and F. Wopfner. 1980b. Chromosome studies on lymphocytes of patients under cytostatic therapy: II. Studies using the BUDR-labelling technique in cytostatic interval therapy. Hum. Genet. 56:157-167.

Gibson, J.F., P.J. Baxter, R.B. Hedworth-Whitty, and D. Gombertz. 1983. Urinary mutagenicity assay: a problem arising from the presence of histidine associated growth factors in XAD-2 prepared urine concentrates, with particular relevance to assays carried out using the bacterial fluctuation test. Carcinogenesis 4:1471-1476.

Gibson, J.F., D. Gompertz, and R.B. Hedworth-Whitty. 1984. Mutagenicity of urine from nurses handling cytotoxic drugs. Lancet i:100-101.

Hedenstedt, A., D. Jenssen, B.-M. Lidesten, C. Ramel, U. Rannug, and R.M. Stern. 1977. Mutagenicity of fume particles from stainless steel welding. Scand. J. Work Environ. Health 3:203-211.

Hirst, M., D.G. Mills, S. Tse, L. Levin, and D.F. White. 1984. Occupational exposure to cyclophosphamide. Lancet i:186-188.

IARC (International Agency for Research on Cancer). 1981. Some Antineoplastic and Immunosuppressive Agents. IARC Monographs on the Evaluation of the Carcinogenic Risk of Chemicals to Humans, Vol. 26, International Agency for Research on Cancer: Lyon, France. p. 441.

IARC (International Agency for Research on Cancer). 1982. Chemicals, Industrial Processes and Industries Associated with Cancer in Humans. Volumes 1 to 29. IARC Monographs on the Evaluation of the Carcinogenic Risk of Chemicals to Humans, Suppl. 4, International Agency for Research on Cancer: Lyon, France, pp. 1-292.

Kolmodin-Hedman, B., P. Hartvig, M. Sorsa, and K. Falck. 1983. Occupational handling of cytostatic drugs. Arch. Toxicol. 54:25-33.

Legator, M.S., E. Bueding, R. Batzinger, T.H. Connor, E. Eisenstadt, M.G. Farrow, G. Ficsor, A. Hsie, J. Seed, and R.S. Stafford. 1982. An evaluation of the host-mediated assay and body fluid analysis. Mutat. Res. 98:319-374.

Maxild, J., M. Andersen, P. Kiel, and R.M. Stern. 1978. Mutagenicity of fume particles from metal arc welding of stainless steel in the Salmonella/microsome test. Mutat. Res. 56:235-243.

National Institutes of Health. 1982. Recommendations for the Safe Handling of Parental Antineoplastic Drugs. Publication No. 83-2621, National Institutes of Health: Washington, DC.

Ng, L.M. 1970. Possible hazards of handling antineoplastic drugs. Pediatrics 46:648-649.

Nikula, E., K. Kiviniity, J. Leisti, and P.J. Taskinen. 1984. Chromosome aberrations in lymphocytes of nurses handling cytostatic agents. Scand. J. Work Environ. Health 10:71-74.

Norppa, H., M. Sorsa, H. Vainio, P. Gröhn, E. Heinonen, L. Holsti, and E. Nordman. 1980. Increased sister chromatid exchange frequencies in lymphocytes of nurses handling cytostatic drugs. Scand. J. Work Environ. Health 67:229-301.

Plewa, M.J. 1982. Maize as a monitor for environmental mutagens. In: Environmental Mutagens and Carcinogens. T. Sugimura, S. Kondon, and H. Takebe, eds. Tokyo University Press: Tokyo. pp. 411-419.

Schimberg, R., E. Skyttä, and K. Falck. 1981. Belastung von Eisengiessereiarbeitern durch mutagene polycyclische aromatische Kohlenwasserstoffe. Staub 41:421-424.

Skyttä, E., R. Schimberg, and H. Vainio. 1980. Mutagenic activity in foundry air. Arch. Toxicol. 4:68-72.

Sorsa, M., K. Hemminki, and H. Vainio. 1982a. Biologic monitoring of exposure to chemical mutagens in the occupational environment. Teratogen. Carcinogen. Mutagen. 2:137-150.

Sorsa, M., H. Norppa, and H. Vainio. 1982b. Induction of sister chromatid exchanges among nurses handling cytostatic drugs. In: Indicators of Genotoxic Exposure. Banbury Report 13, Cold Spring Harbor Laboratory: Cold Spring Harbor, NY. pp. 341-350.

Staiano, N., J.F. Gallelli, R.H. Adamson, and S.S. Thorgeirsson. 1981. Lack of mutagenic activity in urine from hospital pharmacists admixing antitumor drugs. Lancet i:615-616.

Vainio, H., and M. Sorsa. 1983. Application of cytogenetic methods for biological monitoring. Ann. Rev. Pub. Health 4:403-407.

Vainio, H., M. Sorsa, and K. Falck. (in press). Bacterial mutagenicity assay in monitoring exposure to mutagens and carcinogens. In: Monitoring of Human Exposure to Carcinogenic and Mutagenic Agents. A. Berlin, M. Draper, K. Hemminki, and H. Vainio, eds. IARC Scientific Publication No. 59, International Agency for Research on Cancer: Lyon, France.

Vainio, H., M. Sorsa, J. Rantanen, K. Hemminki, and A. Aitio. 1981. Biological monitoring in the identification of the cancer risk of individuals exposed to chemical carcinogens. Scand. J. Work Environ. Health 7:241-251.

Venitt, S., C. Cropton-Sleigh, J. Hunt, V. Speechley, and K. Briggs. 1984. Monitoring exposure of nursing and pharmacy personnel to cytotoxic drugs: Urinary mutation assays and urinary platinum as markers of absorptions. Lancet i:74-77.

Waksvik, H., O. Klepp, and O. Brogger. 1981. Chromosome analyses of nurses handling cytostatic agents. Cancer Treat. Rep. 65:607-610.

Zitting, A., K. Falck, and E. Skyttä. 1980. Mutagenicity of aerosols from the oxidative thermal decomposition of rigid polyurethane foam. Int. Arch. Occup. Environ. Health 47:47-52.

AN INTEGRATED APPROACH TO HUMAN MONITORING

Vincent F. Garry,[1] Robert A. Kreiger,[1] John K. Wiencke,[2] and Richard L. Nelson[1]

[1]Environmental Pathology Laboratory, Department of Laboratory Medicine and Pathology, University of Minnesota, Minneapolis, Minnesota 55455, and [2]Laboratory of Radiobiology and Environmental Health, University of California, San Francisco, California 94143

INTRODUCTION

Our laboratory devotes much effort to the documentation of human effects from exposure to xenobiotics. In general, the approach we use follows the barrier systems concept (Garry et al., in press), which simply states that for a xenobiotic to exert a toxic effect, it must bypass or alter protective physiologic mechanisms present in the affected species. To test this hypothesis, we integrate clinical observations of human effects with *in vivo* and *in vitro* test system information. To provide an example of the various levels and patterns of integration we have utilized, we will review some of our recent findings dealing with ethylene oxide (EO) and formaldehyde.

MATERIALS AND METHODS

Ethylene Oxide

Our first encounter with EO began with a group of workers in the supply and sterilization section of a local hospital (Garry et al., 1979). At the time (Figure 1), these persons reported to us a series of upper respiratory and neurologic complaints. Inspection of the nursing staff's daily report showed periodic absences of the workers due to respiratory complaints. Further investigation indicated that these absences correlated with the visits of the EO sterilizer repair person. These pieces of information suggested to us that health-damaging EO exposure may have occurred (Gross et al., 1979; Dolovich and Bell, 1978; McDonald et al., 1973). Since EO was known to be a mutagen in several species (Glaser, 1977; Flotz and Fierst, 1974; Embree et al., 1977; Pfeiffer and Dunkelberg, 1980), we decided to biologically evaluate the workers' exposure by employing the chromatid exchange assay on samples of peripheral blood. First, we studied those persons who had neurologic symptoms, because this group was

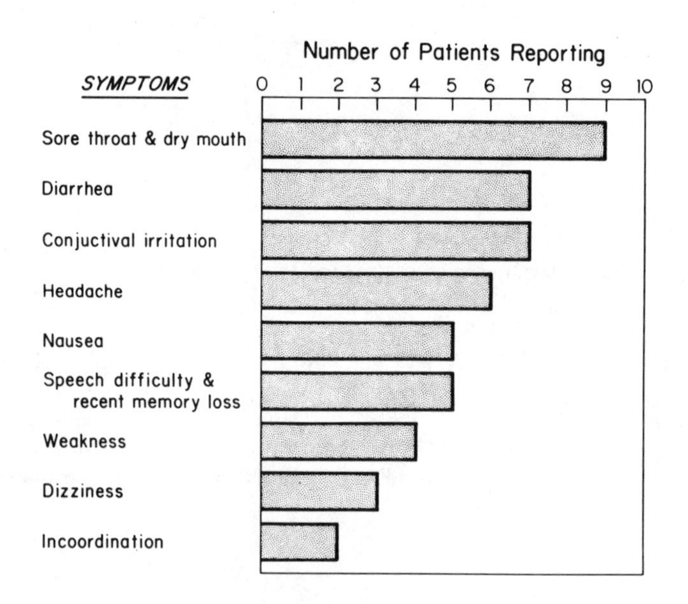

Figure 1. Pooled data from the nursing staff daily report detailing illness among all 15 employees at the sterilization facility. Reproduced with permission from Garry et al. (1979).

thought to have received the greatest exposure (Table 1). Elevated sister chromatid exchanges (SCEs) occurred at three weeks and again at eight weeks after the last exposure. Measurement of ambient EO levels showed 36 ppm in the supply room and 1500 ppm in the rear of the sterilizer adjacent to the purge drain for the expended gas (Figure 2). Persons with respiratory complaints and persons incidentally exposed to the toxicant gas also showed elevated SCEs. One individual who received a single exposure to high EO levels showed an equivocally elevated SCE response. These data suggested that a single high level burst of the toxicant gas may induce abnormal SCE levels. To test this hypothesis and evaluate the sensitivity of the assay to EO, we decided to study the response of human lymphocytes exposed to the gas *in vitro*. In order to achieve this goal, we developed the technology to evaluate a burst dose of EO *in vitro* (Garry et al., 1982). It is based on the passive diffusion of EO through variably sized membranes under nonequilibrium conditions into media containing cultured lymphocytes. Cells contained in four-plex dishes are covered with EO porous membranes and placed in an environmental chamber. EO levels in the chamber are monitored by an infrared analyzer connected with the chamber, forming a closed loop apparatus. EO levels in the media are monitored by gas chromatography. After a 20-min exposure to 100- to greater than 300-ppm ambient EO concentrations, dose/response relationships were observed in the level of SCE induction at the given cumulative media dose (Figure 3) under the conditions mentioned. Significant increases in SCEs occurred at concentrations greater than 10 to 15 μg/ml EO accumulated over the exposure period. These data suggest that high level bursts of the toxicant gas can induce abnormal SCE events. The data from our EO studies represent a relatively simple example for human effects data integration. A single genotoxic chemical exposure and an acute

Table 1. Sister Chromatid Exchanges in EO-exposed Persons Reporting Neurologic Symptoms[a]

Exposed			Controls	
Subject Number	SCE/Metaphase (3 wk after last exposure)	SCE/Metaphase (8 wk after last exposure)	Subject Number	SCE/Metaphase
1	10.68	12.05	5	6.36
2	8.85	10.45	6	5.70
3	9.75	12.15	7	5.85
4	9.70	6.70	8	5.70
			9	4.90
			10	7.00
			11	5.30
			12	7.05

[a] The data are recorded per metaphase; 20 metaphase cells were examined per individual culture. The difference between exposed and unexposed subjects is significant at the $P < 0.005$ level (three weeks) and (eight weeks) $P < 0.01$ by the Mann-Whitney test.

symptomatic clinical response provide the necessary elements from which *in vivo* and *in vitro* laboratory data may be joined. This is particularly important, since bioassays such as SCE, although sensitive, are not specific for any one genotoxic chemical. The more complex situation, formaldehyde exposure, will be discussed in the following section.

Formaldehyde

Formaldehyde is a naturally occurring chemical, a normal metabolite of the human organism (NIOSH, 1976a; Griesemer, 1980). It is also a contributor to indoor air pollution.

We had the opportunity to study persons who were exposed to formaldehyde in their homes (Garry et al., 1980). Sources of exposure range from emissions from chipboard to formaldehyde foam insulation. Adults, children, and infants were reported (Figure 4) to have a variety of upper respiratory complaints consistent with an irritant response. By further investigation, symptoms were found to occur most frequently in the home and during the summer. These factors are consistent with the

Figure 2. Photograph depicting EO sterilizer drain line break. Ethylene oxide at 1500 ppm was measured at this site by a Miran 1a infrared analyzer.

temperature/humidity gradient emission from formaldehyde sources in the home (Andersen et al., 1975). Measurement of ambient air levels (Figure 5) by season and age of the home support the symptomatic responses observed (NIOSH, 1976a). Because of the association of formaldehyde with mutagenic and possibly carcinogenic effects in animals (Auerbach et al., 1977; Swenberg et al., 1980; Hatch et al., 1983), we again used the SCE bioassay on exposed persons. In this relatively small pilot study, we found that there appeared to be a trend between the level and length of exposure in the home and marginally elevated SCEs (Table 2). To begin to address this issue, we turned to the *in vitro* approach. In these studies, human lymphocytes were cultured in the presence of different doses of formaldehyde (Kreiger and Garry, 1983). Commercial solutions containing formaldehyde and formaldehyde prepared from paraformaldehyde were examined. In each case, cytotoxicity by trypan blue dye exclusion, mitotic index, and, finally, SCEs were measured after 72 h of culture at 37°C. The data show (Figure 6) that SCE increase occurs with formaldehyde at concentrations that are highly cytotoxic to the lymphocytes. Since acute toxicity seemed to be an important factor in the generation of SCE, we decided to test the notion that low dose chronic exposure *in vitro* may provide a better parallel to the *in vivo* condition. These studies as well as a study of the interaction of formaldehyde with infectious agents are currently underway in the laboratory.

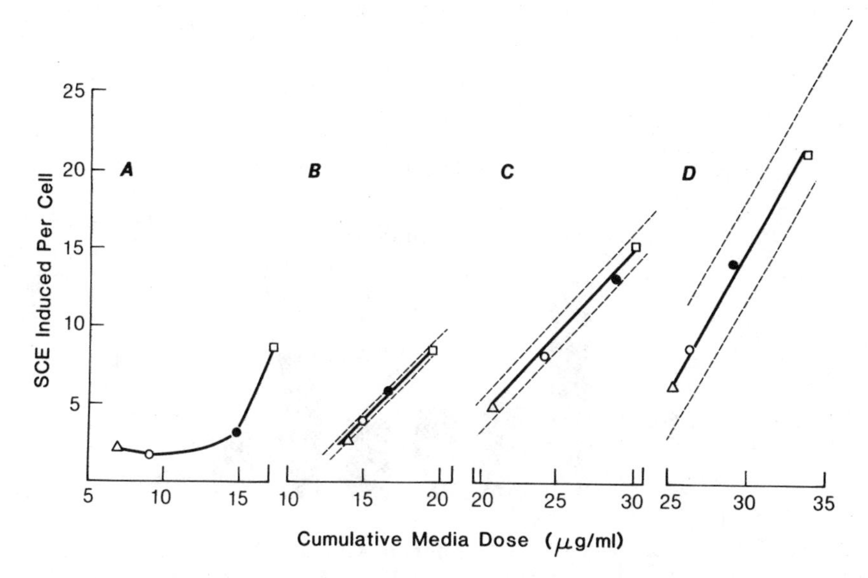

Figure 3. Ethylene oxide-treated cultures show the relationship between EO-induced SCE frequencies and cumulative media dose. Each point on the graphs represents the SCE score and individual EO determination from separate cultures. The symbols indentify the membrane area studied. The ambient dose for each panel is indicated: (A) 100, (B) 131, (C) 218, and (D) 306 ppm. The values ($\pm$16%) recorded in panel B are the mean of three samples from different blood sources. Above 100 ppm, the data follow first order regression. (-----) indicates 95% confidence limits for the regression model. The symbols on the graph refer to the dose/membrane area. (Δ) 71; (o) 160; (●) 388; and (□) 641 mm^2. Reproduced with permission from Garry et al. (1982).

The aforementioned example suggests the multifactorial nature of formaldehyde effects: as a metabolite, a cytotoxic agent, a mutagen, and, potentially, an agent that may alter viral infectivity. The genotoxic effects observed are but a part of the complex biologic activities of a single chemical agent. Formaldehyde is only one among several common home indoor air pollutants. Indeed, cohesive understanding of the home environment and health is an important topic of continuing interest.

SUMMARY

The integration of *in vitro* data with *in vivo* observations in human subjects in the studies reviewed here allows exploration of the multifaceted health effects of xenobiotics in the work environment and the home. It provides a basis for understanding the pathophysiology of noxious biologic effects of xenobiotics in the human organism.

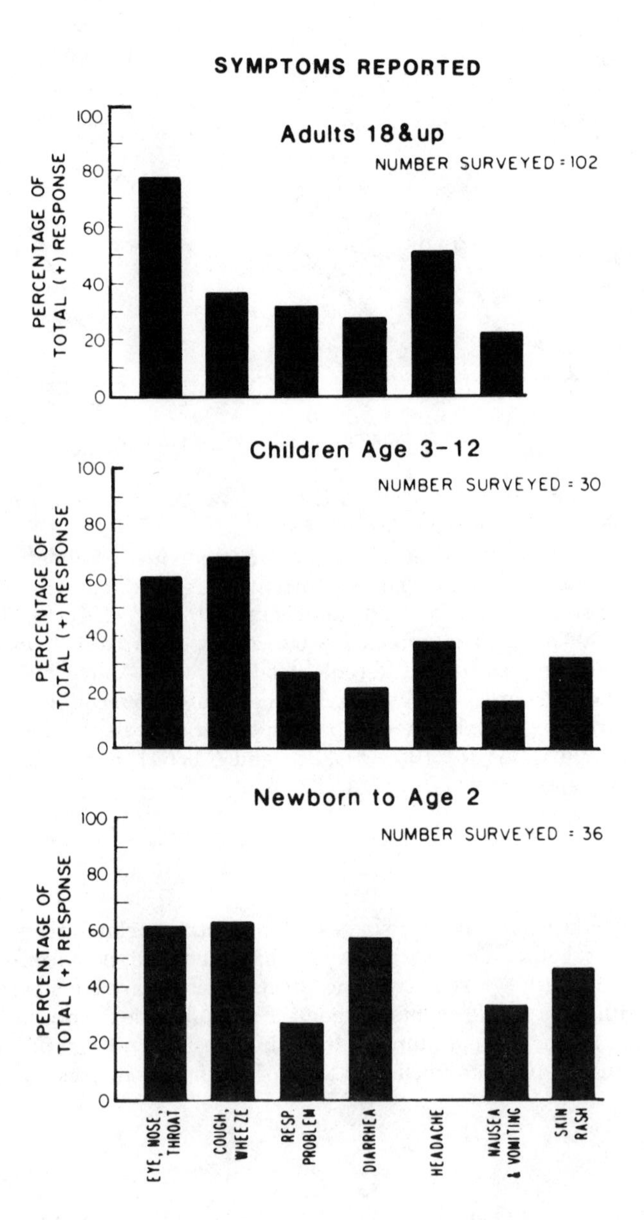

Figure 4. The frequency of symptomatology in the study population. Reproduced with permission from Garry et al. (1980).

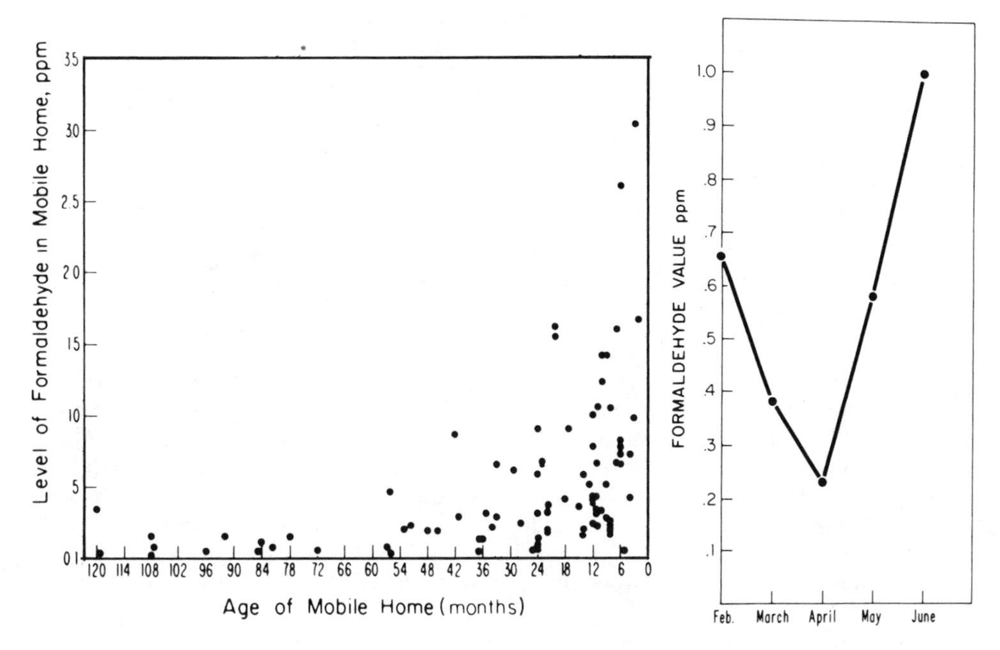

Figure 5. Ambient air levels of formaldehyde in mobile homes of various ages (left) and mean values per month (right) from February through June 1979. All measurements were made by the NIOSH (1976b) impinger method. Reproduced with permission from Garry et al. (1980).

Table 2. SCE Data Recorded from Replicate Cultures Obtained from Mobile Home Residents[a]

Ambient Formaldehyde (ppm)	Residence Time (months)	Formaldehyde Factor Dose (time weighted)	SCE/Cell
0	-	0	6.7 ± 1.3
0.33	18	5.9	8.7 ± 1.3
0.57	17	9.7	7.8 ± 1.3
0.67	16	9.7	8.0 ± 1.3
0.69	14	10.7	8.1 ± 1.3
0.80	18	14.4	9.3 ± 1.3

[a]Forty second divison metaphase cells were counted. Ambient levels of formaldehyde were measured by the impinger method (NIOSH, 1976b). The ambient formaldehyde levels recorded are one-time measurements x time (months) of residence in the mobile home.

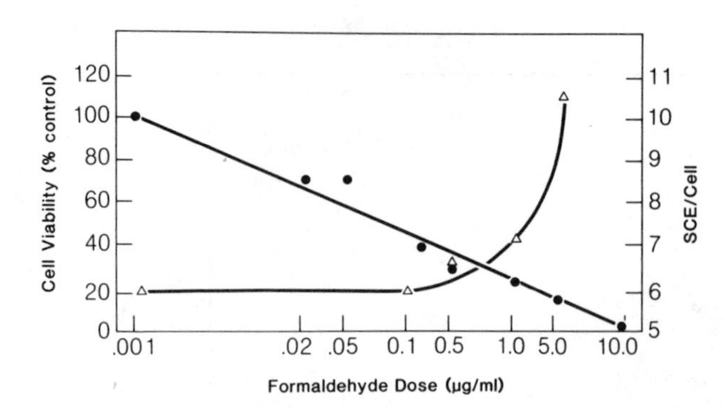

Figure 6. Interrelationship of formaldehyde-induced cytotoxicity and SCE induction. This composite plot of SCE induction and cytotoxic response was obtained from tandem cultures of the same blood sample. Cytotoxicity was measured by trypan blue dye exclusion. The methodology for the chromosomal studies is the same as in Table 2. Sister chromatid exchange induction corresponding to increasing formaldehyde dose is read from the right ordinate axis. ●, toxicity data points; Δ, SCE-level data points. Reproduced with permission from Kreiger and Garry (1983).

REFERENCES

Andersen, I., G.R. Lindqvist, and L. Moelhave. 1975. Indoor air pollution due to chipboard used as a construction material. Atmos. Environ. 9:1121-1127.

Auerbach, C., M. Moutschen-Dahmen, and J. Moutschen. 1977. Genetic and cytogenetical effect of formaldehyde and related compounds. Mutat. Res. 39:317-367.

Dolovich, J., and B. Bell. 1978. Allergy to a product(s) of ethylene oxide gas. J. Allergy Clin. Immunol. 62:30-32.

Embree, J.W., J.P. Lyon, and C.H. Hine. 1977. The mutagenic potential of ethylene oxide using the dominant lethal assay in rats. Toxicol. Appl. Pharmacol. 40:261-267.

Flotz, V.C., and R. Fierst. 1974. Mutation studies with Drosophilia melanogaster exposed to four fluorinated hydrocarbon gases. Environ. Res. 7:275-285.

Garry, V.F., J. Hozier, D. Jacobs, R.L. Wade, and D.G. Gray. 1979. Ethylene oxide: Evidence of human chromosomal effects. Environ. Mutagen. 1:375-382.

Garry, V.F., L. Oatman, R. Pleus, and D. Gray. 1980. Formaldehyde in the home: Some environmental disease perspectives. Minnesota Med. 63:107-111.

Garry, V.F., C.W. Opp, J.K. Wiencke, and D. Lakatua. 1982. Ethylene oxide induced sister chromatid exchange in human lymphocytes using a membrane dosimetry system. Pharmacology 25:214-221.

Garry, V.F., J.K. Wiencke, and R.L. Nelson. (in press). Ethylene oxide and some factors affecting mutagen sensitivity of SCE in humans. In: Sister Chromatid Exchange; 25 Years of Experimental Research. R.R. Tice and A. Hollaender, eds. Plenum Press: New York.

Glaser, Z. 1977. Use of Ethylene Oxide as a Sterilant in Medical Facilities. Pub. No. 77-200, National Institute of Occupational Safety and Health: Washington, D.C.

Griesemer, R.A. 1980. Report of the Federal Panel on Formaldehyde. National Toxicology Program: Research Triangle Park, NC.

Gross, J., M. Haas, and J. Swift. 1979. Ethylene oxide neurotoxicity: A report of four cases and review of the literature. Neurology 29:978-983.

Hatch, G.G., P.M. Conklin, C.C. Christensen, B.C. Casto, and S. Nesnow. 1983. Synergism in the transformation of hamster embryo cells treated with formaldehyde and adenovirus. Environ. Mutagen. 5:49-57.

Kreiger, R.A., and V.F. Garry. 1983. Formaldehyde induced cytotoxicity and sister chromatid exchange in human lymphocyte cultures. Mutat. Res. 120:51-55.

McDonald, T.O., K. Kasten, R. Hervey, S. Gregg, A.R. Borgmann, and T. Murchison. 1973. Acute ocular toxicity of ethylene oxide, ethylene glycol, and ethylene chlorohydrin. Bull. Parenter. Drug Assoc. 27(4):153-164.

NIOSH (National Institute of Occupational Safety and Health). 1976a. Occupational Exposure to Formaldehyde. U.S. Department of Health, Education, and Welfare: Washington, D.C. pp. 1-110.

NIOSH (National Institute of Occupational Safety and Health). 1976b. Occupational Exposure to Formaldehyde, Appendix II. U.S. Department of Health, Education, and Welfare: Washington, D.C. pp. 130-146.

Pfeiffer, E.H., and H. Dunkelberg. 1980. Mutagenicity of ethylene oxide and propylene oxide and of the glycols and halohydrins formed from them during the fumigation. Food Cosmet. Toxicol. 18:115-118.

Swenberg, J.A., W.D. Keyes, R.J. Mitchell, E.J. Gralla, and K.L. Pavkov. 1980. Induction of squamous cell carcinoma of the nasal cavity by inhalation exposure to formaldehyde vapor. Cancer Res. 40:3398-3402.

HUMAN CYTOGENETIC MONITORING: OCCUPATIONAL EXPOSURE TO ETHYLENE OXIDE

S.M. Galloway,[1,5] P.K. Berry,[1] W.W. Nichols,[2,5], S.R. Wolman,[3] K.A. Soper,[4] and P.D. Stolley[4]

[1]Litton Bionetics, Inc., Kensington, Maryland 20895; [2]Institute for Medical Research, Camden, New Jersey 08103; [3]Department of Pathology, New York University School of Medicine, New York, New York 10016; and [4]Clinical Epidemiology Unit, Department of Medicine, University of Pennsylvania Medical School, Philadelphia, Pennsylvania 19104; [5]Present address: Merck Institute for Therapeutic Research, West Point, Pennsylvania 19486

INTRODUCTION

Concern about environmental/occupational exposure to DNA-damaging agents has led to widespread interest in establishing reliable methods to monitor potentially exposed people. Measurement of damage to chromosomes in blood lymphocytes is one of the few monitoring methods available and can be a useful biological dosimeter. Although the health risks associated with chromosome damage are not known, it is possible that exposure to DNA-damaging agents may increase the occurrence of cancer and of heritable genetic changes in the population.

Classical enumeration of chromosome aberrations has been used to study potentially exposed groups for many years. Aberrations are a useful biological dosimeter for ionizing radiation (reviewed by Sasaki, 1983) and can be sensitive to low cumulative doses, e.g., in the nuclear dockyard workers studied by Evans et al. (1979). In cases of testing for chemical exposures, aberrations have often given conflicting results (for examples see Galloway and Tice, 1982). Discrepancies may arise because of the study design, especially from lack of accurate dose estimates for the individuals concerned or from failure to include suitable controls. In well-designed studies, however, aberrations are useful to measure, as they are known to be associated with mutagenic events (e.g., Russell, 1971; Thacker and Cox, 1983) and are usually considered to be undesirable. Since its first development, the newer technique of analyzing sister chromatid exchanges (SCEs) has been acclaimed for its extreme sensitivity to chemicals, based on experiments in tissue culture (e.g., Kato, 1974; Latt, 1974; Perry and Evans, 1975). Sister chromatid exchange analysis was also found to be a very sensitive indicator of exposure in animals (Allen and Latt, 1976; Vogel and Bauknecht, 1976; Schneider et al., 1976) and in humans given cytotoxic drugs (e.g., Perry and Evans, 1975; Raposa, 1978; Nevstad, 1978; Musilova et al., 1979). The use of

SCEs for occupational monitoring has been questioned because the mechanism and significance of SCEs are unknown. Nevertheless, because a positive correlation exists between the ability of a chemical to induce SCEs and its mutagenic capacity in other assays (reviewed by Abe and Sasaki, 1982), the induction of increases in SCEs in a group of individuals certainly implies potential for associated DNA damage.

A wide range of surveys of SCE frequencies has yielded useful results, although conflicting data still arise that are often attributable to interlaboratory differences in experimental protocol, subject selection, and statistical analysis. A well-designed study should include dose/exposure information on the subjects, sufficient subjects and large enough samples of cells scored to allow meaningful statistical analysis, a standard protocol for cytogenetic analysis throughout, and analysis of suitable concurrent controls under "blind code" (Carrano and Moore, 1982; Bloom, 1981).

Sister chromatid exchanges have proven useful in an occupational setting in our study of people exposed to ethylene oxide (ETO). We have found not only that SCEs are increased by ETO exposure in a dose-related fashion, but that an elevated SCE frequency may persist for at least two years.

Mutagenicity of Ethylene Oxide

Ethylene oxide gas is an effective sterilizing agent, widely used for medical supplies not suited to steam/heat treatment. Although the greatest industrial use of ETO is in chemical syntheses (e.g., polyester and polyethylene glycol), the large volumes in such production are usually liquids and are handled in sealed systems, often outdoors. The most intensive known occupational exposures occur in the use of ETO gas for sterilization, especially in hospitals (NIOSH, 1981; Glaser, 1977). There was cause for concern about potential harmful effects of ETO exposure because ETO is a powerful alkylating agent and mutagen (reviewed by Wolman, 1979; NIOSH, 1981). Much information on mutagenicity existed when the present study began in 1980. When given intraperitoneally to mice, ETO induced heritable translocations and dominant lethal mutations (Generoso et al., 1980; Embree et al., 1977), while inhalation exposure to ETO reportedly induced micronuclei and metaphase aberrations in rodent bone marrow (Appelgren et al., 1978; Embree and Hine, 1975). All these effects were observed at doses of ETO close to the lethal range. For exposed humans, elevations in chromosome aberrations (Ehrenberg and Hallstrom, 1967) and in SCEs (Garry et al., 1979) were reported. Preliminary results of one carcinogenicity study reported tumors at the injection site in mice (Dunkelberg, 1979), but the significance of this observation to human risk from ETO inhalation is unknown. In humans, a possible increase in leukemia incidence in exposed individuals was reported (Hogstedt et al., 1979). These results were inconclusive, though, because this study was small and there had been exposure to additional agents.

STUDY DESIGN

Our objective was to determine whether detectable increases in frequencies of SCEs and of chromosome aberrations occurred in lymphocytes of individuals thought to have been exposed to ETO, compared with controls. As part of a large study of employees engaged in sterilization of medical supplies, we carried out a "pilot study" of 114 individuals at three sites (Stolley et al., 1984). The study was designed with the aim of detecting any dose-related response to ETO.

At the time this study was begun, the legal limit determined by the Occupational Safety and Health Administration for air levels of ETO in the workplace was 50 ppm, expressed as a "time weighted average" (TWA) over the 8-h workday. The three worksites in our study had different histories of air levels of ETO (Table 1). Data on air sampling over several years were used, and during the study, individual exposures were also estimated from personal samplers (Qazi-Ketcham charcoal tubes) worn on the lapel during the work shift the day of blood sample collection. Site I had good engineering and exposure controls, and historical air sampling routinely detected average concentrations of 0.5 ppm ETO (8-h TWA), with peaks that did not exceed 5 ppm. At Site II, the mean was 5-10 ppm, and there were some short-term measurements of up to 50 ppm, with rare peaks of 100 to 300 ppm (on one occasion a short-term peak of 532 ppm was noted). After April of 1980, no peaks of over 50 ppm were found by area sampling. (The first blood samples were taken in the autumn of 1980.) Later in the study, during 1980-1982, the mean TWA at site II was less than 1 ppm and respirators were probably worn during peak potential exposure times. At Site III, the historical average was 5-20 ppm, but during March 1980, levels as high as 200 ppm were found and a leak was corrected. Levels of 5-10 ppm (TWA) were found on three occasions during the remainder of 1980. Use of ETO at site III was completely stopped almost immediately after collection of the first blood samples for cytogenetic analysis, so that no further ETO exposure occurred at this site.

Results could thus be compared among three plants known as low (I), medium (II), and high (III) potential exposure sites. Within sites, individuals were also classed as

Table 1. Average Estimated ETO Exposure Levels Prior to May 1980

Worksite	8-h TWA[a] (ppm)
I	0.5
II	5-10
III[b]	10-20
	(50-200)[c]

[a]TWA = time weighted average.
[b]All ETO use ceased after initial blood sampling in September 1980.
[c]Temporary; measured in March 1980. Engineering faults were corrected.

low potential exposed (LPE) or high potential exposed (HPE). For simplicity, classification was solely on the basis of job category at the time the study began; the HPE group was composed of sterilizer operators, and all other potentially exposed employees were in the LPE group. Note that the HPE and LPE groups at the three sites were not equivalent, e.g., a sterilizer operator (HPE) at site I might well have less ETO exposure than a nonoperator (LPE) at site III. Control individuals were randomly selected at the same sites from individuals not exposed to ETO and were matched for sex and age (within five years) for each subject. About half the study population was male, and just under half were smokers. The numbers in each group are shown in Table 2. The employees were all informed about the nature and purpose of the study, and there was a very good rate of continued participation. Blood samples were taken at time 0 (the beginning of the study) and generally 6, 12, and 24 months later. At the time of each blood sample collection, a brief questionnaire was filled out for each employee in consultation with a specially trained nurse. The questions pertained to factors known or suspected to influence cytogenetic results, including history of smoking; consumption of alcohol, drugs, or large doses of vitamins; viral infections; diagnostic or therapeutic radiation; and other potential exposures, e.g., to paints or solvents in previous occupations or outside working hours. Routine hematology and clinical chemistry studies were carried out at the same time as the blood sampling for cytogenetic analysis.

Cytogenetics Protocol

Heparinized blood samples were put into culture usually less than 24 h and never more than 72 h after collection. The samples were coded so that the subject type (control versus potentially exposed) was never known to the cytogenetics laboratory staff. Separate whole blood cultures were used for aberration and for SCE analysis. Two or three replicates of each type were used to ensure that an adequate number of cells would be obtained. For aberration analysis, cultures without 5-bromodeoxyuridine (BrdUrd) were incubated for 48 to 51 h, and 100 cells per person were scored. The aberration results will be the subject of a subsequent report. For SCE analysis, the volume of blood was adjusted to give a standard inoculum of lymphocytes (two million per 10 ml culture) in RPMI 1640 medium with 15% fetal bovine serum, antibiotics, BrdUrd (100 μM), and 1% phytohemagglutinin-M (PHA). The appropriate amount of blood per culture was calculated from the total white cell count (WBC) and the proportion of lymphocytes from a Schilling differential analysis. Occasionally, the total WBC or lymphocyte count was low so that a large volume (e.g., over 2 ml) of blood was required. In these cases, we used only 1 ml of blood per 10 ml of culture because larger amounts of blood interfere with growth, probably due to the excess of red blood cells. Cultures were protected from light throughout growth. Cells were treated with colcemid (0.1 μg/ml) for the final 2.5 to 3 h of culture and were fixed at 68-74 h of incubation. Air-dried chromosome preparations were stained by a modification of the fluorochrome-plus-Giemsa method (Perry and Wolff, 1974; Goto et al., 1978). BrdUrd was present throughout the culture period to avoid any misleading staining patterns in cells that had undergone partial DNA synthesis periods in the presence of BrdUrd. Eighty cells were scored per subject in most cases.

SCE Scoring

To ensure that scoring of SCEs was satisfactory, a standard set of slides scored in the main laboratory was checked by cytogeneticists at two outside laboratories (referee readers). Samples of slides at various times during the study were also re-read by the referee readers in an attempt to ensure consistent scoring throughout (see Discussion).

RESULTS

The results are summarized in Table 2 and in Figures 1 to 3. The statistical analysis has been described in detail by Stolley et al. (1984) and Soper et al. (1984). Within each worksite, the mean SCE frequencies for all groups were compared using an overall analysis of variance. Pairwise comparisons of potentially exposed groups against controls were also done. The data shown in Figures 1 to 3 are presented as the logarithm of the mean SCE; data were analyzed both with and without this logarithmic transformation, and the same overall conclusions were reached by both methods. The P values in Table 2 are those obtained using the logarithmically transformed data.

At site I, no significant differences among groups were detected at times 0, 12, or 24 months, except for a barely significant difference between LPE and control groups (10.5 versus 8.8 SCE/cell) at the initial sampling ($P = 0.043$). This result is probably a "false positive" because the controls were somewhat lower than at the other two sampling times and because the HPE mean was not elevated. At site II, the group mean for the LPE group was never significantly elevated, but the mean for the four HPE individuals was significantly increased at 0, 12, and 24 months. There was an overall reduction in SCE frequencies for which we do not have an explanation at the six-month follow-up period for this site. At that time, there was a slightly higher SCE mean in the HPE group than in controls, but this was not statistically significant. At site III, the LPE group SCE mean was significantly elevated over the worksite controls at zero and six months. The HPE group mean was significantly elevated at all four sampling times. Although there are only two individuals in the HPE group, the results are convincing because the mean SCE frequencies are extremely high (36.1 and 43.6 SCE/cell), and repeated samples from these individuals showed similar high frequencies.

Another notable observation at site III was that the mean SCE level for the worksite controls at the initial sampling, 11.4 SCE/cell, was higher than the control levels at the other sites (means of 8.8 and 10.8 SCE/cell). This raised the possibility that individuals not directly involved in operations associated with ETO might be inadvertently exposed. A second control group ("community controls") was established at site III by selection of peers by the onsite controls. Members of this new group were matched for age and sex as before, but they had never worked in an establishment where ETO gas was handled. This group was incorporated into the 6- and 24-month follow-up studies. It was clear that on both occasions, the onsite controls at site III had higher SCEs on average than the community controls, whose group mean SCE was comparable to the controls at sites I and II. On further investigations at site III, it was clear that many "controls" were in fact potentially exposed because of the physical layout of the plant and because several of the office staff selected as controls spent a significant amount of their time in areas in which ETO was probably present.

Table 2. Mean SCE/Cell for All Groups at Each Follow-up Period

Site	Group[b]	Mean SCE/Cell (Number of Individuals)[a]			
		Initial	6 mo	12 mo	24 mo
I	HPE	10.0 (8)	-	9.5 (8)	10.5 (8)
	LPE	10.5[c] (5)	-	9.3 (5)	10.8 (5)
	WC	8.8 (12)	-	9.6 (13)	10.0 (11)
II	HPE	14.8[d] (4)	10.3 (4)	14.9[d] (4)	12.5[d] (4)
	LPE	10.2 (18)	9.1 (16)	9.9 (18)	10.3 (15)
	WC	10.8 (19)	8.7 (19)	10.1 (21)	10.2 (18)
III	HPE	32.3[e] (2)	35.2[e,f] (2)	21.4[e] (2)	21.1[e,f] (2)
	LPE	14.7[g] (24)	15.1[f,g] (24)	13.5 (23)	12.5[f] (22)
	WC	11.4 (22)	12.2[f] (23)	11.8 (20)	11.8[f] (20)
	CC	-	8.7 (29)	-	9.8 (28)

[a]Values are marked if SCE frequency is significantly different from control mean, $P < 0.05$.
[b]HPE = high potential exposed; LPE = low potential exposed; WC = worksite controls; CC = community controls.
[c]Significantly greater than WC; $P = 0.043$.
[d]Significantly greater than WC; $P = 0.011$; $P < 0.001$; and $P = 0.018$ at 0, 12 and 24 months, respectively.
[e]Significantly greater than WC; $P \leq 0.002$.
[f]Significantly greater than CC; $P < 0.001$.
[g]Significantly greater than WC; $P = 0.024$ and $P = 0.008$ at 0 and 6 months, respectively.

DISCUSSION

It is clear that ETO exposure is associated with a dose-related increase in SCEs in blood lymphocytes. It is also clear that with good engineering controls and enforcement of industrial safety measures such as use of appropriate respirators, exposure (and its consequences) can be reduced. In plant I, where low levels of ETO (below about 1 ppm) were routinely maintained, no increase in SCEs was detected, even in sterilizer operators. One measure of potential exposure may be obtained from the records kept of the number of sterilization cycles and the number of door openings of the sterilizer unit. Despite the fact that the operators at site I had been present at more door openings than those at site III, they did not have a significant increase in

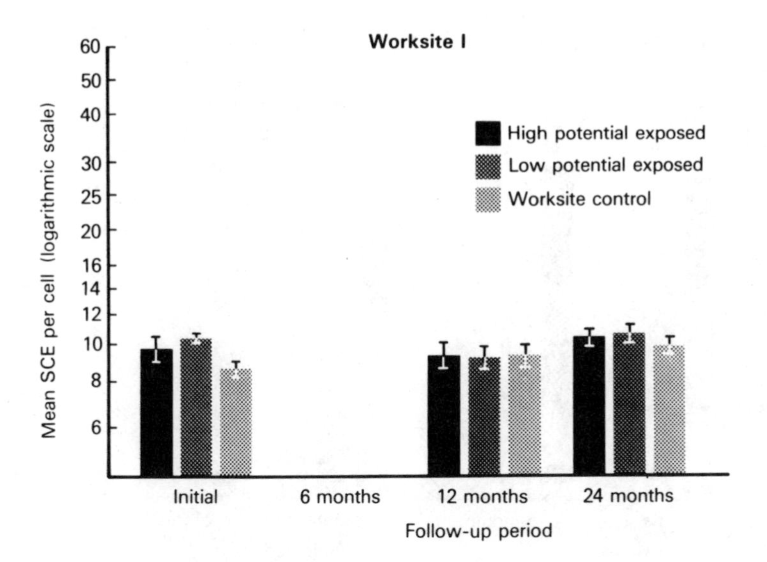

Figure 1. Worksite I. Mean SCE per cell by exposure group and follow-up time, with standard error of the mean (from Stolley et al., 1984).

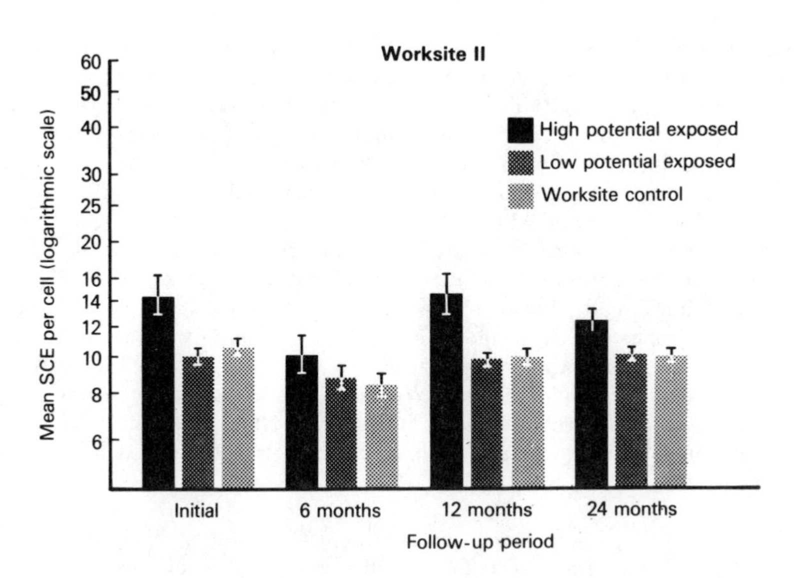

Figure 2. Worksite II. Mean SCE per cell by exposure group and follow-up time, with standard error of the mean (from Stolley et al., 1984).

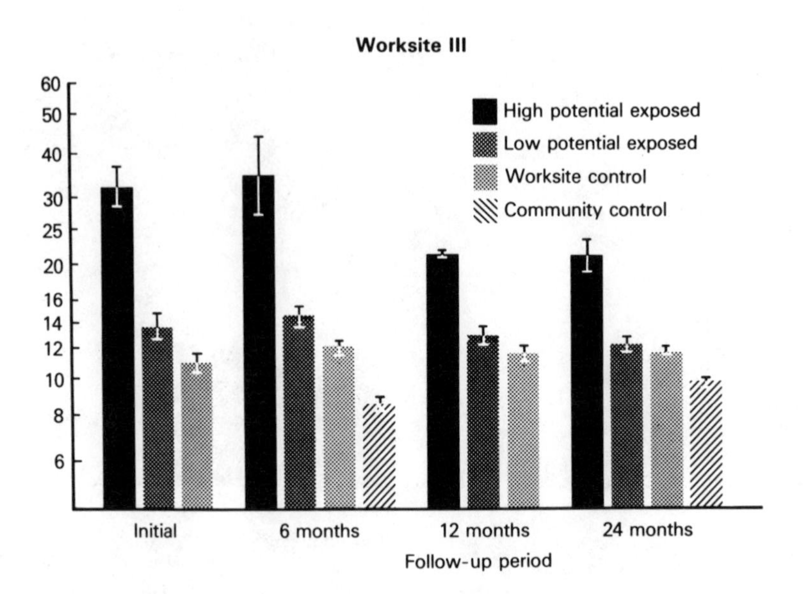

Figure 3. Worksite III. Mean SCE per cell by exposure group and follow-up time, with standard error of the mean (from Stolley et al., 1984).

SCEs. We cannot rule out any potentially hazardous exposure at site I because of the limitations of the sample size and of the SCE technique, but we can at least demonstrate that a detectable increase in SCEs may be avoided.

Perhaps the most striking observation of this study in addition to the high SCE frequencies in the two operators at site III is the apparent persistence of the elevations in SCE frequencies in many individuals. For the LPE group at site III, there was only a slight decrease in SCEs from the initial to the fourth sampling two years after all use of ETO had ceased at that site. In the two site III sterilizer operators, the data shown in Figure 3 suggest that the mean had decreased by 24 months, but results from two further samples taken at 36 and 41 months (Stolley et al., 1984) suggest that the mean SCE frequency may be decreasing in one individual (36.1, 26.7, 21.6, 23.4, 19.3, and 19.4 SCE/cell in samples from 0 to 41 months) but not in the other (28.6, 43.6, 21.1, 18.8, 27.9, and 32.9 SCE/cell). Because of large sampling variation, we cannot yet draw conclusions about trends in SCE frequencies in these individuals. Among the 80 cells scored from each of these samples, there was a wide range of SCE frequencies, including cells with 90 or more SCEs. We may, therefore, expect some random variation in the mean SCE frequency among sampling times in addition to variation caused by possible selective growth of cells in culture or metaphase selection by the reader during scoring.

We have found that reader variation is a significant source of variability in the SCE results. In analysis of the results for the entire control population we have studied (479 control individuals including those for this pilot study but excluding site III onsite controls), significant effects of smoking, age, and gender have been identified; further, almost 20% of the variability among controls can be accounted for by differences among readers (Soper et al., 1984). This variation occurred despite a common training period for all readers. Analysis of SCEs has a reputation for ease and speed, but we have found that it requires very careful training and is not always faster than aberration analysis. We now have a program of repeated restandardization in the laboratory. It is important that new trainees learn to karyotype first and then be thoroughly trained in SCE analysis. However, despite the same initial training, readers tend to become relatively "high" or "low". A standard set of photographs is useful for training, and one method for frequent reassessment of readers is to select a slide at random at frequent intervals, circulate it to all readers, and compare results.

The culture method, with a standard lymphocyte number and rather high BrdUrd concentration, is not always optimal, especially for individuals having low white counts so that a large blood volume is needed. Also, BrdUrd at 100 µM can retard culture growth, although we did obtain results on a very high percentage of the samples. The decision to use this method was based on observations by Carrano et al. (1980) and was intended to reduce intralaboratory variability. Although it may be desirable to minimize the background level of SCEs by using the lowest possible dose of BrdUrd, we decided to use 100 µM because of evidence that at this concentration variations in cell proliferation among subjects would have less effect on the SCE level. This was based on a report that the SCE frequency reflected the relative proportions of growing cells and of BrdUrd molecules in the culture (Mazrimas and Stetka, 1978). At higher concentrations of BrdUrd, the influence of a change in cell number was reportedly less because a plateau was thought to occur in the dose relation of SCE to BrdUrd concentration (Carrano et al., 1980). It now seems, however, that since we cannot control the proliferative response of individuals to PHA, the extra effort involved in standardizing the cell inoculum may not always be worthwhile. There is evidence that at lower concentrations of BrdUrd, e.g., 10 to 33 µM, variations in lymphocyte inoculum do not correlate with the observed SCE frequency (F. Angelosanto, personal communication; Littlefield et al., 1982). Also, data from other laboratories do not always support the observation by Carrano et al. (1980) that the dose relation of SCEs versus concentration of BrdUrd tends to reach a plateau (Galloway et al., unpublished observations). Overall, a lower concentration of BrdUrd may be desirable to avoid any growth retardation. Our control levels, however, are not inordinately high. In the control study the mean was 9.9 SCE/cell, and 95% of the 479 control individuals had means of 13.4 or fewer SCE/cell (Soper et al., 1984). The study design, with 80 cells scored per person and sufficient numbers of individuals in the groups, allowed us to detect quite small increases in SCEs (e.g., a 23% increase over controls in the HPE group at site II at the 24-month follow-up and a 20% increase in the site III worksite controls compared with community controls at the 24-month follow-up (Table 2).

Since the beginning of our study, several other reports have confirmed the ability of ETO to cause cytogenetic changes. Chromosome aberrations and SCEs are induced by ETO in cultured cells (Poirier and Papadopoulo, 1982; Garry et al., 1982), and there are several reports on animals exposed by inhalation and on humans working in sterilization facilities (Table 3). Also, Generoso et al. (1983) have extended their results to demonstrate dominant lethal mutation induction by inhaled ETO.

An important question that remains to be answered is how the type of exposure affects the response; for example, are chromosome changes more marked after low, cumulative exposure over long periods or after brief but very high exposures? We do not yet have the information to answer this question. Garry et al. (1979) reported a slight increase in SCEs in an individual accidentally exposed for a brief period to at least 1500 ppm of ETO, and Ehrenberg and Hallstrom (1967) observed aberrations in lymphocytes of an individual reportedly exposed to ETO for only 2 h. In the inhalation experiments performed by Kligerman et al. (1983), a significant increase in SCEs was induced by exposing rats to about 450 ppm of ETO for one day, but at 50 ppm, three days of exposure were required to achieve an increase in SCEs in blood lymphocytes. In contrast, in a longer-term inhalation study on rabbits (Yager and Benz, 1982), no significant increase was observed in SCEs until seven weeks of continuous exposure had passed, even at 500 ppm of ETO. Although there was no significant increase in mean SCEs during 12 weeks of exposure to 10 ppm ETO, a few cells had higher SCE values than those found in the control cultures (Yager and Benz, 1982).

Several reports on smaller groups confirm our observations in humans (Table 3). It is interesting that a cumulative effect appeared over several years in the study of Laurent et al. (1983), although these results must be treated with caution because of the small group size. Other work in humans (Abrahams, 1980) and in monkeys (NIOSH, 1982) also demonstrated increased SCEs in ETO-exposed individuals. Hogstedt et al. (1983) saw increases in chromosome aberrations and micronuclei in ETO-exposed people but claimed SCEs were not elevated, although in their data at least one individual had a mean SCE frequency (25) that seems unusually high. Thiess et al. (1981) interpreted their own results on chromosome aberrations as negative, but in individuals exposed for over 20 years, they found that 3.5% of the cells had aberrations, compared with 1-1.4% in two control groups. Because the length of the culture period has a marked effect on aberration frequencies (e.g., Buckton and Pike, 1964), the 72-h incubation time in both these studies (Hogstedt et al., 1983; Thiess et al., 1981) may have resulted in reduced numbers of detectable aberrations.

Our study demonstrates persistence of increased SCE levels over a longer period (two years) than any previously observed. Yager and Benz (1982) showed that SCEs in rabbits were slightly elevated over controls 15 weeks after the last ETO exposure. In humans, Garry et al. (1979) reported that SCEs remained elevated up to 8 weeks after exposure terminated. They also noted slight increases with time in three of four individuals, although these increases were very small and were based on only 20 cells scored per person so that they are likely to be within the range of culture variation.

Overall, the present study not only has shown that SCEs can be useful to measure occupational exposure, but it has also raised an unexpected series of questions about persistence of chromosome changes. Although the existence of long-lived lymphocytes (e.g., 4-30 years) is known and indeed was implied from the persistence of chromosome aberrations in lymphocytes of radiation-exposed individuals (e.g., Buckton et al., 1967, 1978; Awa, 1983), SCEs were not thought to persist for long periods of time (reviewed by Lambert et al., 1982). In rabbits exposed to repeated doses of mitomycin C, Stetka et al. (1978) had demonstrated that the SCE frequency returned towards control levels much more slowly than after a large acute dose and that the SCEs were still above control levels more than four months after the end of treatment. There were also some reports of long-lasting (two to three months) increases in SCEs in patients given

Table 3. Effect of Inhaled Ethylene Oxide on Sister Chromatid Exchanges

			Exposed				Controls			
Species	Concentration BrdUrd (μM)	Cells/ Individual	Duration	Dose (ppm)	Number of Individuals	SCE/Cell Mean (Range)	Number of Individuals	SCE/Cell Mean (Range)	Comments	Reference
Rabbit	10	50	12 wk	10	3	7.9	3	7.3	Increase built up during exposure. Not significant until 7 wk. SCEs slightly above control 15 wk after end of exposure.	Yager and Benz, 1982
				50	3	9.5				
				250	3	13.2				
Rat	2	50	1 d	48	3	8.8	4	7.8	Significant increase after 1 d (6 h) exposure at ~450 ppm.	Kligerman et al., 1983
				136	3	9.0				
				451	4	10.4				
			3 d	50	3	9.1	4	7.5		
				140	3	10.3				
				444	4	13.6				

(continued)

Table 3. Continued

			Exposed				Controls			
Species	Concentration BrdUrd (μM)	Cells/Individual	Duration	Dose (ppm)	Number of Individuals	SCE/Cell Mean (Range)	Number of Individuals	SCE/Cell Mean (Range)	Comments	Reference
Human	15	20		36[a]	4[b] 4[c] 8[d]	9.7 (8.9-10.7) 10.3 (6.7-12.2) 8.7[e]	8	6.0 (4.9-7.1)		Garry et al., 1979
Human	100	20			2 ns[f] 3 s[h]	16.1 21.1	63[g] ns 45[g] s	13.9 16.4		Lambert et al., 1982
Human	~33	50-125	0-3 yr 3-6 yr 6-10 yr		25[d]	11.6 (9.6-14.0) 13.1 (9.7-14.0) 14.5 (10.9-17.6) 13.0[e] (9.6-17.6)	10	7.9 (7.0-8.5)		Laurent et al., 1983
Human	20	50		"low" "high"		7.7 (6.2- 9.6) 10.7 (8.0-13.0)	13	7.6 (5.5-8.9)	All exposures said to be <50 ppm.	Yager et al., 1983

[a]Recent. Open floor drain had 1500 ppm.
[b]Reported symptoms of exposure.
[c]Same four as in footnote b, 8 wk after last exposure. Three of four showed slight increase.
[d]Total.
[e]Mean.
[f]ns = non-smoker.
[g]Reference population for that laboratory.
[h]s = smoker.

cytotoxic drugs (e.g., Lambert et al., 1979; Musilova et al., 1979; Aronson et al., 1982). In all these cases, however, clear downward trends were noted, even though the SCE frequency remained above baseline for some time.

A preliminary examination of our data on ETO does not show a similar trend. One would expect a diminution in mean SCE by dilution of cells in the bloodstream with fresh cells from lymphopoietic sites. Thus, two populations of cells would exist: older lymphocytes with high SCEs, and newly produced cells with normal SCE levels. We might, therefore, expect to see changes with time in the relative proportions of "high SCE" and "normal SCE" cells. The expected dynamics of such changes are not known. In patients treated with high doses of X-rays, the frequencies of "unstable" chromosome aberrations in lymphocytes dropped off rapidly during the first four years after the end of treatment. This was thought to reflect death of the damaged cells and repopulation with fresh cells from the lymphopoietic sites (Buckton et al., 1967). Sister chromatid exchanges in individuals exposed to ETO do not appear to behave in the same way. However, in irradiated patients, there is a great deal of lymphocyte turnover because of the initial X-ray-induced suppression in blood lymphocyte count. Similarly, many of the observations that SCEs rapidly returned to normal levels after treatment were made in patients given cytotoxic drugs that cause drastic reductions in lymphocyte count. The situation with occupational ETO exposure may not be directly comparable because there are varied lengths and intensities of exposure, and acute toxicity to lymphocytes may not be characteristic in the exposed individuals. We might, therefore, expect slower lymphocyte turnover and slower changes in the SCE level in ETO-exposed people compared with SCEs after chemotherapy, or aberration changes in X-irradiated patients. Indeed, the reduction in unstable aberration frequency progressed much more slowly after the recovery of the lymphocyte population to normal levels about four years after X-ray treatment (Buckton, 1983). From the aberration studies, Buckton et al. (1967) estimated that the average half-life of the PHA-responsive lymphocytes in their patients, after the recovery of the lymphocyte population, was about 4.5 years. There were also cells that had survived without dividing for 30 years (Buckton, 1983). Longer follow-up of the ETO-exposed individuals will be necessary to study changes with time in the proportions of cells with low and high SCE frequencies.

It is also possible that repopulation does occur, but the fresh cells produced from stem cells have high SCE levels, so that no decrease in the mean SCE in the blood cells is apparent. If this is so, one might postulate not only a long-lived lesion in stem cells, but a lesion that survives several cycles of DNA synthesis and cell division during lymphocyte production. These characteristics are not expected for alkylation of DNA: SCE-inducing DNA adducts in dividing cells are thought to be removed during one to three cell cycles, depending on the cell type and chemical agent (e.g., Wolff, 1982; Connell and Medcalf, 1982; Conner et al., 1983; Littlefield et al., 1983). In some cases, repair occurs even in nondividing cells (Jostes, 1981), although DNA crosslinks induced by a combination of psoralen and ultraviolet radiation were poorly removed in nonproliferating fibroblasts (Bredberg and Lambert, 1983). Thus, in cells that are quiescent or lack repair capacity (Wolff, 1982), DNA lesions capable of inducing SCEs may persist. However, stem cells must proliferate to produce mature lymphocytes and are not likely to have deficient repair in most people. It is possible that ETO directly or indirectly induces some unknown type of alteration that persists through several cell cycles; in mice treated with ethyl carbamate, such persistent lesions have been postulated (Conner and Cheng, 1983). An alternate hypothesis is that ETO exposure might alter normal cellular repair capability. There is evidence for suppression of lymphocyte DNA repair by ETO exposure *in vitro* and *in vivo* (Pero et al., 1981, 1982). Also, incorporation of BrdUrd for SCE analysis could affect the cell's response to ETO effects, but even if this gave rise to an artifact in lymphocyte culture, it would not explain why cells with the potential to develop high SCE frequencies in culture should persist in the body for so long.

In addition to introducing some fascinating scientific questions about SCEs and about the nature of ETO effects on cells, this observation of persistence demonstrates that SCE analysis may be more useful in certain occupational or environmental exposure settings than was widely believed at the time of the initial observations that increases in SCEs were transitory.

ACKNOWLEDGMENTS

We thank P. Unguris, W. Fistere, J. Adams, W. Williams, M. Ehrlich, F. Newman, and T. Hynson for patient and excellent work in data collection and checking; J. Smith for designing the computer systems for data handling; and the following devoted laboratory staff for many hours of work: S. Berg, J. Butler, T. Clement, J. Cole, R. Curvey, D. Drabkowski, H. Duff, J. Edwards, M. George, S. Gianninni, R. Knight, S. Kumar, K. Lavappa, A. Matsukevitch, H. Murli, M. Murtha, R. Nandi, and D. Quann. We also thank P. Archer for expert statistical advice.

REFERENCES

Abe, S., and M. Sasaki. 1982. SCE as an index of mutagenesis and/or carcinogenesis. In: Sister Chromatid Exchange. Progress and Topics in Cytogenetics 2. A.A. Sandberg, ed. Alan R. Liss: New York. pp. 461-514.

Abrahams, R.H. 1980. Recent studies with workers exposed to ethylene oxide. In: The Safe Use of Ethylene Oxide: Proceedings of the Educational Seminar, HIMA Report No. 80-4. J.F. Jorkasky, ed. Health Industry Manufacturers Association: Washington, DC. pp. 27-53.

Allen, J.W., and S.A. Latt. 1976. Analysis of sister chromatid exchange formation in vivo in mouse spermatogonia as a new test system for environmental mutagens. Nature (London) 260:449-451.

Appelgren, L-E., G. Eneroth, C. Grant, L-E. Landstrom, and K. Tenghagen. 1978. Testing of ethylene oxide for mutagenicity using the micronucleus test in mice and rats. Acta Pharmacol. Toxicol., 43:69-71.

Aronson, M.M., R.C. Miller, R.B. Hill, W.W. Nichols, and A.T. Meadows. 1982. Acute and long-term cytogenetic effects of treatment in childhood cancer. Sister-chromatid exchanges and chromosome aberrations. Mutat. Res. 92:291-307.

Awa, A. 1983. Chromosome damage in atomic bomb surviors and their offspring--Hiroshima and Nagasaki. In: Radiation-Induced Chromosome Damage in Man. Progress and Topics in Cytogenetics 4. A.A. Sandberg, ed. Alan R. Liss: New York. pp. 433-453.

Bloom, A.D., ed. 1981. Guidelines for Studies of Human Populations Exposed to Mutagenic and Reproductive Hazards. March of Dimes Birth Defects Foundation: New York. pp. 1-35.

Bredberg, A., and B. Lambert. 1983. Induction of SCE by DNA cross-links in human fibroblasts exposed to 8-MOP and UVA irradiation. Mutat. Res. 118:191-204.

Buckton, K.E. 1983. Chromosome aberrations in patients treated with X-irradiation for ankylosing spondylitis. In: Radiation-Induced Chromosome Damage in Man. A.A. Sandberg, ed. Alan R. Liss: New York. pp. 491-511.

Buckton, K.E., and M.C. Pike. 1964. Chromosome investigations on lymphocytes from irradiated patients: effect of time in culture. Nature (London) 202:714-715.

Buckton, K.E., P.G. Smith, and W.M. Court-Brown. 1967. The estimation of lymphocyte lifespan from studies on males treated with X-rays for ankylosing spondylitis. In: Human Radiation Cytogenetics. H.J. Evans, W.M. Court-Brown, and A.S. McLean, eds. North Holland: Amsterdam. pp. 106-114

Buckton, K.E., G.E. Hamilton, L. Paton, and O. Langlands. 1978. Chromosome aberrations in irradiated ankylosing spondylitis patients. In: Mutagen-Induced Chromosome Damage in Man. H.J. Evans and D.C. Lloyd, eds. Edinburgh University Press: Scotland. pp. 142-150.

Carrano, A.V., and D.H. Moore. 1982. The rationale and methodology for quantifying sister chromatid exchange in humans. In: New Horizons in Genetic Toxicology. John Heddle, ed. Academic Press: New York. pp. 268-304.

Carrano, A.V., J.L. Minkler, D.G. Stetka, and D.H. Moore II. 1980. Variation in the baseline sister chromatid exchange frequency in human lymphocytes. Environ. Mutagen. 2:325-337.

Connell, J.R., and A.S.C. Medcalf. 1982. The induction of SCE and chromosomal aberrations with relation to specific base methylation of DNA in Chinese hamster cells by N-methyl-N-nitrosourea and dimethylsulfate. Carcinogenesis 3:385-390.

Conner, M.K., and M. Cheng. 1983. Persistence of ethylcarbamate-induced DNA damage in vivo as indicated by sister chromatid exchange analysis. Cancer Res. 43:965-971.

Conner, M.K., J.E. Luo, and O. Guttierrez de Gotera. 1983. Induction and rapid repair of sister-chromatid exchanges in multiple murine tissues in vivo by diepoxybutane. Mutat. Res. 108:251-263.

Dunkelberg, H. 1979. On the oncogenic activity of ethylene oxide and propylene oxide in mice. Br. J. Cancer 39:588-589.

Ehrenberg, L., and T. Hallstrom. 1967. Haematologic studies on persons occupationally exposed to ethylene oxide. In: International Atomic Energy Agency Report SM 92/96. IAEA. pp. 327-334. Communicated by L.O. Kallings.

Embree, J.W., and C.H. Hine. 1975. Mutagenicity of ethylene oxide. Toxicol. Appl. Pharmacol. 33:172-173 (Abstract).

Embree, J.W., J.P. Lyon, and C.H. Hine. 1977. The mutagenic potential of ethylene oxide using the dominant-lethal assay in rats. Toxicol. Appl. Pharmacol. 40:261-267.

Evans, H.J., K.E. Buckton, G.E. Hamilton, and A. Carothers. 1979. Radiation-induced chromosome aberrations in nuclear-dockyard workers. Nature (London) 277:531-534.

Galloway, S.M., and R.R. Tice. 1982. Cytogenetic monitoring of human populations. In: Genotoxic Effects of Airborne Agents. Environmental Science Research 25. R.R. Tice, D.L. Costa, and K.M. Schaich, eds. Plenum Press. New York. pp. 463-488.

Garry, V.F., J. Hozier, D. Jacobs, R.L. Wade, and D.G. Gray. 1979. Ethylene oxide: evidence of human chromosomal effects. Environ. Mutagen. 1:375-382.

Garry, V.F., C.W. Opp, J.K. Wiencke, and D. Lakatua. 1982. Ethylene oxide induced sister chromatid exchange in human lymphocytes using a membrane dosimetry system. Pharmacology 25:214-221.

Generoso, W.M., K.T. Cain, M. Krishna, C.W. Sheu, and R.M. Gryder. 1980. Heritable translocation and dominant-lethal mutation induction with ethylene oxide in mice. Mutat. Res. 73:133-142.

Generoso, W.M., R.B. Cumming, J.A. Bandy, and K.T. Cain. 1983. Increased dominant-lethal effects due to prolonged exposure of mice to inhaled ethylene oxide. Mutat. Res. 119:377-379.

Glaser, Z.R. 1977. Ethylene oxide: toxicology review and field study results of hospital use. J. Environ. Pathol. Toxicol. 2:173-208.

Goto, K., S. Maeda, Y. Kano, and T. Sugiyama. 1978. Factors involved in differential Giemsa-staining of sister chromatids. Chromosoma 66:351-359.

Hogstedt, C., O. Roheln, B.S. Berndtsson, O. Axelson, and L. Ehrenberg. 1979. A cohort study of mortality and cancer incidence in ethylene oxide production workers. Br. J. Ind. Med. 36:276-280.

Hogstedt, B., B. Gullberg, K. Hedner, A-M. Kolnig, F. Mitelman, S. Skerfving, and B. Widegren. 1983. Chromosome aberrations and micronuclei in bone marrow cells and peripheral blood lymphocytes in humans exposed to ethylene oxide. Hereditas 98:105-113.

Jostes, R.F. 1981. Sister chromatid exchanges but not mutations decrease with time in arrested Chinese hamster ovary cells after treatment with ethylnitrosurea. Mutat. Res. 91:371-375.

Kato, H. 1974. Induction of sister chromatid exchanges by chemical mutagens and its possible relevance to DNA repair. Exp. Cell Res. 5:239-247.

Kligerman, A.D., G.L. Erexson, M.E. Phelps, and J.L. Wilmer. 1983. Sister-chromatid exchange induction in peripheral blood lymphocytes of rats exposed to ethylene oxide by inhalation. Mutat. Res. 120:37-44.

Lambert, B., U. Ringborg, and A. Lindblad. 1979. Prolonged increase of sister-chromatid exchanges in lymphocytes of melanoma patients after CCNU treatment. Mutat. Res. 59:295-300.

Lambert, B., A. Lindblad, K. Holmberg, and D. Francesconi. 1982. The use of sister chromatid exchanges to monitor human populations for exposure to toxicologically harmful agents. In: Sister Chromatid Exchange. S. Wolff, ed. John Wiley and Sons: New York. pp. 149-182.

Latt, S.A. 1974. Sister chromatid exchanges, indices of human chromosome damage and repair: detection by fluorescence and induction by mitomycin C. Proc. Natl. Acad. Sci. U.S.A. 71:3162-3166.

Laurent, Ch., J. Frederic, and F. Marechal. 1983. Increased sister chromatid exchange frequency in workers exposed to ethylene oxide. Ann. Genet. 26:138-142.

Littlefield, L.G., S.P. Colyer, A.S. Sayer, and R.J. DuFrain. 1982. Lack of correlation between hematological parameters and baseline sister chromatid exchanges in lymphocyte cultures from control subjects. Mammal. Chromo. Newsletter 23:27.

Littlefield, L.G., S.P. Colyer, and R.J. DuFrain. 1983. SCE evaluation in human lymphocytes after G_0 exposure to mitomycin C. Lack of expression of MMC-induced SCEs in cells that have undergone greater than two in vitro divisions. Mutat. Res. 107:119-130.

Mazrimas, J.A., and D.G. Stetka. 1978. Direct evidence for the role of incorporated BUdR in the induction of sister chromatid exchanges. Exp. Cell Res. 117:23-30.

Musilova, J., K. Michalova, and J. Urban. 1979. Sister-chromatid exchanges and chromosomal breakage in patients treated with cytostatics. Mutat. Res. 67:289-294.

Nevstad, N.P. 1978. Sister chromatid exchanges and chromosomal aberrations induced in human lymphocytes by the cytostatic drug adriamycin in vivo and in vitro. Mutat. Res. 57:253-258.

NIOSH, 1981. Current Intelligence Bulletin 35: Ethylene Oxide. DHHS (NIOSH) Publication No. 81-130. U.S. Department of Health and Human Services, Public Health Service, Centers for Disease Control, National Institute for Occupational Safety and Health. U.S. Government Printing Office: Washington, D.C. 22 pp.

NIOSH, 1982. Lynch, D.W., et al. Chronic Inhalation Toxicity of Ethylene Oxide and Propylene Oxide in Rats and Monkeys--A Preliminary Report. Presentation at the 21st Annual Meeting of the Society of Toxicology.

Pero, R.W., B. Widegren, B. Hogstedt, and F. Mitelman. 1981. In vivo and in vitro ethylene oxide exposure of human lymphocytes assessed by chemical stimulation of unscheduled DNA synthesis. Mutat. Res. 83:271-289.

Pero, R.W., T. Bryngelsson, B. Widegren, B. Hogstedt, and H. Welinder. 1982. A reduced capacity for unscheduled DNA synthesis in lymphocytes from individuals exposed to propylene oxide and ethylene oxide. Mutat. Res. 104:193-200.

Perry, P., and H.J. Evans. 1975. Cytological detection of mutagen-carcinogen exposure by sister chromatid exchange. Nature (London) 258:121-125.

Perry, P., and S. Wolff. 1974. New Giemsa method for the differential staining of sister chromatids. Nature (London) 251:156-158.

Poirier, V., and D. Papadopoulo. 1982. Chromosomal aberrations induced by ethylene oxide in a human amniotic cell line in vitro. Mutat. Res. 104:255-260.

Raposa, T. 1978. Sister chromatid exchange studies for monitoring DNA damage and repair capacity after cytostatics in vitro and in lymphocytes of leukaemic patients under cytostatic therapy. Mutat. Res. 57:241-251.

Russell, L.B. 1971. Definition of functional units in a small chromosomal segment of the mouse and its use in interpreting the nature of radiation-induced mutations. Mutat. Res. 11:107.

Sasaki, M.S. 1983. Use of lymphocyte chromosome aberrations in biological dosimetry: possibilities and limitations. In: Radiation-Induced Chromosome Damage in Man. Progress and Topics in Cytogenetics 4. A.A. Sandberg, ed. Alan R. Liss: New York. pp. 585-604.

Schneider, E.L., J. Chaillet, and R.R. Tice. 1976. In vivo BrdU labelling of mammalian chromosomes. Exp. Cell Res. 100:396-399.

Soper, K.A., P.D. Stolley, S.M. Galloway, J.G. Smith, W.W. Nichols, and S.R. Wolman. 1984. Sister chromatid exchange (SCE) report of control subjects in a study of occupationally exposed workers. Mutat. Res. 129:77-88.

Stetka, D.G., J. Minkler, and A.V. Carrano. 1978. Induction of long-lived chromosome damage as manifested by sister-chromatid exchange, in lymphocytes of animals exposed to mitomycin C. Mutat. Res. 51:383-396.

Stolley, P.D., K.A. Soper, S.M. Galloway, W.W. Nichols, S.A. Norman, and S.R. Wolman. 1984. Sister chromatid exchanges in association with occupational exposure to ethylene oxide. Mutat. Res. 129:89-102.

Thacker, J., and R. Cox. 1983. The relationship between specific chromosome aberrations and radiation-induced mutations in cultured mammalian cells. In: Radiation-Induced Chromosome Damage in Man. Progress and Topics in Cytogenetics 4. A.A. Sandberg, ed. Alan R. Liss: New York. pp. 235-276.

Thiess, A.M., H. Schwegler, I. Fleig, and W.G. Stocker. 1981. Mutagenicity study of workers exposed to alkylene oxides (ethylene oxide/propylene oxide) and derivatives. J. Occup. Med. 23:343-347.

Vogel, W., and T. Bauknecht. 1976. Differential chromatid staining by in vivo treatment as a mutagenicity test system. Nature (London) 260:448-449.

Wolff, S. 1982. Chromosome aberrations, sister chromatid exchanges, and the lesions that produce them. In: Sister Chromatid Exchange. S. Wolff, ed. John Wiley and Sons: New York. pp. 41-58.

Wolman, S.R. 1979. Mutational consequences of exposure to ethylene oxide. J. Environ. Pathol. Toxicol. 2:1289-1302.

Yager, J.W., and R.D. Benz. 1982. Sister chromatid exchanges induced in rabbit lymphocytes by ethylene oxide after inhalation exposure. Environ. Mutagen. 4:121-134.

Yager, J.W., C.J. Hines, and R.C. Spear. 1983. Exposure to ethylene oxide at work increases sister chromatid exchanges in human peripheral lymphocytes. Science 219:1221-1223.

DIRECT MUTAGEN RISK ASSESSMENT: THE DEVELOPMENT OF METHODS TO MEASURE IMMUNOLOGIC AND GENETIC RESPONSES TO MUTAGENS

Gary H.S. Strauss

Health Effects Research Laboratory, U.S. Environmental Protection Agency, Research Triangle Park, North Carolina 27711

Accepting society's addiction to the products of industrial technology, it is our responsibility to try to ascertain the adverse effects of chemical and physical agents. An escalating awareness of the potential and manifest severities of mutagen-related health effects confers a sense of urgency upon the problem of producing and utilizing relevant data for sound risk-versus-benefit predictions and decisions. The purpose of this disquisition is to review our efforts to develop methods with which to evaluate the *in vivo* effects of mutagen exposure in cells taken directly from the body.

The direct testing approach offers several important advantages for evaluating the consequences of mutagen exposure in humans: (1) all relevant metabolic and pharmacokinetic processes are intact and active; (2) individuals (and families) may be assessed for differences in susceptibility to effects of various agents; (3) spontaneous (background) mutation frequencies may be determined and monitored so that the consequences of exposure to complex mixtures of mutagens and other relevant environmental factors or to extraordinary exposures may be studied; and (4) mutagenicity test results may be compared directly with those of conventional clinical studies of the same individuals and with epidemiological findings from their peers.

A primary goal of direct mutagenicity tests should be to detect and quantitate genotypic variance. The detection methods should be based upon an unambiguous phenotype at the single-cell level, so that mutant cells can be distinguished from normal cells. The specimens should be easily obtainable tissues containing cells of polyclonal origin, taken directly from the body. Currently, the most promising approaches, though they remain in developmental stages, are those designed to detect mutations at specific gene loci. Among these are two monospecific antibody methods. One detects human red cells containing variant (sickle cell, HvS) hemoglobin (Stamatoyannopoulos et al., 1980) and the other detects an altered sperm-bearing antigen (lactodehydrogenase-X) that differs from the normal form by just one amino acid (Ansari et al., 1980). Severe deficiencies will exist in these new systems until processes underlying the development of variant cells can be more fully characterized (Malling, 1981). The third evolving approach in this category is the Strauss-Albertini test, with which one can demonstrate that the drug-resistant variant cells are indeed products of mutant genotypes (Strauss and Albertini, 1977, 1979).

This method has been formulated and modified with attention to ideal criteria for identifying cells as mutant on the basis of phenotype (Thompson and Baker, 1973; Clements, 1975; Chu and Powell, 1976):

1. Occurs randomly and at low frequency;

2. Remains stable in the absence of selection;

3. Increases in frequency (and, in some cases, in reversion rate) after exposure to mutagens;

4. Can be associated with an altered gene product (usually a protein);

5. Is attributable to a specific chromosomal region and behaves in a Mendelian fashion;

6. Allows viability;

7. Can be isolated and identified unambiguously at the single-cell level;

8. Exists in the haploid state, obviating dominance-recessive relationships. This would require an X- or Y-chromosomal location, although an autosomal locus will suffice for cells in the heterozygous condition prior to testing or having the appropriate small deletion on the homologous chromosome;

9. Two possible causes of the altered phenotype, chromosome deletions and specific gene mutations, can be distinguished from each other;

10. Is transferable as a marker to other cells;

11. Is detectable in the presence of a vast majority of normal cells under selection and can be propagated for further characterization.

The Strauss-Albertini test is the descendant of human cell *in vitro* test systems that rely upon the ability to select, quantify, and isolate mutant drug-resistant cells occurring spontaneously or as a consequence of exposure to test mutagens (for example: DeMars, 1971; Thilly et al., 1976).

The marker phenotype most commonly used in these systems is purine analogue resistance, conferred as a result of DNA damage to the X-chromosomal gene (Hpt locus) responsible for the purine salvage enzyme hypoxanthine-guanine phosphoribosyltransferase (HGPRT). Virtually all cells from males afflicted with the germinally inherited X-linked recessive Lesch-Nyhan (LN) syndrome are HGPRT deficient and therefore survive, using the *de novo* pathway for purine base processing in DNA synthesis, in the presence of the purine analogue 6-thioguanine (TG) at concentrations grossly cytotoxic to normal cells (Kelley, 1968; Demars, 1974).

By design, the Strauss-Albertini test avoids many questions of metabolic realism and target relevancy to which the *in vitro* mutagen exposure systems are subject. Peripheral blood lymphocytes (PBLs) are easily sampled, receive exposure to conditions extant throughout the body, are easily triggered to DNA synthesis, and have been well characterized as to their behavior under conditions of short-term culture and various structural and functional properties. The basic approach, therefore, is to screen PBLs for the presence of LN-like cells. Variants that incorporate tritiated thymidine ([^{3}H]-TdR) *in vitro* in response to phytohemagglutinin (PHA) despite the presence of TG can be counted by light microscopy after autoradiography.

The method used to perform the earliest applications of the Strauss-Albertini test can be found elsewhere (Strauss and Albertini, 1977, 1979; Strauss et al., 1979, 1980a). A revised procedure (Strauss, 1982b, d; Albertini, 1982), presented in Figure 1, takes into account the recent finding that PBLs cycling at the time of biopsy (or at least early in the culture period) can appear as TG-resistant (TG^{r}) variants. These cells, known as "phenocopies," may have resulted in overestimations of variant frequencies (Vfs) determined for normal individuals (as indicated in Figure 4) and those at risk. Preliminary results indicate that the phenocopies can be eliminated by freezing the PBL specimens before they are tested (Albertini, 1982).

The culture protocol was designed to evaluate the products of *in vivo* events, providing sufficient time for vigorous DNA synthesis while denying cell division *in vitro*. Nuclei were freed from PHA-agglutinated cells for enumeration, and it was shown that all persist to be counted.

The selective concentration 2×10^{-4} M TG was chosen on the basis of comparative dose-response studies wherein nearly all PBLs from LN males were resistant (Vf = 7.5×10^{-1}) and PBLs from normal individuals were sensitive (Vf = 9.0×10^{-5}). TG^{r} PBLs from LN heterozygotes were observed at an average Vf of 2.0×10^{-3}. The Strauss-Albertini test was presented as a means to diagnose LN mutation carrier status in females presumed to be at risk (Strauss et al., 1980a). The absolute sensitivity of the assay was estimated to be at least 10^{-5} by evaluating cultures containing artificial mixtures of minority LN PBLs in majority populations of PBLs from a normal individual. These results suggested that efficiency of recovery was not affected by any *in vitro* process of selection against the mutant phenotype.

In a series of 98 experiments, TG^{r} PBL frequencies were determined in 63 healthy, non-LN individuals ranging in age from 11 to 75 yr. The median Vf was 1.0×10^{-4} (the mean was 1.3×10^{-4}, and the 10th and 90th percentiles were 5×10^{-5} and 1.5×10^{-4}, respectively). No correlation between age of donor and Vf was found. We suggested that though many of the pertinent T-lymphocytes are long-lived, failure of TG^{r} PBLs to accumulate with time may be explained, in part, by *in vivo* selection against the well-differentiated cell. Evans and Vijayalaxmi (1981) have reported an age-dependent increase in 8-azaguanine-resistant lymphocytes that they conclude may represent single Xq deletions. This issue obviously requires further inquiry. In the

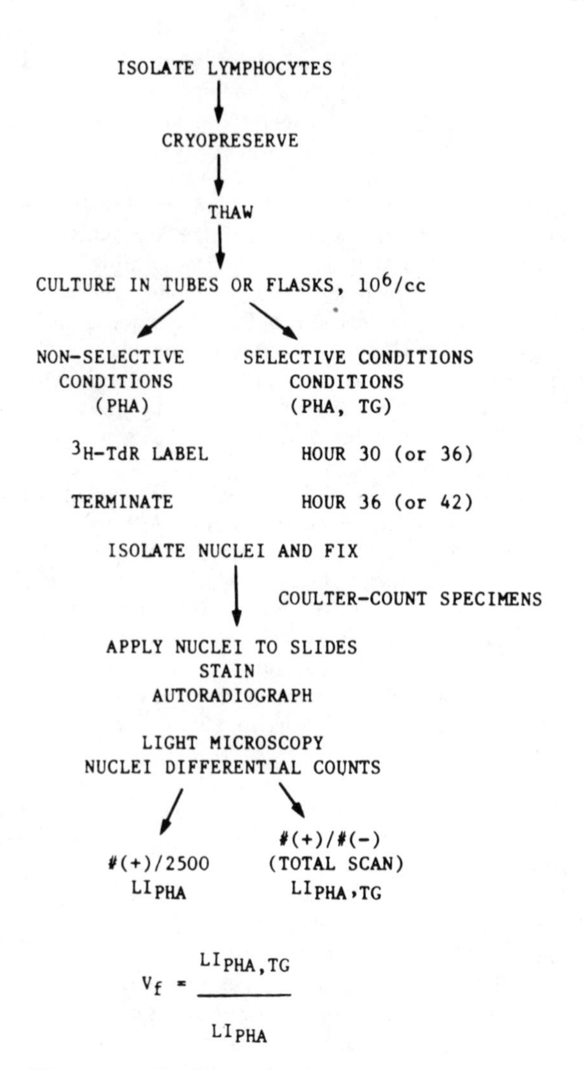

Figure 1. The Strauss-Albertini test procedure.

case of the LN heterozygotes, the finding of mosaicism at less than 50% is assumed to result from negative selective pressure against the mutants both at the stem-cell level and at the level of PBLs at advanced stage of maturity.

We conducted a cross-sectional study of a group of cancer patients heterogeneous in terms of tumor type and treatment regimen. Results from patients with hematological malignancies are excluded from data presented here. They received a variety of combination cytotoxic therapies, including antineoplastic drugs and X irradiation, over widely variable time courses. PBLs from 20 normal controls and 11 treated cancer patients were tested at 2×10^{-4} M TG, while cells from 24 controls and 42 patients were studied at 2×10^{-3} M TG. As Figure 2 shows, the average Vf at the higher

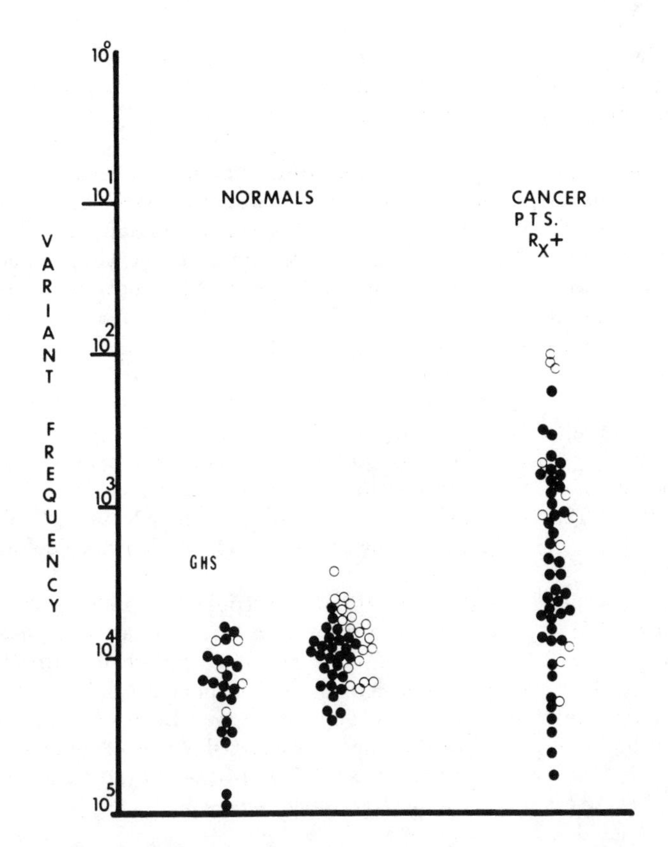

Figure 2. Frequency of TG^r PBLs in a heterogeneous group of cancer patients (right column), a group of normal controls (center column), and a single healthy individual (left column), determined at 2×10^{-4} M TG (o) and 2×10^{-3} M TG (●). Each point represents the results from one or more studies on an individual.

TG concentration for normal controls was 8.7×10^{-5}, and at the lower TG concentration it increased slightly to 1.3×10^{-4}. Values from the single control tested 20 times at the higher and 5 times at the lower TG concentration, except for the outlying low frequencies from two tests, showed variability and a range similar to that for the control group. The distributions of Vfs for treated cancer patients clearly were different from those of the normal controls. Eight of 11 patients tested at 2×10^{-4} TG and 26 of 42 patients tested at 2×10^{-3} M TG showed Vfs higher than the highest seen in a normal person tested at the same TG concentrations.

With reference to the data from this cancer patient study, the relationship between labelling indices of noninhibited PHA-stimulated cultures and Vfs was studied. It was concluded that elevated Vfs were not necessarily correlated with labelling indices lower than in controls. In fact, PBLs from 49 of 53 cancer patients responded normally to PHA within the period of assay. Presently it is not known to what extent phenocopies contributed to these data; however, owing to the secondary leukopoietic effects of cytotoxic drugs, it would not be surprising to find enhanced frequencies of cycling cells in these people.

Preliminary studies (Strauss and Albertini, unpublished results) indicated that untreated breast cancer patients, prior to any treatment, show slightly elevated TG^r PBLs that appear to become greater following surgery and adjuvant chemotherapy. Indeed, one might expect *in vivo* responses to antigens--be they associated with infectious agents, grafts, or tumors--to produce increases in numbers of cells cycling in the peripheral circulation. Cycling cells might well be found to be useful indicators of alterations in test cell subpopulations responding to conditions extant in the body at any given time.

A longitudinal approach was adopted whereby each individual is monitored and assessed in terms of personal baseline (pre-exposure) values. This approach was used with some success in a variety of human, canine, and rodent studies of Vf discussed elsewhere (Strauss, 1982b, c). Efforts to assess some of the effects of PUVA (psoralen and near-ultraviolet light) exposure using this approach are described below.

Though further longitudinal evaluations are required, it appears that the results of Strauss-Albertini test reflect the operation of a brief latency and patency period. Evidently, phenotypic expression of TG^r PBLs is complete within several days following mutagenic exposures. The length of this period (and the magnitude of genotoxic damage observed) probably depends upon the extent of specific and nonspecific killing of PBLs, as well as upon associated states of lymphopoiesis. The persistence of the induced variants (usually days to weeks) must also depend upon these factors plus the operations of negative selection mechanisms.

Further to characterize TG^r PBLs as to their mutant nature, studies were undertaken to measure growth of these cells *in vivo*. The presence of TG^r PBLs in humans is, for the most part, unimportant to health except in LN individuals and in those administered purine analogues therapeutically. Renal graft recipients were evaluated prospectively to monitor the efficacy of azathioprine as an immunosuppressant for transplantation. Vfs increased considerably (as many as 10^{-2} from normal levels) in all 20 patients studied, returning to only moderately increased levels in recipients of well matched kidneys from living, related donors; on the other hand, Vfs continued to climb in those given poorly matched grafts that, ultimately, were rejected (Strauss, 1982b). Again, the degree to which phenocopies may contribute in this situation remains to be determined. There is, however, little doubt that these TG^r cells were generated through a process of antigen-induced proliferation in the face of selection favoring the variant phenotype and, indeed, may mediate rejection. In this context it is interesting to consider that *in vivo* phenocopies may provide a basis for the development of tumors and other somatic cell degenerative proliferative abnormalities.

The growth of TG^r PBLs also was measured *in vitro* through the use of two methods for producing long-term cell lines. PBLs were taken from normal donors and from a renal transplant patient having a Vf of nearly 10^{-2}, and cultures were infected with Epstein-Barr virus B95-8 in the presence and absence of TG. Though PBLs from each donor transformed and grew well in the absence of TG, only the PBLs from the transplant patient developed into an established B-cell line in the presence of TG

(Strauss and Albertini, 1979; Albertini, 1979). In the second case, similar results were obtained using T-cell growth factor (TCGF) to cause the propagation of a TG^r T-cell line (Lane, Strauss, and Albertini, unpublished results). The latter study inspired the development of the clonal assay of lymphocyte mutagenesis (CALM) to isolate, characterize according to the criteria defining somatic cell mutants, and enumerate the TG^r PBLs (Strauss, 1982b; see 1982d for complete details).

TCGF is a general term for a class of naturally occurring lymphokines, individually identified as Interleukins. TCGF-containing supernatants from PHA-stimulated PBLs can be used to establish and support the continuous T-lymphocytes (CTLs) from blood, skin, lymph nodes, spleen, bone marrow, and other tissues. The CTLs are not virus transformed and maintain their morphological and functional integrity (Ruscetti et al., 1977; Gillis et al., 1978). TCGF can be purchased commercially or produced in large volumes in the laboratory from cultures of PHA-stimulated leukopheresed PBLs; its quality easily can be assessed by measuring its influence on the growth of CTLs. CTL densities routinely were maintained between 0.5 to 1.0×10^6 per cubic centimeter by viability counting and refeeding with TCGF-containing medium and T-cell growth medium (TCGM) (each for 2-3 d).

Measurements were performed to compare proliferative responses of CTLs in TCGM from two normals and two LN individuals in the presence of TG at various concentrations. An optimal selective concentration of TG, 2×10^{-5} M, was determined from the kill curves. In one set of experiments, CTLs from normals at densities of 10^5 per culture containing TG were completely eliminated by the eighth day of culture, whereas those from LNs were unaffected. To test the persistence of the TG^r phenotype in CTLs developed from a normal individual in the absence of selection, one set of 10^8 CTLs was challenged with TG. Survivors were raised under these selective conditions for 10 d. By this time, levels of about 10^6 cells were reached and TG^r CTLs were placed in nonselective medium and cultivated for another 10 d. At this stage, 0.5×10^6 TG^r CTLs were planted in selective medium and compared with the nonselected counterparts. The TG^r CTLs were unaffected; the others were killed. The results suggest that if selection against LN-like CTLs from normal individuals indeed occurs *in vitro*, it is far from complete. Figure 3 depicts the procedure used in the early development of CALM. Lymphocytes were diluted in TCGM containing 0.05% x-irradiated sheep red blood cells, which served as filler/feeders; cells were then plated in paired 96-well microtiter plates at 10^5 cells per well in the presence of TG and at 1 cell per well in the absence of TG. Spent media were replaced with fresh media (at 5 and 9 d) and ^{3}H-TdR label was added on day 9. On day 10, activity per well was assessed sequentially by three comparable methods: colorimetry, light microscopy, and liquid scintillation spectrophotometry. By scoring wells positive or negative on the basis of these assays, it was possible to calculate cloning efficiency (CE) assuming an ideal Poisson distribution. For each 96-well plate, the number of cells per well resulting in the percent positive wells was divided by the number of cells per well plated. In this way, for each pair of plates, CEs were determined for cells exposed to TG and cells not exposed to TG such that the ratio $CE_{TCGM,TG}/CE_{TCGM}$ approximated *in vivo* (or, for CTLs, *in vitro*) TG^r PBL mutant frequencies.

Cultures containing serial dilutions of minority PBLs from a LN boy were added to a majority of PBLs from a normal individual before plating, culture, and analysis as described. Expected total mutant frequencies (Mfs) were determined by summing the

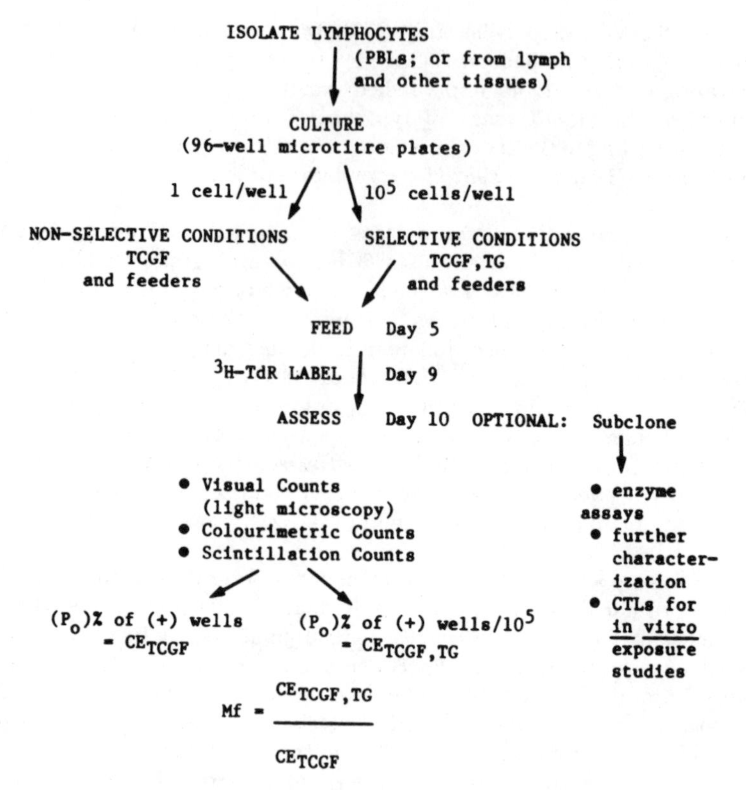

Figure 3. Clonal assay of lymphocyte mutagenesis (CALM) procedure.

observed Mf for the normal individual and the expected Mf of LN PBLs (based on the known CE for these cells grown separately). By this method, efficiency of recovery (EoR) was judged to be rather high, with a test sensitivity of at least 5×10^{-6}. A second set of artificial mixture experiments was performed using minority TG^r and majority TG^s CTLs developed from the PBLs of one normal individual as described earlier. The EoR values again were consistent with the conclusion that *in vitro* selection against rare TG^r lymphocytes probably does not interfere with the performance of this assay. The results of this experiment suggest that cell-mediated cytolysis among alloreactive immunocytes was not problematic in the first EoR experiment (Strauss, 1982a, b, c, d).

The Strauss-Albertini test and CALM were performed in parallel directly to compare *in vivo* and *in vitro* TG^r lymphocyte frequencies among fresh and cryopreserved PBLs and CTLs. Cells from two LNs were tested once, and cells from two normals were tested in four separate occasions; Figure 4 provides the results of this exercise. LNs, as expected, produced Vfs and Mfs near unity in every case. Though they are few, the data seem to support the possibility that cryopreservation eliminates phenocopies from the assay. As this figure also shows, CALM Mfs for CTLs and PBLs from the normal individuals are similar to each other and to Vfs from frozen PBLs (Strauss, 1982d).

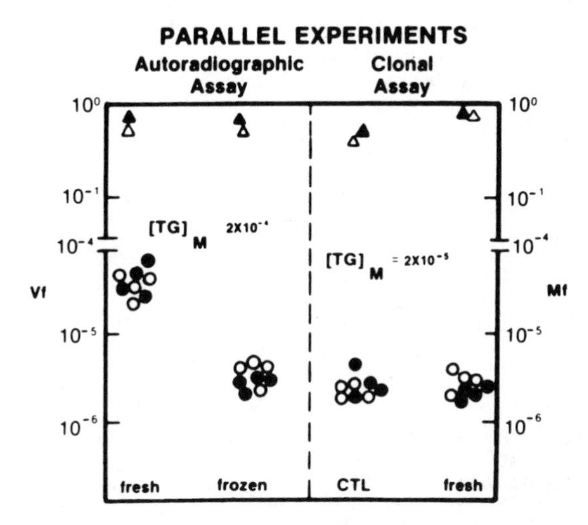

Figure 4. Scattergram of variant and mutation frequencies of PBLs from one parallel test of each of two LN individuals (Δ, ▲) and from four separate parallel tests of two normal individuals (o, ●).

Though results are preliminary and the assay is still under development, it seems safe to suggest that CALM as described here (Strauss, 1982d) and versions now being developed in other laboratories (Albertini et al., 1982; Morley et al., 1983) may aid in verification of the Strauss-Albertini test and possibly may serve as a mutagenicity test system unto itself.

CALM is being adapted to the evaluation of genotoxic damage after *in vitro* exposures to various agents (Figure 5). Individual susceptibility to mutagens and the relevancy of detectable long-term and short-term health risks could then be assessed. In addition, it would be reasonable and valuable to use CALM in mice, both *in vivo* and *in vitro*. The response of target lymphocytes from diverse tissues to a variety of agents could be assayed in mice (and in man to a lesser degree). Such studies would help to develop the means to extrapolate realistically the results from experiments with rodents to man (and back).

Further to answer critical questions concerning the relevancy of mutagenicity test systems, it is necessary to develop mechanisms directly to correlate their results with associated human health effects. New approaches to the study of mutagen-related effects are needed so that the acute, chronic, or latent effects of high-dose-single or low-dose-chronic mutagen exposures can be recognized and quantified. The need to develop new approaches seems especially urgent now, because it is clear that a subtle short-term health effect, either by itself or by interacting with other health effects, can produce long-term ill-health consequences through disturbances in normal regulatory processes. To explore some of these interactions, studies of psoriasis patients and others receiving PUVA (oral psoralen plus whole-body ultraviolet A photochemotherapy) treatments were conducted.

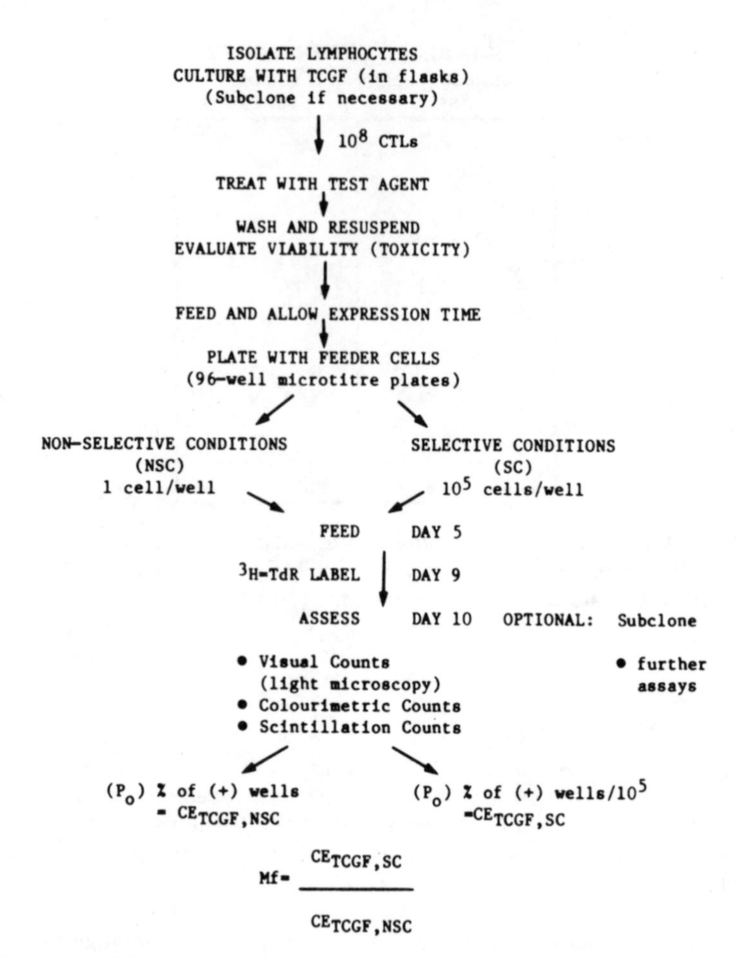

Figure 5. A possible procedure for the adaption of CALM to evaluation of genotoxic damage after *in vitro* exposures to an agent.

Psoralens are known to damage DNA following UVA irradiation by combining covalently with the pyrimidine bases of nucleic acids, which then act as bivalent reagents capable of cross-linking opposite strands of the double helix. PUVA-generated lesions have been detected in a variety of test systems using bacteria, cultured mammalian cells, and human peripheral lymphocytes both *in vitro* and *in vivo* (Bridges, 1978). PUVA treatment protocols are designed to clear the symptoms of generalized psoriasis, mycosis fungoides, urticaria pigmentosa, and vitiligo by producing regulated phototoxicity in the skin (Parrish et al., 1974). Treatments are administered two or three times per week. The actual effective dose of PUVA depends on the interaction of several factors: intensity and duration of UVA exposure, level of active psoralen in the skin at the time of irradiation, and relative quality of protective pigmentation in the skin. The last parameter is described by an index of skin type from I to V in order of increasing ability to produce protective pigmentation in response to light exposure.

PUVA-treated individuals were evaluated using the Strauss-Albertini test. The result of this cross-sectional study are summarized in Figure 6 (for details, see Strauss et al. [1979] and Strauss [1982c]). Although the concurrent control group was small, Vf values fit nicely into the distribution range reported from the standard reference group (mean given here as of 63 normal individuals). After the two groups of psoriasis patients were tested, the additional groups of vitiligo patients were tested. The PUVA-treated vitiligo patients received delivered doses similar to those of the PUVA-treated psoriatics. Vitiligo patients were tested because it was not possible to find untreated, severely diseased psoriatics. Thus, in testing psoriatics, it was not possible to separate the effects of PUVA, conventional therapies (including combinations of UVB, coal tar, steroids, and methotrexate), and possible disease-state effects on PBLs. Vitiligo patients presented a useful alternative, because untreated patients are available, and the test cell (PHA-responsive PBL, a T-cell) does not play a direct role in the pathogenesis of the disease.

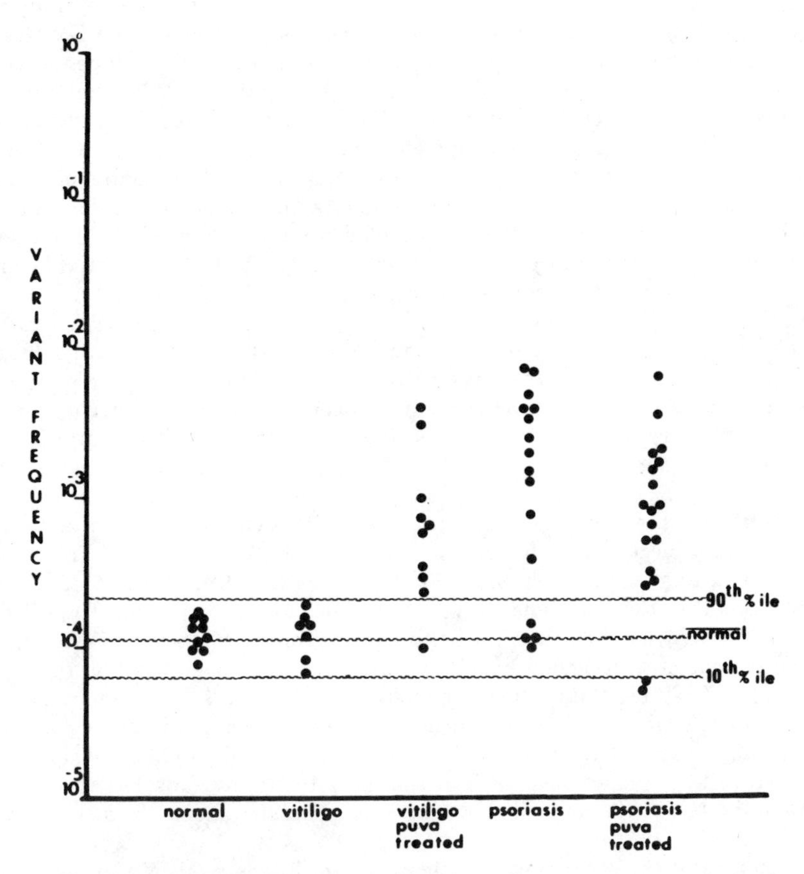

Figure 6. Frequency of TG^r PBLs as a function of treatment group; each point represents one test on one individual. The wavy line is the median variant frequency of 63 normal individuals.

A normal, healthy, 28-yr-old individual volunteered to receive intensive PUVA exposures over a period of 3 mo and to be tested longitudinally for TG^r PBL frequencies. Several tests were performed prior to treatment, and the Vfs were within the normal range. A second normal, untreated individual concurrently tested produced normal Vfs during the entire period for the study. The PUVA-treated subject received the highest tolerable doses twice weekly, and a response was evident within 5 to 6 d, when Vfs climbed out of the normal range. They fell back to near normal when the rate of treatment was dropped to once weekly. Soon after treatment was discontinued, the subject's Vfs returned to the normal range. If changes in Vfs are assumed to be due to changes in the DNA, these results would demonstrate that PUVA can cause mutations and thus presumably initiation of carcinogenesis in exposed tissues. The results demonstrate that the effects of PUVA definitely extend to the capillary beds, with considerable peripheral blood exposure. It is reasonable to expect that the responses of skin cells *in situ* are similar to those of the PBLs (Strauss et al., 1979; Bridges et al., 1981; Strauss, 1982a, c).

PUVA exposure causes the development of cutaneous tumors in shaved and hairless mice (Roberts et al., 1979). In man, the first evidence of a PUVA-related carcinogenic effect was the result of an ill-fated effort to induce protective melanogenesis in four xeroderma pigmentosum (XP) patients. XP is an hereditary disease that, owing to a deficiency in repair of UV-induced damage, is characterized by enhanced sensitivity to the carcinogenic effects of sun exposure. The PUVA-treated XP patients all developed multiple cutaneous neoplasms within months of exposure, and one also presented with lymphatic leukemia. Based on this information, one might expect to find similar tumors in PUVA-treated psoriatics, who should exhibit normal repair capacities following initiation of damage but only after the theoretical lag period of several decades. Several unexpected phenomena were observed: (1) An increased incidence of skin cancer has been reported within 2 yr of starting PUVA treatment; (2) excess tumors were mainly of squamous (not basal) cell origin; and (3) these developed only in patients in whom premalignant foci might be expected due to DNA damage from agents that had produced skin cancer prior to PUVA or from previous therapeutic exposures to X rays, arsenicals, or combination topical coal tar and tar UV (Stern et al., 1979, 1984).

Considering these observations together with evidence that squamous cell carcinomas are subject to immune controls (Walder et al., 1971; Kinlin et al., 1979) and considerable animal data (Morison, 1981; Lynch et al., 1981; Kripke et al., 1981) demonstrating that UV alone and PUVA cause neurological impairment, we (Bridges and Strauss, 1980) hypothesized that the carcinogenic action of PUVA might be functionally that of a pseudo-promoter (permissive carcinogen) rather than (or in addition to) that of an initiator. It seemed to us that if potentially malignant skin cells are subject to surveillance by the cutaneous immune system, then impairment caused by PUVA might promote the growth of such clones. Ironically (and this is a matter of some controversy), the aetiology of psoriasis principally may involve an element of autoimmunity that can be controlled by immunosuppressive agents (Strauss, 1982a, c).

An incidental observation during the period of intensive PUVA exposure of the volunteer subject discussed above provided new insights into both the beneficial and adverse effects (short- and long-term) of PUVA therapy. The subject was involved coincidently with immunologic testing for clinical purposes not related to PUVA. Prior to PUVA exposures he had been sensitized against the skin-test reagent 2,4-dinitrochlorobenzene (DNCB) and repeatedly was found to be a high-normal responder. However, during the entire duration of PUVA treatment he was unresponsive to DNCB. This was discovered serendipitously, and no careful analysis of the phenomenon was attempted at the time. When the treatment period ended, normal reactivity to DNCB returned. The DNCB test measures cell-mediated immunocompetency by evaluating delayed cellular hypersensitivity (DCH) in the skin.

A cross-sectional study of DCH was performed in psoriatics undergoing varying degrees of PUVA therapy. The results are summarized in Figure 7 (for details, including testing and grading methods, refer to Strauss et al. [1980b]). Fifty-five (roughly 54%) of the PUVA-treated psoriatics had sub-normal responses to DNCB sensitization. Under ordinary PUVA treatment conditions there is a striking correlation between skin type and DCH grade. For each skin type the average frequency of sub-normal responses was as follows: type I, 74%; type II, 73%; type III, 65%; type IV, 24%; type V, 0. X^2 analysis of the data from Figure 7 indicates that for all PUVA-treated groups (spanning 1/2-log ranges from 0.07 $J/cm^2/d$), observed DCH grades were dependent upon skin type, and abnormal results are significant ($p < 0.005$). Though they demonstrate that PUVA impairs cutaneous DCH, these studies did not differentiate between failure to respond to sensitization and diminished challenge-response ability.

A second, longitudinal study was designed to examine this question further (Strauss, 1982e). In summary, 20 psoriasis patients were DNCB-sensitized and challenge-tested to determine individual optimal allergic response prior to any PUVA exposure and at 6-wk intervals thereafter for a period of at least 1 yr. Tests determined local DCH reactions at skin sites both exposed to and masked from UVA exposure. A normal, healthy control, who was not PUVA-exposed, was challenged at 8-wk intervals for a period of 2 yr and consistently responded within normal limits. Challenge tests on pre-treatment psoriasis patients revealed DCH response distribution matching or slightly exceeding expectations for normal.

Briefly, the results of this study suggest a direct dose-response relationship between PUVA therapy and immune depression under ordinary treatment conditions and, in some cases, considerable impairment of cutaneous DCH. Sites exposed to UVA are more susceptible to DCH impairment than are nonexposed sites. In general, at moderate doses in individuals with low UVA tolerance and at high doses in those of all skin types, impairment of DCH at exposed sites is evident and can be dramatic. In fact, at exposed sites, 5 of 13 low-skin-type patients became anergic to a DNCB dose (48.0 mg) sufficient to sensitize normal individuals.

Among individuals of high skin type, 3 of 7 developed allergy to 48.0 mg DNCB following PUVA doses far greater than those required to cause the same effect in the low-skin-type group. Further, among low-skin-type individuals under the influence of the most intensive treatments, two individuals were rendered completely anergic to DNCB at both exposed and unexposed sites. In contrast, none of the higher-skin-type group became anergic to DNCB at unexposed sites under any of the treatment conditions. Our findings indicate that PUVA alters both local and systemic immune responsiveness following a short latency period of a day or two, whereas the effect on DCH persists one to two weeks.

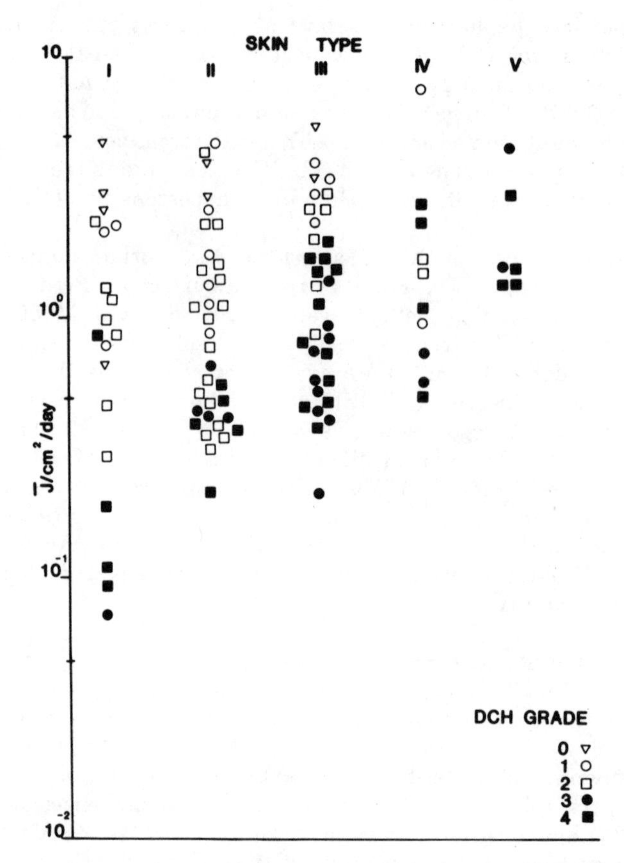

Figure 7. The interaction of a range of delayed cellular hypersensitivity (DCH) reactions (0-4), skin type (I-V), and dose rate ($J/cm^2/d$) in PUVA-treated psoriasis patients.

Studies of guinea pigs using simulated PUVA treatment have provided results similar to those described here (Morison et al., 1979; Vella-Briffa et al., 1981; Morison et al., 1981). When UV-induced skin tumors carrying strong tumor-associated antigens were transplanted to syngeneic mice that had received PUVA, the grafts were accepted; nontreated hosts rejected the tumors immediately (Roberts et al., 1979). This and other animal studies have suggested that alterations in regulatory mechanisms within the immune system (e.g., suppressor T-cell induction) may be responsible for decreased responsiveness to antigenic stimuli (but see Kripke et al. [1981] for a different interpretation of the results). Epidermal Langerhans' cells ordinarily have antigen-presenting functions necessary for contact hypersensitivity to allergens and have been found to be exquisitely sensitive to PUVA. Their disappearance from the skin in mice has rendered the skin insensitive to topical immunizing doses of DNCB; further, it is probable that UV-damaged antigen-presenting cells move to the spleen and become nonfunctional splenic adherent cells (Lynch et al., 1981). In addition to its action on afferent mechanisms required for contact-DCH, PUVA may alter efferent mechanisms by inducing specific suppressor lymphocytes that limit the number of effector cells available to respond to challenge (Roberts et al., 1979; Claman and Miller, 1980; Letvin et al., 1980).

CONCLUSIONS

In conclusion, an anecdotal observation is offered to emphasize the central theme of this review. (Similar findings have been documented more completely; see O'Dell et al. [1980].)

The subject, a normal, healthy, middle-aged man, received moderate sunburn over much of his surface (skin type I) while windsurfing during an unusually pleasant English summer's day. The individual, who in the past had responded normally to sensitization and challenge, was now challenge-tested with three doses of DNCB on each arm. The DNCB was applied to the sunburned side of his right arm and to an area that clearly was not sunburned on his left arm.

Within the test period, normal DCH responses were observed on the left arm but not on the right arm. Evidently, sunlight exposure was sufficient to cause local cutaneous anergy. Whether such immunologic tolerance in the presence of foreign entities (including infectious disease agents and tumor cells) promotes disease remains a matter for speculation. If, for example, the hapten DNCB goes unrecognized, so too might the transplantation antigens of squamous carcinomata. It is very likely, though, that mutagens cause both cytotoxic responses and mutations. These effects may interact over widely ranging time courses to cause disturbances in various normal regulatory processes. A major objective for genetic toxicology in the future should be to identify all relevant short- and long-term health risks from mutagen exposure.

In the near future, direct testing methods should be developed to the point where they can be used to predict manifest disease. Predictions should be based upon a developing understanding of the interactions between exposure and susceptibility. Direct testing methods should be developed so that (1) results from *in vitro* tests can be compared with the results from direct tests, (2) results from mutagenicity tests on nonhumans can be compared with direct test results, (3) results from direct mutagenicity tests on somatic and germ cells can be compared, (4) relevant short-term consequences of mutagen exposures can be recognized and quantified, and (5) individual and population susceptibility to mutagens can be predicted quantitatively.

DISCLAIMER

This manuscript has been reviewed by the Health Effects Research Laboratory, U.S. Environmental Protection Agency, and approved for publication. Approval does not signify that the contents necessarily reflect the views and policies of the Agency, nor does mention of trade names or commercial products constitute endorsement or recommendation for use.

REFERENCES

Albertini, R.J. 1979. Direct mutagenicity testing with peripheral blood lymphocytes. In: Mammalian Cell Mutagenesis: The Maturation of Test Systems. A.W. Hsie, J.P. O'Neill, and V.K. McElheny, eds. Banbury Report 2, Cold Spring Harbor Laboratory Press: Cold Spring Harbor, NY.

Albertini, R.J. 1982. Studies with T-lymphocytes: an approach to human mutagenicity monitoring. In: Indicators of Genotoxic Exposures in Man and Animals. B.A. Bridges, B.E. Butterworth, and I.B. Weinstein, eds. Banbury Report 13, Cold Spring Harbor Laboratory Press: Cold Spring Harbor, NY.

Albertini, R.J., K.L. Castle, and W.R. Borcherding. 1982. T-cell cloning to detect the mutant 6-thioguanine-resistant lymphocytes present in human peripheral blood. Proc. Natl. Acad. Sci. U.S.A. 79:66A-6621.

Ansari, A.A., M.A. Baig, and H.V. Malling. 1980. In vivo germinal mutation detection with "monospecific" antibody against lactate dehydrogenase-x. Proc. Natl. Acad. Sci. U.S.A. 77:7352-7356.

Bridges, B.A. 1978. Possible long-term hazards of photochemotherapy with psoralens and near-ultraviolet light. Clin. Exp. Dermatol. 3:349-353.

Bridges, B.A., and G.H. Strauss. 1980. Possible hazards of photochemotherapy for psoriasis. Nature 283:523-524.

Bridges, B.A., G.H. Strauss, S.P. Hall-Smith, and M.L. Price. 1981. Induction of somatic mutations and impairment of immune capacity by UVA treatment and their relation to skin cancer in man. In: Psoralens in Cosmetics and Dermatology. Pergamon Press: Paris. pp. 287-294.

Chu, E.H.Y., and S.S. Powell. 1976. Selective systems in somatic cell genetics. Adv. Hum. Genet. 7:189-259.

Claman, H.N., and S.D. Miller. 1980. Immunoregulation of contact sensitivity. J. Invest. Dermatol. 75:263-265.

Clements, G.B. 1975. Selection of biochemically variant, in some cases mutant, mammalian cells in culture. Adv. Cancer Res. 21:273-390.

DeMars, R. 1971. Genetic studies of HG-PRT deficiency and the Lesch-Nyhan purine and pyrimidine analogs in relation to mutagenesis detection. Mutat. Res. 24:335-364.

Evans, H.J., and Vijayalaxmi. 1981. Induction of 8-azaguanine resistance and sister chromatid exchange in human lymphocytes exposed to mitomycin C and x-rays in vitro. Nature 292:601-605.

Gillis, S., M.M. Ferm, W. Ou, and K.A. Smith. 1978. T-cell growth factor: Parameters of production and a quantitative microassay for activity. J. Immunol. 120:2027.

Kelley, W.N. 1968. HG-PRT deficiency in the Lesch-Nyhan syndrome and gout. Fed. Proc. 27:1047-1052.

Kinlin, L., et al. 1979. Collaborative United Kingdom/Australian study of cancer patients treated with immunosuppressive drugs. Br. Med. J. ii:1461.

Kripke, M.L., W.L. Morison, and J.A. Parrish. 1981. Differences in the immunologic reactivity of mice treated with UVB and methoxsalen plus UVA radiations. J. Invest. Dermatol. 76:445-448.

Letvin, N.L., I.J. Fox, M. Greene, B. Benacerraf, and R.N. Germain. 1980. Immunologic effects of whole-body ultraviolet (UV) irradiation, II. Defect in splenic adherent cell antigen presentation for stimulation of T-cell proliferation. J. Immunol. 125:1402-1404.

Lynch, D.H., M.F. Gurish, and R.A. Daynes. 1981. Relationship between epidermal Langerhans cell density, ATPase activity and the induction of contact hypersensitivity. J. Immunol. 126:1892-1897.

Malling, H.V. 1981. Perspectives in mutagenesis. Environ. Mutagen. 3:103-108.

Morison, W.L. 1981. Photoimmunology. J. Invest. Dermatol. 77:71-76.

Morison, W.L., M.E. Woehler, and J.A. Parrish. 1979. PUVA and systemic immunosuppression in guinea pigs. J. Invest. Dermatol. 72:273.

Morison, W.L., J.A. Parrish, M.E. Woehler, and J. Bloch. 1981. The influence of ultraviolet radiation on allergic contact dermatitis in the guinea pig, II. Psoralen/UVA radiation. Br. J. Dermatol. 104:165-168.

Morley, A.A., K.J. Trainor, R. Seshardi, and R.G. Ryall. 1983. Measurement of in vivo mutations in human lymphocytes. Nature 302:155-156.

O'Dell, B.L., T.R. Jessen, L.E. Becker, R.T. Jackson, and E.B. Smith. 1980. Diminished immune response in sun-damaged skin. Arch. Dermatol. 116:550-561.

Parrish, J.A., T.B. Fitzpatrick, L. Tanenbaum, and M.A. Patak. 1974. Photochemotherapy for psoriasis with oral methoxsalen and long-wave ultraviolet light. N. Engl. J. Med. 291:1207-1211.

Roberts, L.K., M. Schmitt, and R.A. Daaynes. 1979. Tumor susceptibility generated in mice treated with subcarcinogenic doses of 8-methoxypsoralen and long-wave ultraviolet light. J. Invest. Dermatol. 72:306-309.

Ruscetti, F.W., D.A. Morgan, and R.C. Gallo. 1977. Functional and morphological characterization of human T-cells continuously grown in vitro. J. Immunol. 119:131.

Stamatoyannopoulos, G., P.E. Nute, T.H. Papayannopoulou, T. McGuire, G. Lim, H.F. Bunn, and S. Rucknagel. 1980. Development of a somatic mutation screening system using Hb mutants, IV: Successful detection of red cells containing the human frameshift mutants Hb Wayne and Hb Cranston using monospecific fluorescent antibodies. Am. J. Hum. Genet. 32:484-496.

Stern, R.S., L.A. Thibodeau, R.A. Kleinerman, J.A. Parrish, R.B. Fitzpatrick, and twenty-two participating investigators. 1979. Risk of cutaneous carcinoma in patients treated with oral methoxsalen photochemotherapy for psoriasis. N. Engl. J. Med. 300:809-813.

Stern, R.S., N. Laird, J. Melski, J.A. Parrish, T.B. Fitzpatrick, and H.L. Bleich. 1984. Cutaneous squamous cell carcinoma in patients treated with PUVA. N. Engl. J. Med. 310:1156-1161.

Strauss, G.H.S. 1982a. PUVA: Gambling with a system. Int. J. Dermatol. 21:136-137.

Strauss, G.H.S. 1982b. The Strauss-Albertini test for drug resistant peripheral blood lymphocytes. In: The Use of Human Cells for the Evaluation of Risk from Physical and Chemical Agents. A. Castellani, ed. Plenum Press: New York.

Strauss, G.H.S. 1982c. PUVA therapy--Immunologic and genotoxic approaches to risk evaluation. In: The Use of Human Cells for the Evaluation of Risk from Physical and Chemical Agents. A. Castellani, ed. Plenum Press: New York. pp. 219-235.

Strauss, G.H.S. 1982d. Direct mutagenicity testing. The development of a clonal assay to detect and quantitate mutant lymphocytes arising in vivo. In: Indicators of Genotoxic Exposures in Man and Animals. B.A. Bridges, B.E. Butterworth, and I.B. Weinstein, eds. Banbury Report 13, Cold Spring Harbor Laboratory Press: Cold Spring Harbor, NY.

Strauss, G.H.S. 1982e. Direct Mutagen Risk Assessment: The Development of the Strauss-Albertini Test and Other Methods to Measure Immunologic and Genetic Responses to Mutagens. D.Phil. Dissertation, Sussex University.

Strauss, G.H., and R.J. Albertini. 1977. 6-Thioguanine resistant lymphocytes in human peripheral blood. In: Progress in Genetic Toxicology. D. Scott, B.A. Bridges, and F.H. Sobels, eds. Elsevier/North Holland Press: Amsterdam. pp. 327-334.

Strauss, G.H., and R.J. Albertini. 1979. Enumeration of 6-thioguanine peripheral blood lymphocytes as a potential test for somatic cell mutations arising in vivo. Mutat. Res. 61:353-379.

Strauss, G.H., R.J. Albertini, P.A. Kruzinski, and R.D. Baughman. 1979. 6-Thioguanine-resistant peripheral blood lymphocytes in humans following psoralen long-wave UV light therapy. J. Invest. Dermatol. 73:211-216.

Strauss, G.H., R.J. Albertini, and B.J. Allen. 1980a. An enumerative assay of purine analog resistant lymphocytes in women heterozygous for the Lesch-Nyhan mutation. Biochem. Genet. 18:529-547.

Strauss, G.H., B.A. Bridges, M. Greaves, S.P. Hall-Smith, M.L. Price, and D. Vella-Briffa. 1980b. Inhibition of delayed hypersensitivity reaction in skin (DNCB test) by 8-methoxypsoralen photochemotherapy: Possible basis for pseudo-promoting action in skin carcinogenesis? Lancet ii:556-559.

Thilly, W.G., J.G. DeLuca, I.B.H. Hoppe, and B.W. Penman. 1976. Mutation of human lymphoblasts by methylnitrosourea. Chem.-Biol. Interact. 15:33-50.

Thompson, L.H., and R.M. Baker. 1973. Isolation of mutants of cultured mammalian cells. In: Methods in Cell Biology, Vol. 6. D.M. Prescott, ed. Academic Press: New York. pp. 209-281.

Vella-Briffa, D., D. Parker, N. Tosca, J.L. Turk, and M. Greaves. 1981. The effect of photochemotherapy (PUVA) on cell mediated immunity in the guinea pig. J. Invest. Dermatol. 77:377-380.

Walder, B.K., et al. 1971. Skin cancer and immunosuppression. Lancet ii:1282.

RISK ASSESSMENT OF COMPLEX MIXTURES

Herman J. Gibb and Chao W. Chen

Carcinogen Assessment Group, Office of Health and Environmental Assessment, U.S. Environmental Protection Agency, Washington, D.C. 20460

INTRODUCTION

Risk assessment of suspected carcinogens involves both a qualitative and quantitative evaluation. The qualitative evaluation evaluates the relevant animal, epidemiologic, mutagenesis, and cell transformation studies as to the likelihood that the agent is a human carcinogen.

The quantitative evaluation is an estimate of the carcinogenic potency of the suspected carcinogen. Potency is derived by fitting a mathematical model to the dose-response data from either animal or epidemiologic studies in an attempt to describe what an estimate of the risk would be at low doses. Short-term genetic bioassay data, which are the focus of this symposium, are not used for carcinogenesis risk estimation, although such data have been used in some instances to provide an idea of the comparative carcinogenic potency of different compounds. Several types of risk estimation models have been used for animal dose-response data. Most of the models that have been used for the epidemiologic data are derivatives of the multistage model.

The use of animal data for human risk assessment has several limitations:

1. Animals may respond differently than humans as a result of metabolic or other species-specific differences.

2. The animals are tested at high doses, usually doses that humans would not encounter.

3. The lifetime testing to which animals are subjected is not the equivalent of a human lifetime.

Many limitations are also encountered in the use of epidemiologic data for quantitative evaluation. These include

1. Lack of exposure data for the time period of concern.

2. Small sample sizes and short follow-up periods in the case of cohort studies.

3. Confounding exposures to other carcinogens.

In most cases, epidemiologic or animal data on complex mixtures simply does not exist. In lieu of such data, a comparative potency approach in short-term bioassays, as indicated earlier, has been proposed. By this approach, a unit cancer risk estimate, or in other words the risk at unit dose (e.g., 1 μg/liter), is calculated for a mixture based on its potency in a short-term bioassay relative to that of a mixture for which a unit risk has been calculated. Albert et al. (1983) found that the relative potencies of coke oven emission extracts, roofing tar emission extract, and cigarette smoke condensate determined by skin tumor initiation in SENCAR mice appeared to correlate well with the relative carcinogenic potencies based on epidemiologic data. The limitation of this approach is that potency as determined by a skin tumor initiation bioassay may not correlate well with the potency of the mixture in a cancer bioassay (e.g., the mixture may have both initiation and promotion potential such that the complete carcinogenic potency of the mixture is quite different from its tumor initiation potential). The mathematical implications of the multistage theory with regard to the carcinogenic action of a complex mixture are explored here. Actual data from both epidemiologic studies and animal investigations are used for illustration. Regulatory ramifications of this discussion are also addressed.

RISK ASSESSMENT OF COMPLEX MIXTURES

The main problem associated with doing risk assessments of complex mixtures is that the chemical profiles, and thus the carcinogenic interaction of the mixtures, may vary from source to source. To eventually assess the risk of a population exposed to a complex mixture with a reasonable amount of confidence, it is necessary that we better understand some of the carcinogenic mechanisms involved. Several studies have examined the synergistic and antagonistic effects of chemicals in a mixture with regard to carcinogenicity. Some examples of these studies, both human and animal, are reported in Tables 1 and 2.

In an attempt to explain these phenomena, we will apply the theory of multistage carcinogenesis to interpret and evaluate the data obtained from the animal and human studies. The multistage theory of carcinogenesis, though oversimplified in our example, does offer considerable plausibility for interpreting the dose-response data obtained from animal experiments and epidemiologic studies. However, it is important to understand the underlying assumptions and limitations. For instance, the multistage model assumes that the transition rate from one stage to the next stage is independent of age. While this assumption has been shown by Peto et al. (1975) to be true for benzo[a]pyrene (B[a]P)-induced skin cancer in animals, it may not be true for other carcinogen-induced cancers. For instance, with regard to human breast cancer data, the transition rates may vary with hormone levels that are closely related to the

Table 1. Examples of Synergism with Regard to Carcinogen Response in Human Studies

Agents Involved	Type of Study	Tumor Site	Authors
Smoking and asbestos	Cohort	Lung	Selikoff et al. (1968) Hammond et al. (1979)
Uranium mining and cigarette smoking	Cohort	Lung	Lundin et al. (1969)
Radiation and smoking	Cohort	Lung	Wanebo et al. (1968)
Arsenic and smoking	Cohort	Lung	Pershagen (1982)

age of the individuals. Another possibility that is not included in the simple multistage theory, although the model can be extended to accommodate it, is that the promotion and/or inhibition activity of environmental factors other than the factor under study may have an effect on the proliferation rates of partially or completely transformed cells.

The simple multistage model assumes that a cell is capable of generating a malignant neoplasm when it has undergone k changes in a certain order. The rate, r_i, of the *i*th change is assumed to be linearly related to D(t), the dose at age t, i.e., $r_i = a_i + b_iD(t)$, where a_i is the background rate and b_i is the proportionality constant for the dose (Figure 1). It can be shown (Crump and Howe, in press) that the probability of cancer by age t is given by

$$P(t) = 1 - \exp[-H(t)]$$

where the cumulative incidence rate by time t, H(t), is given by

$$H(t) = \int_o^t \int_o^{u_k} \ldots \int_o^{u_2} \{[a_1 + b_1 D(u_1)] \ldots [(a_k + b_k D(u_k)]\} du_i \ldots du_k$$

When H(t) or the risk of cancer is small, P(t) is approximately equal to H(t). When only one stage is dose related, all proportionality constants are zero except for the proportionality constant for the dose-related stage. The implications of the model when one stage is carcinogen affected have been summarized by Brown and Chu (1983) as follows:

Table 2. Examples of Synergism and Antagonism with Regard to Carcinogen Response in Animal Studies

Agents Involved	Type of Study	Authors
Synergism		
7,12-Dimethylbenz[a]anthracene (DMBA) and extracts of unburned cigarette tobacco	Mouse skin painting	Bock et al. (1964)
7,12-DMBA and each of the following: catechol, pyrogallol, decane, indecane, pyrene, benzo[e]pyrene, and fluoranthene	Mouse skin painting	Van Duuren and Goldschmidt (1976)
Automobile exhaust condensate without particulate matter and B[a]P	Mouse subcutaneous injection	Grimmer (1977)
Antagonism		
Automobile exhaust condensate with particulate matter and B[a]P	Mouse subcutaneous injection	Grimmer (1977)
B[a]P and 10 different noncarcinogens	Mouse subcutaneous injection	Falk et al. (1964)
B[a]P and esculin, quercetin and squalene, and oleic acid (tobacco smoke components)	Mouse skin painting	Van Duuren and Goldschmidt (1976)

For exposure at a near constant level to a carcinogen, the multistage theory predicts the following patterns of excess risk: (1) For any affected stage, excess risk will increase with increasing level and/or duration of exposure; (2) if only the first stage is affected, for fixed exposure duration, excess risk is independent of age at start of exposure and is an increasing function of time since exposure stopped; and (3) if only the penultimate stage is affected, for fixed exposure duration, excess risk is an increasing function of age at start of exposure and is independent of time since exposure stopped.

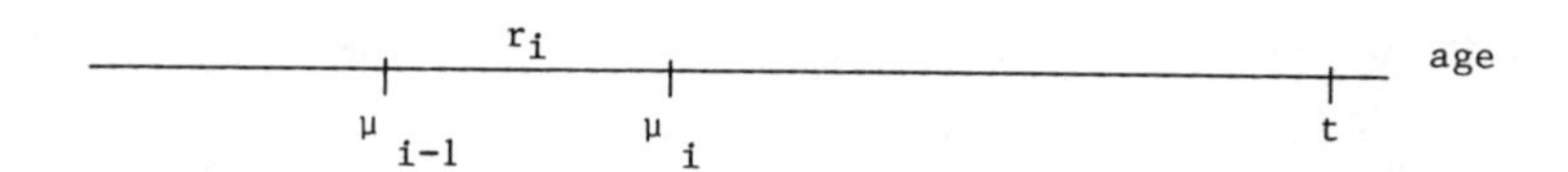

Figure 1. Schematic view of the transition rate of the *i*th change in the simple multistage model. Assumptions: $r_i = a_i + b_iD(u)$, transition rate from $(i-1)$th stage to *i*th stage, where a_i is the background rate and b_i is the proportionality constant for the dose.

Since a complex mixture often contains more than one carcinogen, the likelihood is increased that the mixture will act on more than one stage of the carcinogenic process. Without loss of generality in our discussion, assume that two stages, the *m*th and *n*th $(1 \leq m < n \leq k)$, are dose related. Then,

$$H(t) = H_O + H_1 + H_2 + H_{12}$$

where

$$H_O = (a_1a_2...a_k)t^k/k!$$

$$H_1 = (a_1a_2...a_k)(b_m/a_m)\int_o^t\int_o^{u_k}...\int_o^{u_2} D(u_m)du_1...du_k$$

H_2 is similar to H_1 except that subscript m is replaced by n

$$H_{12} = (a_1a_2...a_k)(b_mb_n/a_ma_n)\int_o^t\int^{u_k}...\int_o^{u_2} D(u_m)D(u_n)du_1...du_k$$

That is, the cumulative incidence function H(t) can be decomposed into four components: H_O is the background cumulative incidence, H_1 and H_2 are cumulative incidences when only one stage is dose related, and H_{12} is the multiplicative term related to the multiple of the two dose-related rates of change. When the two stages are affected separately by two different carcinogens (e.g., B[a]P and a nonB[a]P carcinogen in the mixture), the multiplicative term reflects the synergism due to the two carcinogens. Obviously, the multiplicative effect would not exist if one of the two compounds was removed. When the same stage is affected by different agents, the synergistic effect does not occur under the simple multistage theory, but antagonism may occur due to the competition of carcinogens for the partially transformed cells of a particular stage. These conclusions may not hold if the transition rate from one stage to the next is modified due to external influences (e.g., breast cancer associated with hormonal change).

The above theoretical discussion suggests that exposure to a complex mixture may produce synergistic and/or antagonistic effects. Thus, the multistage theory can be used to interpret the synergistic and antagonistic observations in humans and animals described earlier.

An illustrative example is that of Doll and Hill's (1956, 1964) dose-response data for lung cancer and cigarette smoking in British doctors. These data were analyzed by Doll (1971a) and were presented before the Royal Statistical Society in December, 1970. Doll found that the age-specific lung cancer mortality rate for smokers is approximately proportional to the 5th power of duration since the start of exposure and is linearly related to the amount smoked. This implies that smoking affects an early stage of lung carcinogenesis. Following Doll's presentation, Armitage (Doll, 1971a) raised the issue of whether smoking affects the early or late stage of carcinogenesis. Armitage stated:

> In this connection, I have always been somewhat puzzled about the effect of cigarette smoke as a carcinogen. The dose-response relationship seems to be linear, which suggests that the carcinogen affects the rate of occurrence of critical events at one stage, and one only, in the induction period.... On the other hand, the halt in risk quite soon after smoking stops suggests that a late stage is involved....

The Armitage view can best be seen from Figure 2, which was reported by Doll (1971b) after his presentation of the earlier paper. The fact that the lung cancer rate for the ex-smokers decreased and then increased again approximately 15 yr after smoking stopped suggests than an early stage and a late stage are affected by the cigarette smoke. In a subsequent analysis, Doll and Peto (1978) suggested that a quadratic dose-response relationship seems to be preferred to the linear dose-response relationship, suggesting that more than one stage is dose affected, as was observed by Armitage 14 years ago.

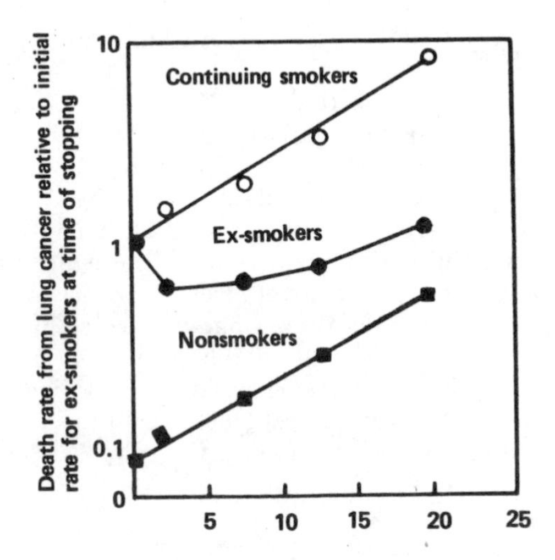

Figure 2. The rate of lung cancer among people who have stopped smoking cigarettes, those who continue to smoke, and those who have never smoked.

Pershagen (1982) recently found that cigarette smoking and exposure to arsenic had a synergistic effect with regard to carcinogenesis. Brown and Chu (1983) concluded from studies of smelter workers that arsenic is a late-stage carcinogen in the multistage model. Brown and Chu found that excess lung cancer risk among smelter workers was an increasing function of age at the start of exposure, and for individuals greater than or equal to 55 yr of age the risk was independent of the time since exposure stopped. This follows the pattern for a late-stage carcinogen, as discussed earlier. It is theorized that older individuals are at a greater risk of lung cancer mortality from exposure to a late-stage carcinogen, since they have had time to accumulate more cells in the earlier stages of the cancer process; such cells are particularly susceptible to a late-stage carcinogen such as arsenic. Since a late-stage carcinogen cannot increase the number of cells in the early stages of carcinogenesis, the individual's risk remains constant after cessation of exposure. Following the simple multistage model, we would then explain the synergistic effect of cigarette smoke and arsenic observed by Pershagen as an interaction between the effect of arsenic on a late stage of carcinogenesis and the effect of components of cigarette smoke on an early stage of carcinogenesis.

The simple multistage model that we have discussed would conclude

1. That carcinogenic synergism in mixtures is a result of constituents of the mixture acting on separate stages of the multistage process of carcinogenesis.

2. That if all constituents of the mixture act on a single stage of the multistage process of carcinogenesis, there will be no synergism. However, an antagonistic effect could result due to the availability of partially transformed cells.

CONCLUSION

Obviously, the simple model cannot explain all of the synergistic or antagonistic carcinogenic effects observed in animal or human studies. As stated earlier, the model does not consider changes in transition rates between stages that may be brought on by age or environmental factors. It is a mathematical model, however, that could certainly explain some of the data. Thus, we feel that implications from the multistage model should be considered in the design of future animal studies or even short-term bioassay studies. In regard to animal bioassays, we would suggest that mixtures be fractionated and administered to the animals, varying the age at which the dose is given and perhaps the duration of the dose. In addition, epidemiologic data should be reported whenever possible to facilitate analysis with regard to the affected carcinogenic stage or stages on which the complex mixture may be acting. The data reported by Doll (1971a) and Brown and Chu (1983) have provided insight with regard to the effects of carcinogens on different stages.

Perhaps one final point should be offered in regard to the understanding of mixtures and components of mixtures by their carcinogenic stage of action. The effects of a late-stage carcinogen would be seen in a relatively short period of time, whereas the effects of an early-stage carcinogen may take many years to be detected. These effects may affect the way we regulate complex mixtures and certainly, we hope, should affect the way in which we study such mixtures.

DISCLAIMER

The views expressed in this article are those of the authors and not necessarily those of the U.S. Environmental Protection Agency.

REFERENCES

Albert, R.E., J. Lewtas, S. Nesnow, T.W. Thorslund, and E.L. Anderson. 1983. Comparative potency method for cancer risk assessment: Application to diesel particulate emissions. J. Risk Analysis 3(2):101-117.

Bock, F.G., S.K. Crouch, and G.E. Moore. 1964. Tumor-promoting activity of extracts of unburned tobacco. Science 145:831-833.

Brown, C.C., and K.C. Chu. 1983. Implications of the multistage theory of carcinogenesis applied to occupational arsenic exposure. J. Natl. Cancer Inst. 70(3):455-463.

Crump, K., and R. Howe. (in press). The multistage model with a time-dependent dose pattern: Applications to carcinogenic risk assessment. J. Risk Analysis.

Doll, R. 1971a. The age distribution of cancer: Implications for models of carcinogenesis (with discussion). J. Royal Stat. Soc. Series A. 134:133-166.

Doll, R. 1971b. Cancer and aging: The epidemiologic evidence. In: Oncology. Tenth International Cancer Congress. Vol. V. R.L. Clark, R.W. Cumley, J.E. McCay, and M.M. Copeland, eds. Chicago Year Book Medical Publishers: Chicago, IL. pp. 1-28.

Doll, R., and A. Hill. 1956. Lung cancer and other causes of death in relation to smoking. Br. Med. J. 2:1071-1076.

Doll, R., and A. Hill. 1964. Mortality in relation to smoking: Ten year's observation of British doctors. Br. Med. J. 1:1399-1410, 1460-1467.

Doll, R., and R. Peto. 1978. Cigarette smoking and bronchial carcinoma: Dose and time relationship among regular smokers and lifelong non-smokers. J. Epidemiol. Community Health 32:303-313.

Falk, H.L., P. Kotin, and S. Thompson. 1964. Inhibition of carcinogenesis. Arch. Environ. Health 9:169-179.

Grimmer, G. 1977. Analysis of automobile exhaust condensate. In: Air Pollution and Cancer in Man. U. Mohr, D. Schähl, and L. Tomatis, eds. Proceedings of the Second International Carcinogenesis Meeting, Hanover, Germany,

October 22-24, 1975. IARC Scientific Publication No. 16. International Agency for Research on Cancer: Lyon, France. pp. 29-39.

Hammond, E.C., I.J. Selikoff, and H. Seidman. 1979. Asbestos exposure, cigarette smoking and death rates. Ann. N.Y. Acad. Sci. 330:473-490.

Lundin, F.E., Jr., J.W. Lloyd, E.M. Smith, V.E. Archer, and D.A. Holaday. 1969. Mortality of uranium miners in relation to radiation exposure, hard rock mining and cigarette smoking--1950 through Sept. 1967. Health Phys. 16:571-578.

Pershagen, G. 1982. Arsenic and lung cancer with special reference to interacting factors--epidemiological and experimental evidence. Submitted as a doctoral thesis at the Karolinska Institute, Stockholm, Sweden.

Peto, R., F. Rose, P. Lee, L. Levy, and J. Clack. 1975. Cancer and aging in mice and men. Br. J. Cancer 32:411-426.

Selikoff, I.J., E.C. Hammond, and J. Churg. 1968. Asbestos exposure, smoking, and neoplasia. J. Am. Med. Assoc. 204:106-112.

Van Duuren, B.L., and B.M. Goldschmidt. 1976. Cocarcinogenic and tumor-promoting agents in tobacco carcinogenesis. J. Natl. Cancer Inst. 56(6):1237-1242.

Wanebo, C.K., K.G. Johnson, K. Sato, and T.W. Thorslund. 1968. Lung cancer following atomic radiation. Am. Rev. Respir. Dis. 98:778-787.

COMPARATIVE POTENCY OF COMPLEX MIXTURES: USE OF SHORT-TERM GENETIC BIOASSAYS IN CANCER RISK ASSESSMENT

Joellen Lewtas

Health Effects Research Laboratory, U.S. Enivronmental Protection Agency, Research Triangle Park, North Carolina 27711

INTRODUCTION

Short-term genetic bioassays are now widely employed to screen both individual chemicals (e.g., pesticides and toxic substances) and complex mixtures (e.g., industrial effluents and automotive emissions) for mutagenic and potential carcinogenic activity. These bioassays are often employed in a tiered or phased approach to prioritize substances for more expensive chronic animal testing (Bridges, 1973; Waters, 1978). The results of both microbial and mammalian cell genetic bioassays are now considered by governmental regulatory agencies and international organizations to provide important suggestive evidence of a substance's carcinogenicity in humans (Interagency Regulatory Liason Group, 1979; International Agency for Research on Cancer, 1983).

The assessment of potential human risk from the introduction of alternative energy sources (e.g., diesel cars and wood stoves) and fuels (e.g., synthetic fuels) presents a somewhat different problem than introducing a new chemical into commerce. Most, if not all, combustion emissions from the conventional energy sources and fuels currently in use contain carcinogenic polycyclic organic compounds that are mutagenic in short-term genetic bioassays in microbial and mammalian cells (Claxton and Huisingh, 1980; Lewtas, 1982; Holmberg and Ahlborg, 1983) and tumorigenic in animals (Nesnow et al., 1982a, b). The primary problem associated with the introduction of new energy sources is whether they will alter the mutagenicity, carcinogenicity, and potential human cancer risk from combustion emissions. New risk assessment methodologies are needed to assess alternative technologies as they are being developed. This will allow the development of new combustion sources, fuels, and control technologies that lower human risk.

In order to assess the relative human cancer risk associated with introducing diesel-powered automobiles as an alternative fuel-efficient energy source for transportation, we have developed a comparative data base (Lewtas et al., 1981; Nesnow and Lewtas, 1981) and risk assessment methodology (Albert et al., 1983). Although this method was initially developed to compare alternative combustion sources (e.g., engines) that are powered by different fuels (e.g., diesel versus gasoline), this risk assessment model and methodology is applicable to the evaluation of various complex mixtures. This paper describes a comparative approach to evaluating alternative energy sources by utilizing short-term genetic bioassays.

COMPARATIVE POTENCY METHOD

The comparative potency method for cancer risk assessment is based on the hypothesis that there is a constant relative potency between two different carcinogens across different bioassay systems. This assumption is implicit in any comparison that utilizes the relative toxicity of two substances in animals to determine which substance would most likely be less toxic to man. This constant relative potency assumption is a testable hypothesis if the relative potency of two carcinogens in humans can be determined and compared to the relative potency in a series of *in vitro* and *in vivo* bioassays.

We have tested this hypothesis for three complex organic emissions from a coke oven, roofing tar pot, and cigarettes by using human lung cancer data from epidemiological studies of humans exposed to these emissions (Albert et al., 1983). The relative potencies for coke oven compared to roofing tar were similar and ranged from 0.2-1.4 across data from humans, mouse skin tumor initiation, and selected short-term genetic bioassays, as shown in Table 1. When the relative potency for cigarette smoke was compared to roofing tar or coke ovens, however, the short-term bioassays showed a much higher relative potency than the human and mouse skin tumor initiation bioassay. A series of diesel and gasoline automotive emissions for which no human data were available were also tested in the battery of *in vivo* and *in vitro* bioassays. The relative potency within the automotive samples across these bioassays was relatively constant, as shown in Table 2. High correlations were observed between the mutagenic activity of these emissions and their tumor-initiating activity (Lewtas, 1982).

This method was initially applied to the estimation of a human lifetime cancer risk (unit risk) for a series of diesel automobiles by comparison to the human cancer unit risk from coke oven, roofing tar, and cigarettes, according to the following equation (e.g, for coke oven emissions): human risk (diesel/coke oven) = relative bioassay potency (diesel/coke oven). The relative potency in the mouse skin tumor initiation bioassay was used to estimate the unit risk of one diesel emission. The average relative potencies in the short-term genetic bioassays were used to estimate the unit risks for the other diesel and gasoline samples. This approach to applying the comparative potency method assumes that when one of the three human carcinogens and automotive emissions are compared, the mouse skin best predicts the relative potency compared to humans, since the relative potencies in mouse skin tumor-initiating activity of the human carcinogens were well correlated with the human lung cancer unit risks. Within the automotive emissions, however, it was assumed that an average relative potency in short-term bioassays can be used to estimate the human risk, since the relative potencies in the short-term bioassays correlated so well with the mouse skin tumor-initiating activity (Albert et al., 1983).

Table 1. Comparison of Relative Potencies of Emission Extracts in Humans and Bioassay Systems[a]

Emission Extracts	Human Lung Cancer	Mouse Skin Tumor Initiation	Mutation in Mouse Lymphoma Cells (+MA)[b]	Mutation in Ames TA98 (+MA)
Coke Oven	1.0	1.0	1.0	1.0
Roofing Tar	0.39	0.20	1.4	0.78
Cigarette	0.0024	0.0011	0.066	0.52
Diesel (Nissan)		0.28	0.24	12.0

[a]The relative bioassay potencies were determined for each of the bioassay systems by dividing the absolute activity (e.g., lung cancer risk, tumor initiation activity, or mutagenicity) of each emission by the activity of the coke oven emission within that bioassay (Albert et al., 1983). The mouse skin tumor initiation data (Nesnow et al., 1982a, b), mouse lymphoma data (Mitchell et al., 1981), and Ames data (Claxton, 1981) are reported and evaluated elsewhere (Lewtas, 1982; Albert et al., 1983). The collection and preparation of emission extracts have been previously reported (Lewtas et al., 1981).

[b]Presence of an Aroclor-induced rat liver S9 metabolic activation system.

The major assumption and main source of uncertainty in this comparative potency method is the validity of the constant relative potency hypothesis. Interspecies differences in absorption and metabolism, which present the most difficult problems in extrapolation from animals to humans, may also affect the relative carcinogenic potency. By developing and evaluating a comparative data base across human, animal, and short-term bioassays for the same substances, these assumptions can be tested. The constant relative potency model does not assume *a priori* that the carcinogens being compared are chemically similar. The coke oven, roofing tar, cigarette smoke, and automotive emission samples are similar in that they all contain polycyclic organic compounds that cause frameshift mutations in *Salmonella typhimurium* (Claxton and Huisingh, 1980). These four emissions differed in their metabolic activation requirements (Claxton, 1981), and upon chemical fractionation and characterization, significant differences were observed in the specific mutagenic/carcinogenic chemicals present (Lewtas, in press). The similarity in chemical composition of the different diesel automobile emissions may be the basis for the constant relative potency observed across the short-term bioassays and animal tumorigenicity for this series of emissions. The mutagens and carcinogens present in these mixtures may differ only in their concentration.

Table 2. Comparison of Relative Potencies of Diesel and Gasoline Emission Extracts in Several Bioassay Systems[a]

Emission Extracts	Mouse Skin Tumor Initiation	Mutation in Mouse Lymphoma Cells		SCE in CHO Cells		Mutation in Ames TA98	
		(−MA)[b]	(+MA)[c]	(−MA)	(+MA)	(−MA)	(+MA)
Diesels							
Nissan	1.0	1.0	1.0	1.0	1.0	1.0	1.0
Volkswagen Rabbit	0.41	0.23	0.25	0.25	0.42	0.34	0.23
Oldsmobile	0.53	0.29	0.45	Neg	0.24	0.19	0.11
Caterpillar	Neg	0.06	0.022	0.037	Neg	0.034	0.023
Gasoline Catalyst							
Mustang II	0.29	0.90	0.38	0.25	NT	0.26	0.26

[a]The relative potencies were determined by dividing the tumorigenicity or mutagenicity (linear slope of the dose response) of each emission by the activity of the Nissan diesel within that bioassay. Several emissions were either negative (Neg) or not tested (NT) due to insufficient amounts of sample. The data for each assay are reported elsewhere (Lewtas, 1982; Nesnow et al., 1982a, b; Albert et al., 1983; Claxton, 1981; Mitchell et al., 1981). The generation, collection, and preparation of these automotive emissions have been previously reported (Lewtas, 1982; Lewtas et al., 1981).

[b]Absence of an Aroclor-induced rat liver S9 metabolic activation system.

[c]Presence of an Aroclor-induced rat liver S9 metabolic activation system.

This comparative potency method provides a framework for evaluating alternative energy sources and fuels by comparing them to the conventional technologies in short-term genetic bioassays. It is possible to extend the method described here for estimating lung cancer unit risks for diesel and gasoline automotive emissions to unit risk estimates for various combustion emissions or other complex mixtures through the data base established that links human, animal, and short-term genetic bioassay data. A simplified comparative approach to evaluating alternative energy sources is to employ parallel bioassay studies of the alternative (a) and conventional (c) source emissions and determine a relative risk by direct comparison, as follows: increased risk (a/c) = relative bioassay potency (a/c). Since there is no one conventional standard petroleum-derived fuel or one standard combustion source, such studies need to consider the range of mutagenic and carcinogenic potency between different conventional sources and fuels. The establishment of such a range could then serve as a guide for evaluating alternative fuels or sources. For example, if the relative potency of the alternative source or fuel emissions fell within the range for the conventional fuels, one could assume no increased human risk. The battery of bioassays to be employed in such a study would depend upon availability of emission sample, resources, and time. As more bioassays are employed, however, confidence in the relative potency estimates should increase.

The range of mutagenic activity (e.g., revertants per microgram) and mutagenic emission rate (e.g., revertants per kilometer) in a battery of short-term genetic bioassays has been compared for a series of petroleum-derived fuels that meet the specifications for No. 2 diesel fuel for automobiles. Within these specifications, however, is a range of fuels from minimum quality fuels (high nitrogen, sulfur, and aromatic content and low cetane value) to premium quality fuels (low sulfur, nitrogen, and aromatic content and high cetane value). Table 3 shows the range in mutagenic emission rates for three short-term bioassays when three No. 2 diesel fuels were combusted in two vehicles. The mutagenic emission rates generally ranged from two to five fold, except for the sister chromatid exchange (SCE) assay for the truck emissions in the absence of S9, which showed a 31-fold range.

The combustion emissions from synfuels with chemical compositions significantly different from petroleum-derived fuels could be evaluated in the same bioassays to determine if the relative risk from these emissions is increased by a factor significantly greater than the differences observed between currently utilized petroleum-derived fuels.

EXPERIMENTAL METHODS

Emission Sample Collection and Extraction Methods

Combustion emissions containing both gaseous and particulate matter are typically collected by filtration or electrostatic precipitation after dilution and cooling (Bradow, 1982; Schuetzle, 1983) or are directly collected by condensation (Grimmer et al., 1973). The automotive emissions described here as well as the residential oil and wood combustion emissions were collected by a modification of the standard dilution tunnel method (Lewtas et al., 1981; Bradow, 1982), in which the exhaust is collected on large 20 in. x 20 in. Teflon-coated Pallflex T60-A20 filters. The collection methods for

Table 3. Range of Mutagenic Emission Rates from No. 2 Diesel Petroleum Fuels[a]

Diesel Vehicle	MA[b]	Mutagenic Activity/km: Mutation in Ames TA98		Mutation in Mouse Lymphoma Cells		SCE in CHO Cells	
Truck	–	6.8 -14.0	(2.1x)	0.91-3.5	(3.8x)	0.21-6.6	(31x)
	+	5.2 -17.0	(3.3x)	0.85-1.4	(1.6x)	1.2 -2.1	(1.7x)
Bus	–	0.22 - 1.1	(5.0x)	-		0.91-2.8	(3.1x)
	+	0.37 - 0.86	(2.3x)	1.3 -4.3	(3.3x)	0.72-1.7	(2.4x)

[a]The mutagenic emission rate, mutagenic activity/km, was determined from the mutagenic activity dose-response slope of the particulate emission extracts, the percent organic extractables, and particulate emission rates for each vehicle. The mutagenic emission rate units for each bioassay are as follows: Ames (revertants/km x 10^5), mouse lymphoma (mutants/10^6 survivors/µg/ml/km x 10^5), and SCE in CHO cells (SCE/cell/µg/ml/km x 10^4). The range in activity between the different fuels is shown in parentheses, e.g., (2.1x), for each bioassay. The truck was a heavy-duty Ford Bob Tail Van (LN-7000) with a Caterpillar 3208 engine. The bus was a General Motors city bus with a Detroit diesel DD8V-71 engine. Both vehicles were operated on the 83 transient cycle with premium (EM-242F), average (EM-239F), and minimum quality (E-241F) petroleum-derived diesel fuels. The specifications and chemical composition of these fuels have been previously reported (Huisingh et al., 1979).

[b]An Aroclor-induced rat liver S9 metabolic activation system.

the coke oven, roofing tar, and cigarette smoke condensate have been described elsewhere (Lewtas et al., 1981). The particulate samples were Soxhlet extracted with dichloromethane (DCM) for 48 h, filtered (0.2-µm pore), and the DCM was removed by evaporation and solvent-exchanged into dimethylsulfoxide (DMSO) for all the bioassays discussed here except for mouse skin tumorigenesis, in which acetone was used as the solvent.

Bioassay Methods

The short-term genetic bioassays in which dose-dependent and reproducible increases in response were obtained for many of the combustion emission samples included the following:

1. Reverse mutation in *Salmonella typhimurium* TA98 (Ames bioassay) (Claxton, 1981)

2. Forward mutation at the thymidine kinase locus in L5178Y mouse lymphoma cells (mouse lymphoma bioassay) (Mitchell et al., 1981)

3. Sister chromatid exchange in Chinese hamster ovary (CHO) cells (SCE bioassay) (Mitchell et al., 1981)

For these bioassays, one or more preliminary range-finding tests were performed to determine the optimum dose range for defining the slope of the dose response. The bioassays were generally performed with duplicate or triplicate treatments at five to seven doses. The bioassays were all performed in the presence and absence of an Aroclor-induced rat liver S9 metabolic activation system (MA). The linear slope of the dose response determined from simple regression analysis was utilized for the comparative potency analysis, as previously described (Albert et al., 1983).

Mouse skin tumorigenesis assays, which have had considerable use in the determination of the tumorigenic and carcinogenic effects of chemicals and complex mixtures, were performed in SENCAR mice, as previously described (Nesnow et al., 1982a), utilizing five single doses (80 mice/dose) and scoring individually at six months for papillomas and at one year for squamous cell carcinomas (Nesnow et al., 1982b). The carcinoma responses in the initiation assay were correlated with the papilloma responses; only the papilloma responses were used in the comparative potency analysis (Albert et al., 1983).

COMPARATIVE ASSESSMENT OF COMPLEX MIXTURES FROM ALTERNATIVE ENERGY

Short-term bioassays have been proven to be of use in evaluating the effects of various engines, fuels, and control technologies on the mutagenicity of emissions. To draw meaningful conclusions from such comparisons, however, the normal day-to-day variations in the emissions were studied for one engine under standard conditions (Claxton and Kohan, 1981). The coefficients of variation for the particulate emission rates, percent organic extractables, and mutagenic activity ranged from 0.07 to 0.11. To compare the mutagenicity of particulate emissions from different sources or fuels, the percent of organic material extractable from the particles and the vehicle's particulate emission rate must also be included in the analysis. The final rate of mutagenic emissions is determined from all three parameters.

Although pure compounds have been reported to range in carcinogenic and mutagenic potency over six orders of magnitude (Meselson and Russel, 1977), the relative potency of combustion emissions ranged from only a 5-fold difference in the mouse skin tumorigenesis assay to a 150-fold difference in the mouse lymphoma gene mutation assay. The Ames *Salmonella typhimurium* mutagenesis assay showed approximately two orders of magnitude difference (100- to 110-fold) in mutagenic activity from the weakest to the most potent emission sample when measured per mass of extractable organics (e.g., revertants per microgram). The emission rates (revertants per kilometer or revertants per joule) for the particle-bound organics, however, differed in some cases by as much as five orders of magnitude. To compare the mutagenicity of emissions from various sources, it is critical to compare the rates of mutagenic emissions on the same basis (e.g., fuel consumption, distance driven, energy consumption, or yield). To compare rates of mutagenic emissions of mobile sources, we have compared the mutagenic activity per kilometer, as shown in Table 4. The rates of mutagenic emissions from stationary sources are compared on the basis of fuel or energy consumption, as shown in Table 5.

This direct comparative analysis of mutagenic emission rates provides a simple means of evaluating alternative energy sources. It is apparent from this data base that the organic emission rate may vary more than the mutagenicity or carcinogenicity of the emissions; therefore, in the final analysis the organic emission rate has a greater impact on the relative carcinogenic risk from those sources.

COMPARATIVE EVALUATION OF ALTERNATIVE FUELS

Fuel composition may affect the mutagenic and potential carcinogenic activity of combustion emissions by (1) direct contribution of mutagens in the fuel, (2) contribution of compounds that give rise to mutagens when combusted, or (3) influencing the combustion processs so that generation of mutagenic and carcinogenic compounds is altered. The influence of fuel composition on the mutagenicity of diesel emissions was initially examined for five petroleum-derived fuels in two light-duty diesel vehicles (Huisingh et al., 1979). In one vehicle, the emissions from the minimum-quality fuel, with a higher aromatic and nitrogen content, were approximately five times more mutagenic per milligram particulate emission in the Ames mutagenesis bioassay than the emissions from the other fuels. The emissions from these fuels did not significantly differ in mutagenicity, however, when they were burned in the second vehicle. Similar mutagenesis studies were conducted when the same fuels were combusted in the three heavy-duty vehicles (Lewtas, 1983). The minimum-quality fuel produced a higher rate of mutagenic emissions in two of the vehicles; however, this difference was not observed for the third vehicle.

Since aromaticity and nitrogen content were implicated as influencing the microbial mutagenicity of diesel emissions, a study was designed to examine these two parameters by the use of aromatic and nitrogen additives to a base fuel (Sklarew et al., 1982). Nonmutagenic nitrogen fuel additives, hexylnitrate and isoquinoline, had little effect on the mutagenicity of the diesel emissions. In this study, the emissions from the three fuels with a higher aromatic content of 30-35% were more mutagenic than fuels with 3-7% aromatics; however, the differences were small and were within the ranges shown in Table 3. A second study was designed to examine correlations between nitrogen content, aromatic content, and the concentrations of specific polycyclic aromatic hydrocarbons (e.g., fluoranthene, pyrene, and benzo[a]pyrene). This study,

Table 4. Comparative Mutagenic Emission Rates for Transportation Combustion Sources[a]

Vehicle	Fuel Type	Mutagenicity of Organics		Organic Emission Rate (mg/km)	Mutagenic Emission Rates (activity/km)	
		Ames TA98 (+MA)[b]	Mouse Lymphoma (+MA)		Ames TA98	Mouse Lymphoma
Diesel engine						
Car (Mercedes)	Diesel	12.0	1.5	19.2	2.4	0.30
Truck (Ford/Cat)	Diesel	1.7	0.28	302.0	5.3	5.85
Bus (GM)	Diesel	0.1	0.35	357.0	0.37	1.3
Gasoline engine						
Noncatalyst (Ford van)	Unleaded gasoline	32.0	5.7	5.7	1.8	0.32
Catalyst (Mustang II)	Leaded gasoline	3.5	1.1	1.4	0.05	0.02

[a]The mutagenicity of the organics in the Ames bioassay is expressed in revertants/μg and in the mouse lymphoma bioassay as mutants/10^6 survivors/μg/ml (mut. freq./μg/ml). The mutagenic emission rates were determined as described in Table 3. The methods for collecting these emissions and performing the mutagenesis bioassays have been reported elsewhere (Lewtas, 1982; Lewtas et al., 1981; Lewtas, in press).

[b]Presence of an Aroclor-induced rat liver S9 metabolic activation system.

Table 5. Comparative Mutagenic Emission Rates for Stationary Combustion Sources[a]

Residential Combustion Source	Fuel Type	Mutagenicity of Organics Ames TA98 (revertants/μg)	Organic Emission Rate (ng/J)	Mutagenic Emission Rate (revertants/J) (x10^3)
Woodstove (airtight)	Pine	1.3	508.0	660.0
Woodstove (airtight)	Oak	0.9	187.0	168.0
Oil furnace #1	Oil #2	2.0	0.5	1.0
Oil furnace #2	Oil #2	5.1	1.5	7.6
Utility power plants				
Conventional	Coal	3.1	0.01	0.031
Fluidized-bed	Coal	9.4	0.03	0.28

[a]The mutagenic emission rates were determined as described in Table 3. The mutagenicity of these stationary source combustion emissions have been reviewed elsewhere (Lewtas, in press). The mutagenicity of the woodstove and oil furnace emissions is shown in the presence of metabolic activation and the utility coal power plant emissions are shown in the absence of metabolic activation.

using six different petroleum fuels in one diesel engine and five petroleum fuels in a second engine, found the highest correlations between mutagenicity and nitrogen content (Zweidinger, 1982).

These studies suggest that although low-quality fuel, with relatively high aromatic and nitrogen content, can result in increased rates of mutagenic emissions in certain vehicles, such differences are not observed in all vehicles. In general, the observed changes in mutagenic activity of emissions as a function of fuel quality were much smaller than the differences between different types of engines (e.g., diesel versus gasoline) and control technologies (catalyst-equipped versus non-catalyst-equipped vehicles).

The increased or decreased human cancer risk from combustion emissions may be of greater concern than the absolute risk from these emissions when we consider the effect of alternative energy technologies and fuels. The data of greatest value in determining this increased or decreased risk are comparative bioassay data on emissions from the complex mixture being evaluated or comparative bioassay data on the conventional and alternative technology being evaluated. Such studies should be useful in providing direction for engineers and chemists to design alternative energy sources and fuels that result in less mutagenic and potentially carcinogenic emissions.

DISCLAIMER

This paper has been reviewed by the Office of Research and Development, U.S. Environmental Protection Agency, and has been approved for publication. Approval does not signify that the contents necessarily reflect the views and policies of the U.S. Environmental Protection Agency.

REFERENCES

Albert, R.E., J. Lewtas, S. Nesnow, T.W. Thorslund, and E. Anderson. 1983. Comparative potency method for cancer risk assessment: Application to diesel particulate emissions. Risk Anal. 3:101-117.

Bradow, R.L. 1982. Diesel particle and organic emissions: Engine simulation, sampling and artifacts. In: Toxicological Effects of Emissions from Diesel Engines. J. Lewtas, ed. Elsevier Science Publishing Co., Inc.: New York. pp. 33-47.

Bridges, B.A. 1973. Some general principles of mutagenicity screening and a possible framework for testing procedures. Environ. Health Perspect. 6:221-227.

Claxton, L.D. 1981. Mutagenic and carcinogenic potency of diesel and related environmental emissions: *Salmonella* bioassay. Environ. Int. 5:389-391.

Claxton, L.D., and J.L. Huisingh. 1980. Comparative mutagenic activity of organics from combustion sources. In: Pulmonary Toxicology of Respirable Particles. C.L. Sanders, F.T. Cross, G.E. Dagle, and J.A. Mahaffey, eds. CONF-791002, Department of Energy, U.S. Government Printing Office. pp. 453-465.

Claxton, L., and M. Kohan. 1981. Bacterial mutagenesis and the evaluation of mobile-source emissions. In: Short-Term Bioassays in the Analysis of Complex Environmental Mixtures II. M.D. Waters, S.S. Sandhu, J. Huisingh, L. Claxton, and S. Nesnow, eds. Plenum Press: New York. pp. 299-317.

Grimmer, G., A. Hildebrandt, and H. Bohnke. 1973. Investigations on the carcinogenic burden by air pollution in man. II. Sampling and analytics of polycyclic aromatic hydrocarbons in automobile exhaust gas. I. Optimization of the collecting arrangement. Zbl. Bakt. Hyg., I. Abt. Orig. B 158:22-34.

Holmberg, G., and U. Ahlborg. 1983. Consensus report: Mutagenicity and carcinogenicity of car exhausts and coal combustion emissions. Environ. Health Perspect. 47:1-30.

Huisingh, J., R. Bradow, R. Jungers, L. Claxton, R. Zweidinger, S. Tejada, J. Bumgarner, F. Duffield, M. Waters, V.F. Simmon, C. Hare, C. Rodriquez, and

L. Snow. 1979. Application of bioassay to the characterization of diesel particle emissions. In: Application of Short-Term Bioassays in the Fractionation and Analysis of Complex Environmental Mixtures. M.D. Waters, S. Nesnow, J.L. Huisingh, S.S. Sandhu, and L. Claxton, eds. Plenum Press: New York. pp. 381-418.

Huisingh, J.L., D.L. Coffin, R. Bradow, L. Claxton, A. Austin, R. Zweidinger, R. Walter, J. Sturm, and R.H. Jungers. 1980. Comparative mutagenicity of combustion emissions of a high quality no. 2 diesel fuel derived from shale oil and a petroleum derived no. 2 diesel fuel. In: Health Effects Investigation of Oil Shale Development. W.H. Griest, M.R. Guerin, and D.L. Coffin, eds. Ann Arbor Science: Ann Arbor, MI. pp. 201-207.

Interagency Regulatory Liaison Group. 1979. Fed. Regist. 44, 39858.

International Agency for Research on Cancer. 1983. IARC Internal Technical Report No. 83/001. IARC: Lyon, France.

Lewtas, J., R.L. Bradow, R.H. Jungers, B.D. Harris, R.B. Zweidinger, K.M. Cushing, B.E. Gill, and R.E. Albert. 1981. Mutagenic and carcinogenic potency of extracts of diesel and related environmental emissions: study design, sample generation, collection, and preparation. Environ. Int. 5:383-387.

Lewtas, J. 1982. Mutagenic activity of diesel emissions. In: Toxicological Effects of Emissions from Diesel Engines. J. Lewtas, ed. Elsevier Science Publishing Co., Inc.: New York. pp. 243-264.

Lewtas, J. 1983. Evaluation of motor vehicle and other combustion emissions using short-term genetic bioassays. In: Mobile Sources Emissions Including Polycyclic Organic Species. D. Rondia, M. Cooke, and R.K. Haroz, eds. D. Reidel Publishing Co.: Dordrecht. pp. 165-180.

Lewtas, J. (in press). Combustion emissions: Characterization and comparison of their mutagenic and carcinogenic activity. In: Carcinogens and Mutagens in the Environment, Vol. IV. H.F. Stich, ed. CRC Press: Boca Raton, FL.

Meselson, M., and K. Russel. 1977. Comparisons of carcinogenic and mutagenic potency. In: Origins of Human Cancer, Book C. H.H. Hiatt, J.D. Watson, and J.A. Winsten, eds. Cold Spring Harbor Laboratory: New York. pp. 1473-1481.

Mitchell, A.D., E.L. Evans, M.M. Jotz, E.S. Riccio, K.E. Mortelmans, and V.F. Simmon. 1981. Mutagenic and carcinogenic potency of *in vitro* mutagenesis and DNA damage. Environ. Int. 5:393-402.

Nesnow, S., and J. Lewtas. 1981. Mutagenic and carcinogenic potency of extracts of diesel and related environmental emissions: Summary and discussion of the results. Environ. Int. 5:425-429.

Nesnow, S., L. Triplett, and T.J. Slaga. 1982a. Comparative tumor-initiating activity of complex mixtures on mouse skin. J. Natl. Cancer Inst. 68:829-834.

Nesnow, S., C. Evans, A. Stead, J. Creason, T.J. Slaga, and L.L. Triplett. 1982b. Skin carcinogenesis studies of emission extracts. In: Toxicological Effects of Emissions from Diesel Engines. J. Lewtas, ed. Elsevier Science Publishing Co., Inc.: New York. pp. 295-320.

Schuetzle, D. 1983. Sampling of vehicle emissions for chemical analysis and biological testing. Environ. Health Perspect. 47:65-80.

Sklarew, D.S., R.A. Pelroy, S.P. Downey, R.H. Jungers, and J. Lewtas. 1982. Chemical and mutagenic characteristics of diesel exhaust particles from different diesel fuels. In: Diesel Emissions Symposium Proceedings. EPA-600/9-82-014 (NTIS PB 82-244013), U.S. Environmental Protection Agency: Research Triangle Park, NC. pp. 593-597.

Waters, M.D. 1978. A phased approach to the bioscreening of emissions and effluents from energy technologies. In: Proceedings of Symposium on Potential Health and Environmental Effects of Synthetic Fossil Fuel Technologies, September 25-28, 1978, Gatlinburg, Tennessee. First Annual Oak Ridge National Laboratory Life Sciences Symposium. Department of Energy (CONF-780903). pp. 143-152.

Zweidinger, R.B. 1982. Emission factors from diesel and gasoline powered vehicles: Correlation with the Ames test. In: Toxicological Effects of Emissions from Diesel Engines. J. Lewtas, ed. Elsevier Science Publishing Co., Inc.: New York. pp. 83-96.

INDEX